T0250038

Biomechanics in Animal Behaviour

EXPERIMENTAL BIOLOGY REVIEWS

Series advisors:

D.W. Lawlor
AFRC Institute of Arable Crops Research, Rothamsted Experimental Station, Harpenden, Hertfordshire AL5 2JQ, UK

M. Thorndyke
School of Biological Sciences, Royal Holloway, University of London, Egham, Surrey TW20 0EX, UK

Environmental Stress and Gene Regulation
Sex Determination in Plants
Plant Carbohydrate Biochemistry
Programmed Cell Death in Animals and Plants
Biomechanics in Animal Behaviour

Forthcoming titles include:
Cell and Molecular Biology of Wood Formation

Biomechanics in Animal Behaviour

P. DOMENICI
International Marine Centre, Localitá Sa Mardini, 09072 Torregrande (Oristano), Italy

R.W. BLAKE
Department of Zoology, University of British Columbia, 6270 University Boulevard, Vancouver, BC, V6T 2A9, Canada

Taylor & Francis
Taylor & Francis Group

LONDON AND NEW YORK

A CIP catalogue record for this book is available from the British Library.

ISBN 1 85996 162 2

Published by Taylor & Francis
2 Park Square, Milton Park, Abingdon, Oxon, OX14 4RN
270 Madison Ave, New York NY 10016
World Wide Web home page: http://www.taylor&francis.com/

Transferred to Digital Printing 2006

Production Editor: Fran Kingston.
Typeset by Saxon Graphics Ltd, Derby, UK.

Cover image of Spanish ibex kindly supplied by Roberto Travesi.

Publisher's Note
The publisher has gone to great lengths to ensure the quality of this reprint but points out that some imperfections in the original may be apparent

Contents

Contributors

Alerstam, T., Department of Animal Ecology, Lund University, Ecology Building, S-223 62 Lund, Sweden

Alexander, R.McN., School of Biology, University of Leeds, Leeds, LS2 9JT, UK

Batty, R.S., Centre for Coastal and Marine Sciences, Dunstaffnage Marine Laboratory, PO Box 3, Oban, PA34 4AD, UK

Bellwood, D.R., Department of Marine Biology, James Cook University, Townsville, Queensland 4811, Australia

Bels, V.L., Centre Agronomique de Recherches Appliquées du Hainaut, rue Paul Pastur 11, B-7800-ATH, Belgium

Binder, W.J., Department of Organismic Biology, Ecology and Evolution, University of California, Los Angeles, CA 90095–1606, USA

Blake, R.W., Department of Zoology, University of British Columbia, 6270 University Boulevard, Vancouver, BC, V6T 2A9, Canada

Copp, N., Keck Science Center, The Claremont Colleges, Claremont, CA 91711, USA

Domenici, P., International Marine Centre, Localitá Sa Mardini, 09072 Torregrande (Oristano), Italy

Ellers, O., Department of Biology, Bowdoin College, 220A Druckenmiller Hall, Brunswick, ME 04011, USA

Full, R.J., Department of Integrative Biology, University of California, Berkeley, CA 94720, USA

Gerstner, C.L., Department of Conservation, John G. Shedd Aquarium, 1200 South Lakeshore Drive, Chicago, IL 60605, USA

Hughes, R., University of Wales, Bangor, Gwynedd, LL57 2UW, UK

Jamon, M., Laboratoire de Neurobiologie et Mouvements, CNRS, 31 Chemin Joseph Aiguier, 13402 Marseille Cedex 20, France

Kitchener, A.C., National Museums of Scotland, Chambers Street, Edinburgh, EH1 1JF, UK

Lauder, G.V., Museum of Comparative Zoology, Harvard University, 26 Oxford Street, Cambridge, MA 02138, USA

Moore, P., Laboratory for Sensory Ecology, Department of Biological Sciences, and J.P. Scott Center for Neuroscience, Mind and Behavior, Bowling Green State University, Bowling Green, OH 43403, USA

Rayner, J.M.V., School of Biology, University of Leeds, Leeds, LS2 9JT, UK

Schneider, R.W.S., Laboratory for Sensory Ecology, Department of Biological Sciences, and J.P. Scott Center for Neuroscience, Mind and Behavior, Bowling Green State University, Bowling Green, OH 43403, USA

Swaddle, J.P., School of Biological Sciences, University of Bristol, Woodland Road, Bristol, BS8 1UG, UK

Van Valkenburgh, B., Department of Organismic Biology, Ecology and Evolution, University of California, Los Angeles, CA 90095–1606, USA

Vollrath, F., Department of Zoology, South Parks Road, Oxford, OX1 3PS, UK and Department of Zoology, Universitetsparken B135, DK 8000 Aarhus C, Denmark

Wainwright, P.C., Section of Evolution and Ecology, University of California, One Shields Avenue, Davis, CA 95616, USA

Webb, P.W., School of Natural Resources and Environment, University of Michigan, Ann Arbor, MI 48109–1115, USA

Weinstein, R.B., Department of Physiology, University of Arizona, 1501 North Campbell Avenue, Tucson, AZ 85724, USA

Westneat, M.W., Department of Zoology, Field Museum of Natural History, 1400 South Lakeshore Drive, Chicago, IL 60605–2496, USA

Abbreviations

BCF	body and caudal fin
BMR	basal metabolic rate
BPBP	bucco-pharyngeal breathing pump
BSI	beach state index
COT	cost of transport
DAP	display action pattern
FG	fast glycolytic
IL	intensity level
MAP	modal action pattern
MAS	maximum aerobic speed
MPF	median and paired fins
SO	slow oxidative (muscle)
SPL	sound pressure level
STST	selective tidal-stream transport
VBPBV	ventilatory bucco-pharyngeal breathing pump

Preface

The currently prevailing perspective is that biomechanics can be used as a tool to bear upon issues in certain classical divisions of organismal biology. This book investigates the value of integrating biomechanical analysis and animal behaviour for the first time.

Chapter 1 (Domenici and Blake) sets a general context for expanding the scope for integration of biomechanics and behaviour. Whilst biomechanical analyses and perspectives are currently common in studies of locomotion and feeding, they are rare in other areas (e.g., social behaviour, aggression, orientation). In addition, they provide a general discussion of the relationships between behaviour and the mechanical properties of the environment, animal materials and structures. In Chapter 2, Lauder departs from the conventional approach whereby the vast majority of biomechanical studies focus on a single species to understand how behaviour is generated at an intraspecific level, to explore the nascent areas of comparative and historical analysis of behaviour The next two chapters deal with the energetics and behavioural strategies in terrestrial locomotion. Weinstein and Full (Chapter 3) discuss the energetic implications of non-steady-state locomotor behaviour, such as that observed in nature, in crabs and other invertebrates and lower vertebrates. By combining laboratory and field studies, Weinstein and Full show that, for a high work-rate behaviour performed intermittently, the total work accomplished can far exceed the work done during the same behaviour performed continuously at the same average work rate. Alexander (Chapter 4) discusses walking and running strategies in mammals bearing upon speed in relation to gait and optimal routes in relation to risks. The latter is discussed in the context of the nature of the terrain (soft or rough patches, hills) and predator–prey encounters. Swimming biomechanics and behaviour are discussed by Webb and Gerstner, and Blake (Chapters 5 and 6, respectively). Webb and Gerstner employ basic physical principles to bear upon gait choice, choices of how, when and where to swim, and complex behaviours such as predator– prey interactions and social interactions. They point out that the physical properties of propulsion systems are important in habitat partitioning. Blake describes the energetic consequences of intermittent swimming behaviour in aquatic vertebrates. Among other issues, new analyses and perspectives are provided on porpoising and the flight of flying fish. The next two chapters are concerned with flight biomechanics and behaviour. Alerstam (Chapter 7) shows how flight mechanics provides a fundamental basis for evaluating and understanding the possibilities and constraints on bird migration performance. Combining flight mechanics and simple flight vector trigonometry, the speed, range and precision of a bird's migration in relation to the influences of wind and orientation behaviour are discussed. Rayner and Swaddle (Chapter 8) employ aerodynamics analysis to investigate aspects of the annual cycles of birds (breeding, migration, etc.) from the standpoint of the energetics of moult as it bears upon take-off flight behaviour. In addition, they discuss how the energetic cost of moult may be understood in the context of adaptive radiation across avian taxa. The next two chapters are concerned with issues related to orientation behaviour and the physical properties of the environment. Schneider and Moore (Chapter 9) discuss the relationship between the physical features of the environment, chemical signals and hydrodynamics. A principle finding is that habitat-specific hydrodynamics sets constraints on the type of information available to the organisms for locating the source of a chemical signal. The issue of information transfer to organisms in the context of passive orientation is addressed by Ellers in Chapter 10. The particular example involves a mode of locomotion called 'swash-riding' in certain clams. In addition, Ellers discusses the physical characteristics of the wave-swept shore that are relevant to habitat choice in sandy beach invertebrates. The next four chapters are concerned with the issues of feeding and predator–prey interactions. In Chapter 11, Hughes applies a biomechanical perspective to understanding the function of crab claws as tools and weapons. Among other issues, Hughes illustrates how the characterization of biomechanical properties of claws and shelled prey represents an excellent system for investigating the general behavioural phenomenon of skilled transfer. Linking feeding behaviour and biomechanics is further explored by Wainwright *et al.*, in Chapter 12, by reference to the jaw mechanics of fishes. Quantitative mechanical modelling is employed to link the jaw mechanics of fish to prey capture kinematics. Two case studies are discussed to test predictions on the scaling of jaw mechanics for inter-specific differences and to examine the ability of jaw mechanics to account for differences in prey capture. Van Valkenburgh and Binder review the biomechanics and feeding behaviour in terrestrial carnivorous mammals in Chapter 13. The relationships between diet, killing behaviour, cranial and dental morphology are discussed. In addition, a case study, the spotted hyena, is employed to explore the morphological and behavioural development of bite strength and feeding performance. Predator–prey relationships in fish and other aquatic vertebrates is discussed by Batty and Domenici in Chapter 14. They show how the interplay between swimming kinematics and a number of behavioural variables can help elucidate fundamental issues of predator–prey interactions, such as the scaling of attack and escape locomotion, the biomechanical implications of various attack strategies, and temporal (latencies) and spatial (trajectories) variability in escape behaviour. The next three chapters are concerned with display and fighting behaviour. Bels (Chapter 15) addresses biomechanical aspects of display behaviour in tetrapods. In particular, the display-action-pattern in lizards is employed to emphasise the role of biomechanical aspects of display behaviour in tetrapods. A possible scenario for the evolution of throat display is also discussed. Biomechanical and behavioural aspects of the defence response in crayfish are discussed in Chapter 16 by Copp and Jamon. They bear upon the integration of visual, hydrodynamic and tactile stimulation to show how biomechanical studies set in an integrative framework can help understand a particular behaviour from the standpoint of the interplay between the control of muscles and the application of forces. Kitchener (Chapter 17) discusses the biomechanical and behavioural aspects of fighting with horns and antlers in bovids and cervids, in both extant and extinct species. The material properties of the fighting devices are linked to the kinematics of encounters and discussed in relation to fitness-related issues. Finally, in Chapter 18, Vollrath discusses the co-evolution of behaviour and material properties of webs in spiders. Using orb webs as an example, he shows how the interaction of behaviour and material is optimised by evolution with respect to both the behaviour and the material.

The initial grounding for this book was the organization of a symposium 'Biomechanics and Behaviour' held at Heriot-Watt University in 1999, sponsored by the Society for Experimental Biology (SEB). We are grateful to the SEB for their help in bringing together such an eminent collection of contributors. On behalf of all the contributors to this book, we would like to thank the Society for Experimental Biology and BIOS Scientific Publishers Ltd for their support. We also wish to thank S. Ferrari and D. Gonzalez Calderon for editorial assistance.

Paolo Domenici
Robert Blake

Biomechanics in Behaviour

Paolo Domenici and Robert W. Blake

1. Introduction

Biomechanics is the study of the mechanical design of organisms and their movements, integrating physics and biology. A biomechanical approach can therefore be useful to study animal behaviour, as animals are constrained by the laws of physics, be they hummingbirds hovering while feeding on nectar, intertidal organisms withstanding wave forces, migrating whales, or seabirds resting on water. The interactions of animals with their physical world are based on trial and error, and shaped by evolution through maximizing fitness. For example, a cat applies the proper ground force needed to jump up on a windowsill, and a spider builds a web with the mechanical and structural properties that resist the impact of a prey.

Animal behaviour involves evolutionary compromises resulting from multiple, often competing, evolutionary forces. Hence, animal behaviour does not always maximize particular features related to laws of mechanics; similarly, morphology may not simply reflect the needs required by such laws (for example, a bird's tail may be subject to selection for high manoeuvrability in flight and for sexual attractiveness; Thomas and Balmford, 1995). The properties of a spider's web may be due not only to its mechanical requirements for prey capture, but also, for example, to its level of visibility to potential prey (Craig et al., 1996), and a foraging crow may not use the optimal height according to physics (i.e. the height that gives the maximum net energy gain) for dropping a shell to break it open if competitors are present (Plowright et al., 1989). Arguably, knowledge of the laws of mechanics governing certain animal movements or structures is fundamental for understanding most behavioural patterns, regardless of whether the behaviour being investigated has evolved to maximize any given physical (mechanical) parameter or variables of any other nature.

Although the integration of biomechanics and behaviour is a fundamental step in organismal biology, the advantages of this integration have not been thoroughly explored. For example, while current approaches in locomotion and feeding often encompass behaviour and biomechanics, this approach is less common in other areas of animal behaviour such as social behaviour, aggression and orientation. Here, we discuss the role of behavioural biomechanics in the context of organismal biology to

Biomechanics in Animal Behaviour, edited by P. Domenici and R.W. Blake.

understand how fitness is determined within the paradigm, 'design→performance→ behaviour→fitness'. In addition, the relationships between behaviour and the mechanical properties of the environment, animal materials and structures are explored. Finally, an overview of many areas in which behavioural biomechanics can be useful are considered.

2. Behavioural biomechanics in organismal biology

In organismal biology, an integrative approach that accounts for all disciplines pertaining to whole-animal biology is desirable (Wainwright and Reilly, 1994). Past reductionist approaches have given rise to traditional fields, recognizable by programmes and departments in schools, universities and funding agencies. Those dealing with organismic functions are usually found in the fields of physiology, behaviour, ecology and morphology (*Figure 1*). In the past three decades, strong links have been created between some traditional fields (e.g. ecomorphology, physiological ecology, behavioural ecology, neuroethology). Such syntheses have led to rapid advances as a result of seeing problems from new perspectives. Indeed, such syntheses have often been so successful that they too have spawned new fields. For example, disciplines complementary to animal behaviour have resulted in the rise of behavioural ecology and neuroethology. Behavioural ecology was derived from the need to integrate behavioural observations with ecological data, as 'the way in which

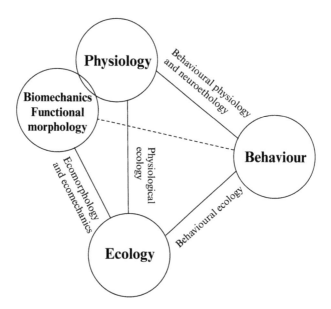

Figure 1. A simplified scheme of the current traditional (circles) and integrative disciplines (lines) in whole-animal biology. The number of disciplines is kept to a minimum for simplicity. Biomechanics and functional morphology are pooled within one area as they are closely related. Similarly, physiology 'touches' biomechanics, as biomechanics spans from it and is present in textbooks of animal physiology. Established integrative disciplines (e.g. behavioural ecology, physiological ecology, etc.) are indicated by continuous lines. The integration of biomechanics and behaviour is indicated by a discontinuous line. The area of evolution is not indicated, as it is related to all disciplines and it is not restricted to whole-animal biology.

behaviour contributes to survival and reproduction depends on ecology' (Krebs and Davies, 1984). Similarly, neuroethology arose from the need to understand the neural basis of behaviour, as 'the reason the nervous systems evolved in the first place was to produce behaviour' (Camhi, 1984).

An integrative approach of this kind can link all traditional disciplines in a new way (*Figure 1*). Any given study of organismal biology may be focused on a given traditional or integrative field while considering the overall framework. Although some studies have focused on both biomechanics and behaviour, the potential for integration within the organismal biology framework has not been explored fully. Biomechanics is currently a 'tool' for investigating a variety of problems in physiology, ecology and evolution (see, e.g. Alexander and Goldspink, 1977; Rayner and Wotton, 1991; Wainwright and Reilly, 1994). Our goal is to demonstrate how biomechanics can also be useful in evaluating, making predictions and understanding behaviour.

2.1 *Historical issues: two lines of thought*

We believe that, to date, the integration of biomechanics and behaviour has not developed to its full potential because of historical reasons. Biomechanics has developed as a discipline spanning comparative physiology and functional morphology, with a strong experimental and laboratory tradition. In contrast, behaviour has developed 'hand in hand' with natural history and ecology on one side and neurophysiology on the other (hence neuroethology) (*Figure 1*). Biomechanicists focus on physiological, morphological or ecological implications but rarely consider behavioural aspects. Similarly, behavioural scientists often consider evolutionary, ecological or neurobiological implications, but rarely biomechanical functional and/or causal implications and explanations of their work.

The worth of an integrated biomechanical–behavioural approach is especially well illustrated by studies of the escape response. In these studies, biomechanicists have focused on locomotor performance, often assuming that escape responses would result in maximum speed and acceleration. On the other hand, from a behavioural and evolutionary perspective, escape performance is measured in terms of the ability to escape a predator attack. The assumption that escape success is directly related to speed (and therefore speed should be maximized) may not always apply (e.g. Domenici and Blake, 1993b; Webb, 1986). In addition, other, non-biomechanical variables influence escape success (e.g. latency and reaction distance; Eaton and Hackett, 1984). Therefore, for biomechanical studies to have relevance in terms of fitness, behaviour needs to be considered.

These considerations may be relatively straightforward for biomechanical studies of a specific behaviour (such as an escape response) that is employed by an animal in a particular situation (escaping from predators). However, not all locomotor patterns are as situation specific as the escape response. Escape responses in fish are triggered by a specific neuronal system (the Mauthner system; Eaton and Hackett, 1984), which can be stimulated in laboratory situations, where precise kinematic analyses can be performed. However, in the case of fish steady swimming, the swimming pattern may not be easily related to any particular behaviour observable in nature. The behavioural context of steady swimming observed in the laboratory, therefore, is not as 'strong' as that of escape responses. However, this is not to say that steady swimming does not imply certain natural behaviours. Fish swim steadily within various behavioural contexts, for example, searching for food, finding mates, migrating, and plankton feeding (see Chapter 5) (*Figure 2*). Therefore, certain characteristics such as speed or locomotor strategies may be associated with specific behaviours. We can predict, for instance, that migration will be accomplished at the velocity with the lowest cost of transport (Weihs, 1973a). Therefore, performance levels can be related to behaviour, even for those locomotor patterns that are not particularly situation specific, such as steady swimming or running. Other performance values that can be measured in the laboratory, such as maximum sustained speed and maximum prolonged speed, give an indication of the endurance capabilities of animals, but are not necessarily used by fish in particular behavioural situations.

Behavioural biologists, on the other hand, can include biomechanical variables and/or use biomechanical tools to investigate a number of behaviours. In studies where animal displacement is the issue, the idea that a biomechanical approach can be fundamental for elucidating behaviour is current (locomotor strategies in walking, Chapters 3 and 4; in swimming, Chapters 5 and 6; in flying, Chapters 7 and 8). Biomechanics can be used also for analysing behaviours that appear more complex than locomotion *per se*, such as feeding and predator–prey relationships (Chapters 2, 11–14), ritualization, fighting and sexual and aggressive displays (e.g. Chapters 15–17), orientation and communication (Chapters 9 and 10) as well as behaviours that imply the use of material and tools, such as nest and spider web building (Chapter 18), and food cracking and opening (Chapter 11).

2.2 *Linking biomechanics with behavioural ecology and neuroethology: proximate and ultimate causes of behaviour*

In the last couple of decades, interactions between traditional areas of whole-animal biology have given rise to relatively independent areas (*Figure 1*), as is the case for

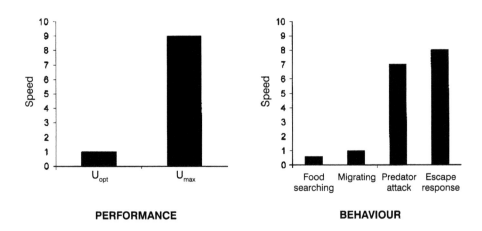

Figure 2. *Theoretical representation of performance and behavioural data on fish locomotion. Values are only indicative (in relative units) and do not correspond to actual data. Optimum speed (the speed with the lowest cost of transport, U_{opt}) and maximum speed (U_{max}) are represented as performance variables, whereas food searching, migrating, predator attack and escape response are represented as behavioural variables. It should be noted that migrating speed is similar to U_{opt}, and predator attack and escape response speeds are slightly lower than U_{max} (see text for explanation).*

behavioural ecology and neuroethology. These two disciplines can also be directly linked to biomechanics. The main aim of behavioural ecology is to find functional (ultimate) explanations for the behaviours of animals; that is, to identify the value of behaviour in terms of lifetime fitness. On the other hand, the scope of neuroethological studies is mainly to investigate the causal (proximate) explanations of behaviour in terms of the internal (physiological) or external (i.e. temperature, light cycle, etc.) factors that cause it.

In terms of solving behavioural questions, both causal and functional questions should be considered at the same time, for they are often complementary (Lorenz, 1960), and asking 'why' questions can often help in solving 'how' questions, and vice versa (Krebs and Davies, 1984). Behavioural biomechanics is within the framework envisioned by Lorenz (1960), because it is not limited to one of the two questions (proximate and ultimate). The purpose of behavioural biomechanics is to explain animal behaviour in relation to physical laws. Such explanations can be functional and/or causal. For example, the functional value of leaping behaviour in dolphins, as an energy-saving mechanism (Au and Weihs, 1980; Blake, 1983b; Chapter 6) (a 'why' question), can be analysed from a biomechanical perspective. Likewise, at least part of the 'how' of the same behaviour (e.g. determination of the forces necessary for leaping out of the water) would also require a biomechanical approach.

Behavioural ecologists will find biomechanics a useful tool for elucidating the functional value of behaviours that are constrained by and/or dependent on mechanics (e.g. locomotion, feeding, tool use, etc.). Also, neuroethologists seeking causal explanations may find biomechanics useful for analysing stereotypic behaviours, reflexes, and the coupling between the neural and the muscular system of animals.

2.3 The 'design → performance → behaviour → fitness' paradigm

The success of any trait, including behaviour, is ultimately measured in terms of fitness. How does the integration of biomechanics and behaviour affect our understanding of how fitness is determined? Arnold (1983) proposed a framework relating morphology (intended as a shorthand for any measurable aspect of structure, physiology or behaviour), performance and fitness. He argued that selection and adaptation can be studied directly by statistically characterizing the relationships between morphology, performance and fitness. Arnold's paradigm was modified by Garland and Losos (1994) to include behaviour as a separate level. They argued that selection acts most directly on what an animal actually does in nature, that is, its behaviour. Their proposed framework consists of morphology (including physiology and biochemistry) → performance → behaviour → fitness.

Arnold's paradigm and its modifications have proved useful because the potential roles of each level of the scheme within an evolutionary context (e.g. see Wainwright and Reilly, 1994) can be specified. Our version (*Figure 3(a)*) of the original paradigm rests heavily on that of Garland and Losos (1994). We consider design (corresponding to 'morphology' *sensu* Arnold) as the first level of this chain. Design is not necessarily equivalent to structure; rather it is a more general concept of structure in relation to function, where this relationship is a complex one that does not imply a tight structure–function matching (Lauder, 1996). This first level is broadly defined as the 'design box' because it includes the characteristics that can determine performance (physiological, biochemical, neural and morphological).

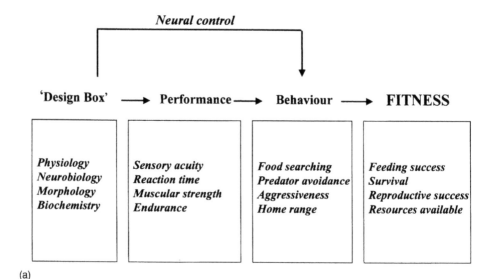

(a)

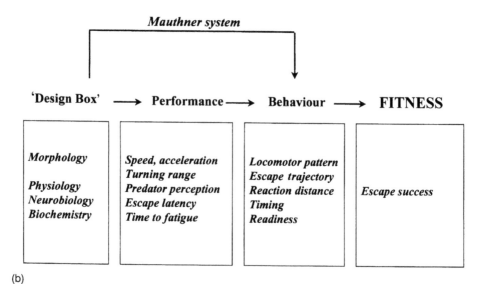

(b)

Figure 3. (a) *Theoretical framework showing some examples of the relationships between design, performance, behaviour and fitness. The design characteristics of each line are mainly related to performance, behaviour and fitness parameter of the same line. For example, physiological characteristics influence sensory acuity (performance level), which in turn is used during food searching (behaviour level). The latter is a fundamental component of feeding success (fitness level). (b) Escape response in fish as an example of the variables involved in the design → performance → behaviour → fitness paradigm. Morphology influences locomotor performance (speed, acceleration, turning), whereas physiological, neurobiological and biochemical characteristics influence predator perception, escape latency and time to fatigue, respectively. Each of these performance types is used in a behavioural aspect of the escape response, resulting in a given escape success (fitness level).*

Performance is intended here as the measured level of given characteristics of a spec-ified action (usually observed in the laboratory) and can be determined by one or more design characteristics. 'Designs' affect performance, for example, in terms of temper-ature tolerance, hypoxia tolerance, sensory acuity, sensory discrimination, metabolic scope, endurance and muscular strength. Biomechanical performance is mainly linked to morphology and muscle mechanics and physiology. In addition, biomechanical performance can be modulated by behaviour (Emerson and Koehl, 1990). In our example (*Figure 3(b)*), speed and acceleration can be related to the morphology of fish. There are, however, cases that show that such a relationship is not so clear-cut; that is, performance cannot necessarily be predicted on the basis of morphological design alone. Knifefish, for example, lack a tail fin and, on the basis of functional morpho-logical arguments, their speed and acceleration performance should be compromised. However, their escape kinematics are peculiar in that escape responses include pitch-and-roll movements that increase 'effective body depth', which mitigates against the absence of a tail blade (Kasapi *et al.*, 1993). In this case, behaviour is probably the main factor modulating speed performance. In our example (*Figure 3(b)*), a number of performance variables contribute to varying extents to the overall escape performance level, which, from a fitness standpoint, corresponds to the likelihood of escaping a predator attack. These performance variables include speed, acceleration performance (typical biomechanical variables), response latency (neural performance, directly related to 'neurobiological design' such as axon diameter, number and type of synapses) and other factors such as acuity of predator perception and turning angles (Domenici and Blake, 1993a).

Behaviour can be seen as a 'filter' between performance and fitness (Garland and Losos, 1994). Behavioural studies are necessary to establish if and to what extent any given performance variable may contribute to fitness. Whereas performance is defined operationally usually by laboratory experiments, and gives a measure of what an animal can do (although it is debatable whether motivational aspects of animals in laboratory settings would be equivalent to those in natural situations; see Blake (1991), for a general discussion), behaviour is what the animal 'chooses' to do in nature, within the context of its performance capabilities. This 'choice' is modu-lated via neural commands (*Figure 3(a)*). In our example (*Figure 3(b)*), maximum speed found in laboratory experiments (at the 'performance' level) may not corre-spond to maximum speed in natural situations, which may vary depending on such factors as the predator type, the prey motivation or the prey's habituation. Lack of data on the fish's actual escape speed in natural situations does not currently allow for a test of these hypotheses. Modulation of behaviour via neural commands (in our example, the Mauthner system, Eaton and Hackett, 1984) implies other charac-teristics of escape behaviour chosen from the pool of those observed at the 'perfor-mance' level, such as response latency and escape trajectories. All of these can contribute to the individual fitness, in the sense that they determine the fish's ability to actually escape a predator's attack.

Unfortunately, the relative contribution to escape success of all the different perfor-mance characteristics expressed at the behavioural level is not known. At first sight, it might be reasonable to assume that maximum speed or minimum latencies would be advantageous for the prey. However, predator–prey interactions are complex; in certain situations escaping too soon or too fast may not pay off. In fact, the speed of escaping fish, measured in the laboratory is not always maximal. Prey speed in fish was often sub-maximal in responses to predator attacks that were not followed by chases (Webb, 1986). Therefore, motivation may play an important role in determining the fast-start performance of escaping prey. Performance is modulated behaviourally to a sub-maximal level that may be optimal in terms of minimization of the energy costs of an escape when the risk is assessed to also be sub-maximal. Considering that energy is often a limiting resource, such behaviour could lead to greater growth and reproductive potential, and lower risk to gape-limited predators. Behaviour can also determine a certain performance level in relation to morphological constraints. Domenici and Blake (1993b) found that large angelfish used sub-maximal speed when escaping at large turning angles, possibly because of flexibility limitations. Such morphological constraints give the animal the potential choice of escaping at small angles and high speed or at large angles and low speed. In this case, therefore, the turning angle chosen behaviourally has a direct effect on distance-derived performance (i.e. speed).

In escape responses, what is 'best' at a given performance level may vary with the prey and/or predator species and with their habitat (e.g. prey in habitats with a high density of hiding places may use different timing from those in habitats with low cover density; Grant and Noakes, 1987). Measuring performance in terms of escape trajectories and reaction distance (variables that have a direct relationship to swimming kinematics; Chapter 14), is even less clear. There is little evidence that one trajectory may be superior to another in predator–prey encounters, and our knowledge of fish escape trajectories is limited to laboratory situations (Domenici and Blake, 1993a). In addition, variability in trajectories (Domenici and Batty, 1997; Domenici and Blake, 1997; Godin, 1997) rather than single trajectories may be the variable linked to different levels of escape performance in different individuals. High trajectory variability can be advantageous for the prey as it prevents predators from learning a fixed pattern of response (Domenici and Blake, 1993a; Godin, 1997). Escape responses constitute a clear case where considerations of behavioural issues can help us determine the adaptive significance of biomechanical performance.

Only a full understanding of the whole scheme (*Figure 1*) can give insights into any one area and allow us to put it within an evolutionary context (Chapter 2). Therefore, to understand the 'fitness' implications of biomechanical performance, it is necessary to know how it is modulated by behaviour. Similarly, from a behavioural standpoint, it is necessary to know the role of biomechanical performance in the significance of a certain movement pattern.

Behaviour not only 'filters' the effect of performance on fitness (by 'choosing' the level of performance), but may also compensate for decrements in ability (e.g. Garland and Losos, 1994; Chapter 5). For example, Garland and Losos (1994) showed that various species of lizards compensate for diminished locomotory abilities either by increasing their reaction distance (e.g. *Anolis* lizard; Rand, 1964) or by using aggressive defence instead of escape responses (e.g. leopard lizard, *Gambelia wislezenii*; Crowley and Pietruszka, 1983).

Although integrating biomechanics and behaviour allows us to study the relationship between performance data and behavioural observations, biomechanics can also be utilized at all the levels of the paradigm (*Figure 3*). It can be useful for measuring the relationship between morphological traits and performance and in identifying the fitness consequences of certain behaviours. For example, biomechanics can be a tool for modelling the energetic advantages of certain locomotor modes

(Chapters 3 and 6). In addition, behavioural biomechanics has great potential in the area of evolution, providing new insights into organismal structure and function (Chapter 2). The increased utilization of biomechanical analyses in the comparative study of behaviour promises to assist in all forms of causal explanation, and hence to enhance greatly our understanding of the evolution of behaviour and its physiological basis (Chapter 2).

This simple scheme (*Figure 3(a)*) can be expanded to include other factors, such as habitat, which can influence behaviour, performance capabilities and morphology (Garland and Losos, 1994). In addition, the resulting fitness of the various morphological, performance and behavioural characteristics is habitat related. For example, body rigidity is advantageous as a drag reduction mechanism for fish, such as tuna, living in a pelagic environment and employing continuous swimming, but it impairs manoeuvrability (Blake *et al.*, 1995), making tuna unsuited for structurally complex environments.

3. Behaviour and the mechanical properties of the environment, certain animal structures and tools

3.1 *Behaviour and the physics of the environment*

Arguably, animal behaviour has evolved as an adaptation to the environment, where environment here is intended to imply both actual habitat as well as lifestyle. At the most basic level, habitat types may be categorized as aquatic, terrestrial or aerial. Within each category, a number of sub-categories can be defined, including structurally complex and open habitats (valid for all primary categories, aquatic, terrestrial and aerial), bottom or surface (aquatic), or composed of different terrain (terrestrial). In addition, some habitats are at the interface between two or more of the primary categories (e.g. intertidal habitats, arboreal habitats), and some animals can be found in a variety of habitats. This simplified view of habitat types introduces the idea that each habitat, with its physical characteristics, requires certain behavioural adaptations. These behavioural adaptations can be seen as hardwired in the neural system in some cases, such as the kinematics necessary for steady locomotion in each primary habitat (swimming, walking, running and flying). Locomotor patterns at all levels of complexity can be subject to behavioural modulation. For example, animals have evolved specific locomotor strategies for different habitats (e.g. see Chapters 3, 4 and 6) whereby the locomotor pattern is adjusted behaviourally to minimize energy consumption.

The advantages of certain locomotor patterns may be twofold, that is, minimum cost of locomotion and minimum sensory background noise, caused by the animal's own motion, which could decrease the signal-to-noise ratio for the perception of predators or prey (Denton and Gray, 1993). The kinematics of steady swimming in herring showed yaw movements in agreement with a physical model predicting kinematics minimizing both the energy expenditure as well as the stimulation that a fish's own movement would give to the receptor organs of the lateral line system (Denton and Gray, 1993; Lighthill, 1993; Rowe *et al.*, 1993). In this case, biomechanics has proved essential for interpreting the evolutionary advantages of a kinematic pattern.

In addition, locomotion may require behavioural modulation when subject to modification of the physical properties of the environment and/or to modification in the relative position of a target (Chapter 4). For example, in terrestrial locomotion on different surfaces, human runners adjust their leg stiffness to accommodate changes in surface stiffness (Ferris *et al.*, 1998). The locomotory output is regulated via feedback at each stride, in such a way as to adjust both the force applied and the trajectory of the limb (Zehr *et al.*, 1998). Such feedback implies behavioural adjustment to the physical properties of the terrain, and it can be refined by learning. Similarly, a continuous feedback is necessary in those movements that can be considered as taxis in the broadest sense, that is, any movement directed towards (positive taxis) or away from (negative taxis) a stimulus. In these cases, the animals have to behaviourally adjust their goal with continuous feedback on the physical properties of the environment and the position of the goal, which changes over time in relation to the animal's own movement (Chapter 16); this may imply learning as well. For example, crayfish trained to reach a shelter along an imposed curved trajectory overcompensated, at the beginning of each walking sequence, for the leg angle drift that such a walking manoeuvre requires (Domenici *et al.*, 1998).

The above examples illustrate situations where locomotion is modulated while taking into account the physical characteristics of the environment. The physics of the environment has, however, implications other than locomotor behaviour. For example, animals living in wave-swept environments must withstand water motion associated with breaking waves. These water motions are accompanied by large hydrodynamic forces that tend to push organisms downstream (drag and acceleration reaction) and to pull them away from the substratum (lift) (Denny, 1994). While these forces are repetitive, animals in the wave-swept zone manage to accomplish all the basic requirements for living, including feeding and reproducing. Acorn barnacles utilize internal fertilization, requiring that the penis of one individual be manoeuvred into the opercular opening of a neighbour. Denny (1994) hypothesized that, until the penis is firmly inserted into the neighbour's operculum, the imposition of flows above a critical velocity may prevent copulation. Therefore, barnacles must have a certain combination of mechanical and behavioural capabilities to be able to copulate in such conditions. In Chapter 10, Ellers shows that behavioural modulation as well as optimal (i.e. maximizing the distance travelled) utilization of the forces of waves is accomplished by donacid clams.

We have shown that basic and complex locomotor behaviours as well as some sexual behaviours have biomechanical implications. Further relatively complex behaviours such as social interactions and the hydrodynamic and aerodynamic advantages of group behaviours can be analysed biomechanically. Animals in a group may experience energetic advantages relative to solitary individuals, by positioning themselves in certain ways to take advantage of the dynamic forces of their neighbours. There are two main mechanisms that can account for this, draughting and active vortex shedding. The first mechanism involves pace-lining and accounts for reduction of drag (e.g. bicycle riders in a group, Kyle, 1979; spiny lobsters migrating, Bill and Herrnkind, 1976; formation swimming in ducks, Fish, 1995). The second involves an animal positioning at certain angles relative to its neighbours, to take advantage of the vorticity actively (i.e. by their oscillating foils) shed by others. This vorticity transports momentum within the fluid and affects its relative velocity, which can lead to a reduction in the required force for propulsion at any given speed (e.g. energetic advantages in group flying, Kshatriya and Blake, 1992; Lissaman and Schollenberger, 1970; in group swimming, Belyayev and Zuyev, 1969; Herskin and Steffensen, 1998; Weihs,

1973b). In addition, the three-dimensional structure of a school was found to be behaviourally modulated in the presence or absence of predators (Abrahams and Colgan, 1985). Abrahams and Colgan (1985) found that fish adopted two alternative strategies of three-dimensional school structure. In the absence of predators, the school structure was consistent with the requirements for hydrodynamic advantage. In the presence of predators, visual monitoring of the predator by all or most members of the school explained the observed school structure. The interaction between grouping and the physical environment has also been investigated for sessile animals (e.g. bryozoans, Grunbaum, 1995, 1997), where hydrodynamic interactions among colony members can affect filtering rates, and flow is affected by defence structures (spines) and colony spatial organization (Grunbaum, 1995, 1997).

The physical properties of the environment can be 'utilized' by animals, as in the case of the burrows of prairie dogs (Vogel et al., 1973). These animals are colonial and live in underground tunnels that they build. The tunnels are simple U-shaped passages with a mound built at either end. How are these underground passages ventilated? The mounds were originally hypothesized to serve as lookout points or flood barriers. However, a closer inspection revealed that they are of different shapes and heights. Vogel et al. (1973) provided an explanation based on the Bernoulli effect. Simply stated, the different heights of the two mounds, associated with the gradient of wind speed from the ground (i.e. the wind is faster at the entrance with the high mound compared with the entrance with the low mound) cause air to be sucked into the entrance with the low mound and out of the entrance with the high mound. This ventilates the tunnel.

3.2 Behaviour and the biomechanics of animal structures such as teeth, beaks, claws, horns and antlers

Certain animal structures (e.g. jaw, teeth, beaks, claws, antlers and horns) can be considered as tools and/or weapons with given physical characteristics, and the analysis is of the types of forces they can exert can bear upon their use (e.g. Chapters 11–13 and 17). These forces can be used for manipulating food, for breaking it open or tearing it (e.g. claws; see Chapter 11) for various feeding modes (suction feeding, ram feeding and manipulation using the jaw apparatus; see Chapter 12), for biting, cutting skin, cracking bones and killing (e.g. teeth in carnivorous mammals; see Chapter 13). Animal structures that can be considered weapons are those involved in fighting, such as antlers and horns (Chapter 17) and claws (Chapter 11).

Kinematic analyses of the movement of these animal structures (e.g. the jaw movements in feeding) can reveal motor correlates of learning behaviour (Galis et al., 1994; Wainwright, 1986). Biomechanical and energetic considerations on the usage of animal structures such as the shell-smashing feeding appendages of stomatopods (Gonodactylus bredini) can be used for investigating prey selection (Full et al., 1989). Further examples include the spoonbill, which uses its bill as a hydrofoil in gathering food particles by swinging it in midwater at the appropriate angle (Weihs and Katzir, 1994), and killer whales, which use underwater tail slaps at a speed around 14 m s^{-1} as weapons to stun schools of herring (Chapter 14; Domenici et al., 2000). In all these cases, knowledge of the kinematics and the dynamics of animal structures aids in understanding their functional role in behaviour. In extinct animals, analysis of the biomechanical properties of fossilized morphological structures has given important insights into behaviour (Chapters 13 and 17).

The characteristics of some animal structures may change over time, as a result of development or of certain cycles such as moulting. It is well known that these variations affect behaviour; for example, moulting birds are more vulnerable than are non-moulting birds. Knowledge of the aerodynamic implications of moulting in birds' flying feathers is fundamental for understanding the related behavioural strategies (Chapter 8).

3.3 *Any morphological adaptation related to behaviour (e.g. sexual, courtship, defence) has an impact on the relationship between the animal and the environment according to physical laws*

There has been debate on whether tail elongation in birds such as the barn swallow evolved mainly as a sexual character (Moller *et al.*, 1998) or as an aerodynamic one (e.g. Balmford *et al.*, 1993; Thomas and Balmford, 1995). Similarly, it has been debated whether the massive antlers of the Irish elk were used for fighting (Kitchener, 1987) or display (Gould, 1974). The rostrum of the swordfish may have evolved as a prey capture device (McGowan, 1988) or as a hydrodynamic adaptation to delay flow separation and thereby reduce drag (Aleev, 1977). Our point here is not to take a position on these issues, rather to suggest that, regardless of the main evolutionary forces that may have led to the development of such structures, they undoubtedly have physical implications for the animals that bear them (which could imply costs or benefits), and such implications should be investigated using an integrative approach including biomechanics and behaviour.

3.4 *Behaviour and the structural and material biomechanics of tools and other objects*

Animals interact with objects (live or inanimate) that may provide food and/or shelter *et cetera*. This involves a variety of behaviours that may involve tools but not necessarily 'tool use' in the strict sense. Tool use can be defined as 'the use of an external object as a functional extension of the body' (McFarland, 1987). However, although there is a behavioural distinction between dropping a shell on rocks to break it and using a rock to break a shell (tool use properly defined), biomechanically both cases can be analysed to obtain information on behavioural strategies. Dropping shells to break them is a common behaviour in some birds (e.g. Corvidae and Laridae). Zach (1979) suggested that crows drop shells from the optimal height (i.e. the height that minimizes the total height summing all the ascents required to break one shell). Further studies (Plowright *et al.*, 1989) incorporating the handling costs have suggested, however, that the birds ascend to a lower height than is predicted by all optimality models considered, using various currencies (i.e. total height currency, net energy gain, time required to break a shell, net rate of energy gain). As birds may not use the optimal height from a simple energetics point of view, certain behavioural modulations of 'dropping behaviour' have been suggested, such as those related to height constraints because of the possibility of losing the shell, or because of the presence of competitors (Plowright *et al.*, 1989). Additional biomechanical modelling of dropping behaviour, including a more complete accounting of all aspects of flying to various dropping heights, including energy expenditure for take-off (not considered in earlier studies) and biomechanical analysis of the breaking point of shell, would certainly help clarify this issue.

An example of the use of objects in relation to the properties of the physical environment is the station-holding behaviour in some fish (Chapter 5). Wakes can be caused by flow around habitat protuberances such as rootwads, submerged branches, coral heads and substratum protuberances such as ripples. These structures can be used by fish to avoid swimming or to reduce costs of swimming, termed flow refuging or entrainment.

Animal building is another area of behaviour that biomechanical analyses can bear upon (see Chapter 18). Sources of building material may be external (as in bird nests) or internal (e.g. spider webs). Biomechanical analyses of both nest and web material can lead to the understanding of building and architectural characteristics (Chapter 18). In the case of nest building, choices are made on materials to use, and knowledge of material properties can be instrumental in understanding behavioural choice.

Other examples include tool use for capturing food, which can imply such precise manoeuvres as in the spitting behaviour of the archer fish (Dill, 1977), specialized tools such as the cactus needles used by the Galapagos woodpecker finches to capture insect larvae from crevices when foraging, and complex tool use involving high-level learning as in sea-otters and in primates. The mechanical principles involved in choosing and manipulating tools have not been studied. Such behaviours would benefit from integrated behavioural and biomechanical analyses.

4. Conclusions

The above examples illustrate the advantages of a biomechanical approach in providing a basis for understanding animal behaviour. Biomechanics can be effective in understanding behaviours pertaining to locomotion (e.g. *predator–prey relationships*; Chapter 14, and *locomotor strategies*, Chapters 3–8) and feeding (from *feeding mechanisms* to *food choice*; Chapters 11–13). Whereas these are the fields in which biomechanics and behaviour are currently often integrated, there are other areas of behaviour that can benefit from biomechanics, as follows.

Habitat selection. Habitat selection is related to morphological design, performance and behaviour. Within this framework, biomechanical studies of functional morphology and performance give insights into the adaptive value of habitat selection (Garland and Losos, 1994). For example, in Chapter 10, Ellers shows how physical principles can aid in predicting the distribution of various invertebrates on sandy beaches.

Orientation. Chapters 9 and 10 show how the physical properties of the flow are fundamental to understanding passive orientation in sandy beach invertebrates and chemosensory mediated orientation in a variety of animals, respectively.

Crypsis and mimicry. Crypsis is often associated with immobility, but this is not always the case. Some animals remain inconspicuous during locomotion. For example, the seahorse's body remains rigid and cryptic while swimming. Blake (1983a) hypothesized that the seahorse's low locomotor efficiency when compared with other undulating fin swimmers may be adaptive in that, while the fins move at a high frequency, such a frequency exceeds the flicker fusion frequency of the eyes of its potential predators. In addition, kinematic analysis of such behaviours as well as those related to locomotor mimicry (e.g. where two or more sympatric butterfly species show similarities in wing motion as a mimetic signal; Srygley, 1999), can provide useful insights into the behavioural mechanisms at their basis.

Communication. Certain aspects of animal sound production (e.g. Pfau and Koch, 1994; Westneat *et al.*, 1993; Young and Bennet-Clark, 1995) and vibratory communication (e.g. Dierkes and Barth, 1995) have been investigated using a biomechanical approach. In terms of chemical communication, the fluid dynamics of the odour plume can be considered a biomechanical tool in analysing chemical signals as outlined in Chapter 9 (for a study on the hydrodynamic effects on chemosensory-mediated predation, see also Weissburg and Zimmer-Faust (1993)).

Aggression. A biomechanical approach is of value in studying aggression especially in terms of the kinematics of display, where kinematic patterns can be useful for identifying the neural correlates of behaviour (Chapters 15 and 16), and mechanics of fighting, where the mechanical properties of antlers and horns can be related to fight strategies (Chapter 17).

Sexual behaviour. A biomechanical approach can be used especially in terms of display, ritualization and physical constraints on fertilization in animals, and for assessing the cost of sexual morphological traits.

Social behaviour. For example, although the most widely accepted view for schooling and flocking behaviour is for antipredator, feeding and reproductory advantages, these behaviours have fluid-mechanical implications.

Animal building. The integration of biomechanics and behaviour permits the analysis of the coevolution of behaviour and material in the spider's web (Chapter 18) and has great potential for studying the relationship between the mechanical properties of building material and material choice.

Learning. Biomechanics can be a valuable tool in providing accurate measurements of the kinematics of certain behaviours to identify and evaluate motor correlates of learning (Galis *et al.*, 1994; Wainwright, 1986).

In conclusion, many of the aspects traditionally treated within animal behaviour can benefit from a biomechanical approach. Biomechanics has a fundamental role at all the levels modelled by the paradigm *Design → performance → behaviour → fitness.* Biomechanics can be useful for measuring the relationship between design and performance, and in identifying the performance limits and the fitness correlates of behaviour. The dynamic relationship between animals and their environment, be it the substrate on which they walk or the objects they manipulate, requires biomechanical analysis. Within this framework, we have illustrated a wealth of examples and areas of behaviour that would benefit from a biomechanical approach. Behavioural biomechanics is a key step for further development of a holistic approach to the study of organisms, and future studies integrating biomechanics and behaviour promise to answer fundamental questions regarding the proximate and/or the ultimate significance of behaviour.

Acknowledgements

We thank George Lauder, Roger Hughes and Paul Webb for valuable comments on this paper.

References

Abrahams, M.V. and Colgan, P. (1985) Risk of predation, hydrodynamic efficiency and their influence on school structure. *Env. Biol. Fish.* **13**: 195–202.

Aleev, Y.G. (1977) *Nekton*. W. Junk, The Hague.

Alexander, R.McN. and Goldspink, T. (1977) *Mechanics and Energetics of Animal Locomotion*. John Wiley, New York.

Arnold, S.J. (1983) Morphology, performance and fitness. *Am. Zool.* **23**: 347–361.

Au, D. and Weihs, D. (1980) At high speed dolphins save energy by leaping. *Nature* **284**: 348–350.

Balmford, A., Thomas, A.L.R., and Jones, I.L. (1993) Aerodynamics and the evolution of long tails in birds. *Nature* **361**: 628–631.

Belyayev, V.V. and Zuyev, G.V. (1969) Hydrodynamic hypothesis of schooling in fishes. *J. Ichthyol.* **9**: 578–584.

Bill, R.G. and Herrnkind, W.F. (1976) Drag reduction by formation movement in spiny lobsters. *Science* **193**: 1146–1148.

Blake, R.W. (1983a) *Fish Locomotion*. Cambridge University Press, Cambridge.

Blake, R.W. (1983b) Energetics of leaping in dolphins and other aquatic animals. *J. Mar Biol. Assoc. UK* **63**: 61–70.

Blake, R.W. (1991) On the efficiency of energy transformations in cells and animals. In: *Efficiency and Economy in Animal Physiology* (ed. R.W. Blake). Cambridge University Press, Cambridge, pp. 13–31.

Blake, R.W., Chatters, L.M. and Domenici, P. (1995) The turning radius of yellowfin tuna (*Thunnus albacares*) in unsteady swimming manoeuvres. *J. Fish Biol.* **46**: 536–538.

Camhi, J.M. (1984) *Neuroethology*. Sinauer, Sunderland, MA.

Craig, C.L., Weber, R.S. and Bernard, G.D. (1996) Evolution of predator–prey systems: spider foraging plasticity in response to the visual ecology of prey. *Am. Nat.* **147**: 205–221.

Crowley, S.R. and Pietruszka, R.D. (1983) Aggressiveness and vocalization in the leopard lizard (*Gambelia wislizenii*): the influence of temperature. *Anim. Behav.* **31**: 1055–1060.

Denny, M.W.(1994) Roles of hydrodynamics in the study of life on wave-swept shores. In: *Ecological Morphology: Integrative Organismal Biology* (eds P.C. Wainwright and S.M. Reilly). University of Chicago Press, Chicago, IL, pp. 169–204.

Denton, E.J. and Gray, J.A.B. (1993) Stimulation of the acoustico-lateralis of clupeid fish by external sources and their own movement. *Philos. Trans. R. Soc. Lond. Ser B* **341**: 113–127.

Dierkes, S. and Barth, F.G. (1995) Mechanism of signal production in the vibratory communication of the wandering spider *Cupiennius getazi* (Arachnida, Araneae). *J. Comp. Physiol. A* **176**: 31–44.

Dill, L.M. (1977) Refraction and the spitting behaviour of the archer fish (*Toxotes chatareus*). *Behav. Ecol. Sociobiol.* **2**: 169–184.

Domenici, P. and Batty, R.S.(1997) The escape behaviour of solitary herring and comparisons with schooling individuals. *Mar. Biol.* **128**(1): 29–38.

Domenici, P. and Blake, R.W.(1993a) Escape trajectories in angelfish (*Pterophyllum eimekei*). *J. Exp. Biol.* **177**: 253–272.

Domenici, P. and Blake, R.W.(1993b) The effect of size on the kinematics and performance of angelfish (*Pterophyllum eimekei*) escape responses. *Can. J. Zool.* **71**: 2319–2326.

Domenici, P. and Blake, R.W. (1997) Fish fast-start kinematics and performance. *J. Exp. Biol.* **200**(8): 1165–1178.

Domenici, P., Jamon, M. and Clarac, F. (1998) Curve walking in freely moving crayfish *Procambarus clarkii. J. Exp. Biol.* **201**: 1315–1329.

Domenici, P., Batty, R.S., Simila, T. and Ogam, E. (2000) Killer whales (*Orcinus orca*) feeding on schooling herring (*Clupea harengus*) using underwater tail slaps: kinematic analyses of field observations. *J. Exp. Biol.* **202**: 283–294.

Eaton, R.C. and Hackett, J.T. (1984) The role of Mauthner cells in fast-starts involving escape in teleost fish. In: *Neural Mechanisms of Startle Behaviour* (ed. R.C. Eaton). Plenum, New York, pp. 213–266.

Emerson, S.B. and Koehl, M.A.R. (1990) The interaction of behavioral and morphological change in the evolution of a novel locomotor type: 'flying' frogs. *Evolution* 44: 1931–1946.

Ferris, D.P., Louie, M. and Farley, C.T. (1998) Running in the real world: adjusting leg stiffness for different surfaces. *Proc. R. Soc. Lond., Ser. B* 265: 989–994.

Fish, F.E. (1995) Kinematics of ducklings swimming in formation: energetic consequences of position. *J. Exp. Zool.* 272: 1–11.

Full, R.J., Caldwell, R.L. and Chow, S. (1989) Smashing energetics: prey selection and feeding efficiency of the stomatopods, *Gonodactylus bredini. Ethology* 82: 134–147.

Galis, F., Terlouw, A. and Osse, J.W.M. (1994) The relation between morphology and behaviour during ontogenetic and evolutionary changes. *J. Fish Biol.* 45 (Suppl. A): 13–26.

Garland, T., Jr and Losos, J.B. (1994) Ecological morphology of locomotor performance in squamate reptiles. In: *Ecological Morphology: Integrative Organismal Biology* (eds P.C. Wainwright and S.M. Reilly). University of Chicago Press, Chicago, IL, pp. 240–302.

Godin, J.G.J. (1997) Evading predators. In: *Behavioural Ecology of Teleost Fishes* (ed. J.G.J. Godin). Oxford University Press, Oxford, pp. 191–236.

Gould, S.J. (1974) The evolutionary significance of 'bizarre' structures: antler size and skull size in the 'Irish elk' *Megaloceros giganteus. Evolution* 28: 191–220.

Grant, J.W.A. and Noakes, D.L.G. (1987) Escape behaviour and use of cover by young-of-the-year brook trout, *Salvelinus fontanalis. Can. J. Fish. Aquat. Sci.* 45: 1390–1396.

Grünbaum, D. (1995) A model of feeding currents in encrusting bryozoans shows interference between zooids within a colony. *J. Theor. Biol.* 174: 409–425.

Grünbaum, D. (1997) Hydromechanical mechanisms of colony organization and cost of defence in an encrusting bryozoan, *Membraniphora membranacea. Limnol. Oceanogr.* 42: 741–752.

Herskin, J. and Steffensen, J.F. (1998) Reduced tail beat frequency and oxygen consumption due to hydrodynamic interactions of schooling sea bass, *Dicentrarchus labrax* L. *J. Fish Biol.* 53: 366–376.

Kasapi, M., Domenici, P., Blake, R.W. and Harper, D.G. (1993) The kinematics and performance of the escape response in the knife fish (*Xenomystus nigri*). *Can. J. Zool.* 71: 189–195.

Kyle, C.R. (1979) Reduction of wind resistance and power output of racing cyclists and runners traveling in groups. *Ergonomics* 22: 387–397.

Kitchener, A. (1987) The fighting behaviour of the extinct Irish elk. *Mod. Geol.* 11: 1–28.

Krebs, J.R. and Davies, N.B. (1984) *An Introduction to Behavioural Ecology.* Blackwell Scientific, Oxford.

Kshatriya, M. and Blake, R.W. (1992) Theoretical model of the optimum flock size of birds flying in formation. *J. Theor. Biol.* 157: 135–174.

Lauder, G.V. (1996) The argument from design. In: *Adaptation* (eds M.R. Rose and G.V. Lauder). Academic Press, San Diego, CA, pp. 55–91.

Lighthill, M.J. (1993) Estimates of pressure differences across the head of a swimming clupeid fish. *Philos. Trans. R. Soc. Lond., Ser. B* 341: 129–140.

Lissaman, P.B.S. and Schollenberger, C.A. (1970) Formation flight in birds. *Science* 168: 1003–1005.

Lorenz, K. (1960) Foreword. In: *The Herring Gull's World,* by N. Tinbergen. Harper and Row, New York.

McFarland, D. (1987) *The Oxford Companion to Animal Behaviour.* Oxford University Press, Oxford.

McGowan, C. (1988) Differential development of the rostrum and mandible of the swordfish (*Xiphias gladius*) during ontogeny and its possible significance. *Can. J. Zool.* 66: 496–503.

Moller, A.P., Barbosa, A., Cuervo, J.J., deLope, F., Merino, S. and Saino, N. (1998) Sexual selection and tail streamers in the barn swallow. *Proc. R. Soc. Lond., Ser. B* 265: 409–414.

Pfau, H.K. and Koch, U.T. (1994) The functional morphology of singing in the cricket. *J. Exp. Biol.* **195**: 147–167.

Plowright, R.C., Fuller, G.A. and Paloheimo, J.E. (1989) Shell dropping by Northern crows: a reexamination of an optimal foraging study. *Can. J. Zool.* **67**: 770–771.

Rand, A.S. (1964) Ecological distribution in anoline lizards of Puerto Rico. *Ecology* **45**: 745–752.

Rayner, J. and Wotton, R. (1991) *Biomechanics in Evolution.* Cambridge University Press, Cambridge.

Rowe, D.M., Denton, E.J. and Batty, R.S. (1993) Head turning in herring and some other fish. *Philos. Trans. R. Soc. Lond., Ser. B* **341**: 141–148.

Srygley, R.B. (1999) Locomotor mimicry in *Heliconius* butterflies: contrast analyses of flight morphology and kinematics. *Philos. Trans. R. Soc. Lond., Ser. B* **354**: 203–211.

Thomas A.R.L. and Balmford, A. (1995) How natural selection shapes birds' tails. *Am. Nat.* **146**: 848–868.

Vogel, S., Ellington, C.P. and Kilgore, D.L. (1973) Wind-induced ventilation of the burrows of the prairie dog *Cynomys ludovicianus. J. Comp. Physiol.* **85**: 1–14.

Wainwright, P.C. (1986) Motor correlates of learning behaviour: feeding on novel prey by pumpkinseed sunfish (*Lepomis gibbosus*). *J. Exp. Biol.* **126**: 237–247.

Wainwright, P.C. and Reilly, S. (1994) *Ecological Morphology: Integrative Organismal Biology.* University of Chicago Press, Chicago, IL.

Webb, P.W. (1986) Effect of body form and response threshold on the vulnerability of four species of teleost prey attacked by largemouth bass (*Micropterus salmoides*). *Can. J. Fish. Aquat. Sci.* **43**: 763–771.

Weihs, D. (1973a) Optimal fish cruising speed. *Nature* **245**: 48–50.

Weihs, D. (1973b) Hydrodynamics of fish schooling. *Nature* **241**: 290–291.

Weihs, D. and Katzir, G. (1994) Bill sweeping in the spoonbill, *Plataea leucornia*—evidence for a hydrodynamic function. *Anim. Behav.* **47**: 649–654.

Weissburg, M.J. and Zimmer-Faust, R.K. (1993) Life and death in moving fluids: hydrodynamic effects on chemosensory-mediated predation. *Ecology* **74**: 1428–1443.

Westneat, M.W., Long, J.H., Hoese, W. and Nowicki, S. (1993) Kinematics of birdsongs: functional correlation of cranial movements and acoustic features in sparrows. *J. Exp. Biol.* **182**: 147–171.

Young, D. and Bennet-Clark, H.C. (1995) The role of the tymbal in cicada sound production. *J. Exp. Biol.* **198**: 1001–1020.

Zach, R. (1979) Shell dropping; decision-making and optimal foraging in Northwestern crows. *Behaviour* **68**: 106–117.

Zehr, E.P., Stein, R.B. and Komiyama, T. (1998) Function of sural nerve reflexes during human walking. *J. Physiol. Lond.* **507**: 305–314.

2

Biomechanics and behaviour: analyzing the mechanistic basis of movement from an evolutionary perspective

George V. Lauder

1. Introduction

All organisms face demands imposed by the physical world and exhibit movements that are dependent on the material properties of the media in which they live. Aquatic life, for example, faces a very different set of physical challenges than organisms on land (Denny, 1993), and such physical constraints may set boundary conditions for the design of organisms in media such as air and water. Furthermore, virtually every aspect of organismal design, including patterns of skeletal organization, movement, sensory biology, prey capture, osmoregulation, and circulatory function, is significantly affected by the environment inhabited by organisms. As movement is the major feature of many behaviours, the behaviour of organisms must also be subject to physical constraints imposed by the environment (see also Chapter 1).

As a consequence, the study of behaviour is inextricably linked with understanding how organisms function, and the mechanical and physiological bases of movement. Indeed, the disciplines of functional anatomy, biomechanics, and physiology provide a critical mechanistic underpinning to our understanding of how behaviour is generated. Without the added dimension of a mechanistic analysis, the study of how behaviours evolve has the potential of becoming a largely phenomenological and descriptive enterprise.

To date, the vast majority of biomechanical and physiological studies have focused on single species and have attempted to understand how behaviour is generated at an intraspecific level. This approach has been exceptionally fruitful, and whole disciplines such as neuroethology have arisen to address the causal generation of behaviour. In addition, as is well demonstrated in this volume, biomechanical approaches have allowed considerable progress in understanding how behaviour is generated within species. Several key model behaviours, such as routine locomotion and rapid escape behaviour, have been

Biomechanics in Animal Behaviour, edited by P. Domenici and R.W. Blake.
© 2000 Taylor & Francis, Oxford.

fruitfully explored using diverse approaches including histochemistry, anatomy, kinematics, force plate measurements, and hydrodynamics (e.g. Domenici and Blake, 1997;
Eaton et al., 1988; Full and Weinstein, 1992; Harper and Blake, 1990; Rome et al., 1993).

A key additional issue, and one that will form the main theme of this chapter, is that
broader, comparative approaches promise to provide new insights into the relationship
between biomechanics and behaviour. It is my contention that it is precisely in the area
of the comparative and historical (evolutionary) analysis of organismal behaviour and
biomechanics that these two areas show the greatest joint potential for providing new
insights into organismal structure and function.

2. Phylogeny and the study of biomechanics and behaviour

The relevance of biomechanical, physiological, and functional anatomical studies for
understanding behaviour is readily apparent by considering a set of traits that might be
involved in generating behaviour (*Table 1*; Lauder, 1991; Lauder and Reilly, 1996).
Although many such arrangements of traits are possible, the one presented in *Table 1*
emphasizes a causal chain of mechanistic associations originating in the nervous
system and extending to the topological arrangement of the peripheral musculoskeletal system. Thus, for example, patterns of neuronal activation drive a sequence of
muscle activity, and the effect of muscle activation on the skeleton is determined in
part by the fiber types activated, the activity level of muscle enzymes and various ion
pumps within these muscle fibers, and the mechanical properties (such as elasticity) of
connective tissue elements within the muscle and at attachments to the skeleton. The
particular behaviour generated as a result of muscle activity further depends on the
pattern of arrangement of skeletal elements. Muscles that attach near a joint have a
different behavioural effect at a fixed level of activation than the same muscle attaching
elsewhere. Rearrangement of joints or muscle lines of action without changing muscle
physiological properties will also result in a completely different behaviour.

Although the term 'behaviour' can be applied to a number of possible organismal
traits, for the purposes of this chapter I will refer to kinematic data on skeletal
movement patterns as behaviours. Although 'behaviour' may refer to a broader array
of organismal traits than just kinematics, kinematic information certainly provides
critical data on the behaviour of organisms.

The traits listed in *Table 1* promise nothing new by themselves, and yet the implications for comparative analyses of biomechanics and behaviour are nontrivial. The most

Table 1. *One set of possible characters
(traits) that serve as mechanistic
underpinnings for the analysis of behaviour.*

Type of character

Musculoskeletal topology: anatomy
Tissue mechanical properties
Physiological properties of muscles
Timing of muscle activation: motor pattern
Neuronal circuit activation

intriguing implications are apparent when we consider possible patterns of evolution among biomechanical, physiological and behavioural characters, and how such traits might differ interspecifically. For example, it is possible, as shown in *Figure 1*, that the origin of novel behaviours in species is associated with new characters arising at several underlying hierarchical levels. In the simplest case, the origin of a new behaviour in species A, for example (*Figure 1*), occurs as a result of underlying changes in both morphology and motor activation of muscles. Hence, in *Figure 1* these traits are shown as undergoing a transformation in state from the primitive condition present at the base of the entire clade to a derived condition present along the branch leading to species A. Furthermore, convergent evolution of the novel behaviour shown by species A, in species F, occurs similarly as the result of changes in both morphology and motor pattern. The convergent acquisition of biomechanical novelties in this clade has thus resulted in two species, A and F, both possessing a similar behaviour.

However, there is no reason to suspect that phylogenetic change need be so clear-cut. Because changes in any one of the hierarchical levels that causally underlie behaviour (*Table 1*) can result in changes in behaviour, phylogenetic patterns to biomechanical traits and behaviour could be rather complex. *Figure 2* shows one possible phylogenetic distribution of such characters. A novel motor pattern (diagonal hatched bar) arises by modification from the plesiomorphic state to char-acterize the clade composed of species F and G. A novelty in musculoskeletal anatomy arises in species F and results in a change in behaviour in that species, whereas retention of the plesiomorphic anatomical condition (i.e. that present at the

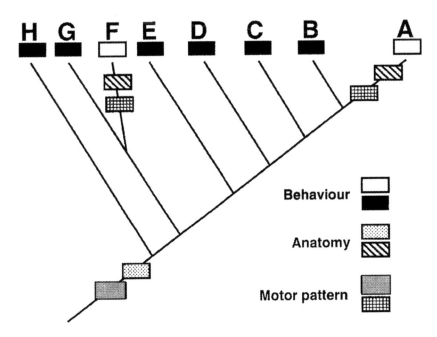

Figure 1. *Phylogenetic pattern showing how motor pattern, anatomical, and behavioural traits may show similar evolutionary histories. The convergently similar behaviours in taxa A and F are generated by similar novelties in anatomy and motor pattern. Only two causal levels from the hierarchy shown in Table 1 are used for clarity, but a similar argument could be made for any of the other levels.*

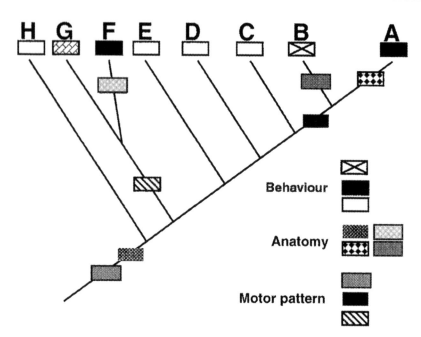

Figure 2. *Phylogenetic pattern showing how muscle activity, anatomical, and behavioural traits may show incongruent evolutionary histories. The convergently similar behaviours in taxa A and F are in this case generated by different combinations of evolutionary novelties in anatomy and motor pattern. Taxa B and G show divergent behaviours because they possess unique combinations of motor pattern and anatomical character states. Other combinations of causal levels from Table 1 could have been selected, but only two levels are shown for clarity.*

base of the clade) in species G interacting with the novel motor pattern produces a new behaviour in G. Further changes in motor pattern and morphology generate changes in behaviour in species A and B, but in this case species A displays a similar (convergent) behaviour to species F. Despite differences in morphology and motor pattern, species A and F have acquired similar behaviours. In other words, these two species have taken different mechanistic paths through the biomechanical hierarchy to arrive at a similar behavioural endpoint.

Although this result as depicted in *Figure 2* may seem unlikely, the simple schematic diagram shown in *Figure 3* illustrates that such relatively complex phylogenetic patterns to behavioural and biomechanical traits may easily be generated biologically. Changes in morphology that arise as a result of alterations in the rate and relative timing of musculoskeletal development (heterochrony, see Alberch, 1980; Fink, 1988; McKinney, 1988), for example, may alter the attachment site of muscles to produce a novel topological configuration relative to outgroup taxa as shown in the upper right panel of *Figure 3*. Alterations in motor pattern from the plesiomorphic condition (muscle M2 becoming activated before M1; vertical hatched bar) generate a novel behaviour. But this same behaviour can also be generated by changes in musculoskeletal arrangement while retaining the primitive motor pattern (*Figure 3*). Fiber type changes, alterations in connective tissue elasticity, or changes in muscle physiological properties such as time to peak tension could also modify the behaviour observed.

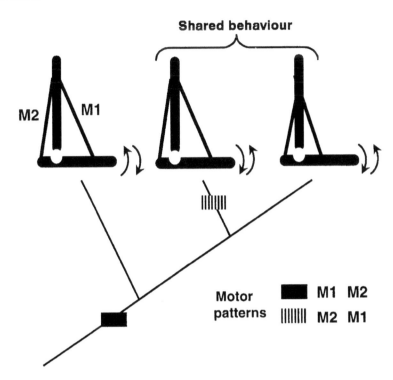

Figure 3. *Schematic diagram to illustrate how a similar behavioural endpoint might be achieved by different combinations of morphology and motor pattern. The derived morphological condition on the right-hand side could occur by changes in the timing of muscle growth during ontogeny. M1 and M2, muscles connecting skeletal elements. The sequence of M1 and M2 to the right of the motor pattern codings shows the order in which the muscles are activated by the nervous system. Arrows show the direction and order of movement of the lower skeletal element (assuming that the vertical element is fixed) as a result of the indicated sequence of muscle activity.*

The general issues raised by a phylogenetic approach to biomechanics and behaviour are ones not typically addressed by intraspecific analyses. How do the causal mechanical systems that underlie behaviour change in evolution? What types of paths through the mechanistic hierarchy are taken by organisms? How correlated are phylogenetic changes in different types of biomechanical traits? These general questions relating to the transformation of behaviour and the origin of behavioural novelty by alteration of biomechanical systems are of key importance for the integrated study of biomechanics and behaviour.

3. Two case studies

The two case studies presented below document the nature of patterns of biomechanical and behavioural transformation in clades of fishes and salamanders. Although neither of these case studies can be considered as by any means complete at present, they do illustrate the considerable complexity that exists in understanding the relationship between biomechanics and behaviour even when relatively simple comparative analyses are undertaken.

3.1 *Chewing behaviour in osteoglossomorph fishes*

The Osteoglossomorpha is a basal clade of teleost fishes characterized by possession of a novel morphological condition: a tongue-bite between the hyoid and the base of the skull (Arratia, 1996; Lauder and Liem, 1983). Most fishes possess two sets of jaws: anterior mandibular jaws, used primarily in the biting and capture of prey, and posterior pharyngeal jaws located near the esophagus and used for subsequent prey maceration and swallowing (Lauder, 1983; Liem, 1970, 1973). Osteoglossomorph fishes possess *three* sets of jaws, with the tongue bite intercalated between the mandibular and pharyngeal jaws. In some taxa such as the mooneye, *Hiodon*, and knifefish, *Notopterus*, the tongue-bite is a formidable structure and has been suggested to function in prey manipulation after capture.

Although the tongue bite has long been known as a morphological feature of osteoglossomorph fishes, little was known about its function in natural situations. However, given the remarkable specialization at the level of musculoskeletal anatomy and the rarity of such a trait within teleost fishes, a reasonable null hypothesis is that such a novel morphological feature might be accompanied by behavioural (kinematic) specialization. Species of osteoglossomorphs possessing a tongue-bite might be expected to exhibit congruent behavioural specialization in a manner similar to the phylogenetic pattern shown in *Figure 1* for taxon A.

Investigation of one species, the knifefish *Notopterus*, confirmed that the tongue-bite consists of novel skeletal specializations that are used after prey capture to disable and puncture prey by the shearing action of teeth on the tongue against those on the base of the skull (Sanford and Lauder, 1989). The effect on the prey of even a few tongue-bites can be dramatic, with prey completely immobilized as a result of perforations and lacerations by tongue teeth. High-speed films were used to study the process of prey capture, use of the tongue-bite, and to quantify all aspects of feeding behaviour. Analysis of the films revealed two distinct intra-oral prey processing behaviours, which we termed 'raking' and 'open-mouth chewing' (Sanford and Lauder, 1989). Measurement of muscle activity patterns in *Notopterus* in comparison with outgroup taxa (*Lepisosteus* and *Amia*) showed that use of the tongue-bite during prey processing involved several novelties in motor pattern. Thus, on the basis of the one species studied, specialization at the hierarchical levels of morphology and motor pattern (*Table 1*) appeared to be causally linked to the presence of novel prey processing behaviours, the phylogenetic pattern depicted for taxon A in *Figure 1*.

Anticipating that other taxa in this clade would show similar kinematic patterns to raking and open-mouth chewing when the tongue-bite was used, we examined two other species, *Osteoglossum* and *Pantodon* (Sanford and Lauder, 1990). The purpose of investigating other species in the osteoglossomorph clade was to test the hypothesis that shared morphological specializations would be congruent with shared behavioural novelties in the pattern depicted for taxon F in *Figure 1*. Surprisingly, statistical analysis showed that the raking and open-mouth chewing behaviours were significantly different in kinematic variables among the taxa despite shared similarities in the tongue-bite morphology (Sanford and Lauder, 1990). Each of the three species studied thus possesses an apomorphic kinematic pattern although sharing a primitive morphological specialization with other members of the osteoglossomorph clade. This must be due to as yet unstudied differences among the species in tissue mechanics or muscle activity patterns.

These results suggest that possession of the structurally specialized tongue-bite has little predictive power at the behavioural level, and that behavioural and derived morphological features may exhibit considerable phylogenetic independence in this clade, similar to the pattern shown in *Figure 2* for taxa A, B, and F.

3.2 *Comparative analysis of prey capture in salamanders*

Methods. The hierarchy of classes of characters shown in *Table 1* and then considered phylogenetically in *Figures 2* and *3* is one dimensional in the sense that individual characters are represented at each level and analyzed discretely by mapping on a phylogeny of relevant taxa. But all levels of the causal biomechanical hierarchy are actually multi-dimensional, as there are many morphological, physiological, mechanical, or behavioural variables that are associated with any given set of species. Indeed, it could be argued reasonably that without a multivariate analysis it is impossible to capture accurately the extent of variation and hence the complexity of the relationship between behaviour and underlying biomechanical variables.

One way to avoid the problems inherent in univariate analyses is to measure multiple variables for each hierarchical category and to conduct a multivariate statistical analysis for each level. *Figure 4* depicts such a scheme in theoretical form, modified from the method outlined by Reilly and Lauder (1992). The behavioural level (*Figure 4*) can serve as an example of this approach. Given five taxa (A–E), locomotor behaviour, for example, of multiple individuals in each species can be quantified by high-speed videos and kinematic measurements made from those video records. Each stride can be quantified by measuring a number of kinematic variables such as angular velocity, stride length, and the maximum angles of limb segment excursions. Together these variables provide a quantitative depiction of the locomotor pattern in one species. By making similar measurements from several species one obtains a data set that captures the interspecific differentiation in locomotor behaviour as quantified using limb kinematics. Such data may then appropriately be analyzed using multi-variate statistical methods such as principal components analysis (PCA) (Chatfield and Collins, 1980; Harris, 1975; Shaffer and Lauder, 1988), where each principal component represents a linear combination of the original variables.

There are a number of advantages of this approach. First, much of the variation in a large data set can be represented in a bivariate plot of components one and two, which capture the majority of variance in the original data. If analysis of additional variation is needed, the other components can be examined also. Second, each principal component is uncorrelated with the others, and so patterns of intercorrelation present among the original univariate variables are accounted for and do not confound interpretation of differences among species. Third, differences among taxa in mean component scores can be assessed statistically using a multivariate analysis of variance (MANOVA, Lauder and Reilly, 1996), which allows quantitative statements about differences among species. In *Figure 4*, for example, taxa whose locations in bivariate PC1–PC2 space are *not* significantly different from each other are encircled, showing that they share common overall characteristics at that level. In *Figure 4*, each species is represented as the mean value of principal components of all the individuals studied, although the PCA and MANOVA are conducted on the entire data set of all individual values.

This procedure is then repeated for other classes of biomechanical traits (*Table 1*) and a multilevel, multivariate analysis is constructed showing how species differ

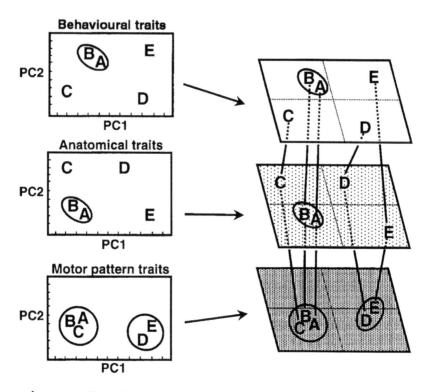

Figure 4. *One possible method of examining the mechanistic and evolutionary relationships among classes of traits that generate behaviour. Panels on the left represent bivariate plots of principal components analyses of behavioural, anatomical, and muscle activity pattern variables measured from individuals in five species. Each letter represents the mean position for that species in principal component space. Taxa enclosed by an ellipse are not significantly different from each other in position. Panels on the right show graphically the pattern of variation among hierarchical levels (Table 1) by connecting the positions of taxa between adjacent planes. Modified from Lauder and Reilly (1996). (See text for further discussion.)*

among different classes of traits (*Figure 4*). To visualize the mappings across levels, it is convenient to connect taxa across the bivariate principal component plots to show how their relative position changes. The relevant statistic is changing patterns of significant relationship to the other taxa, and not simply absolute location on the plots. Thus, in *Figure 4* taxa A and B share behaviour as well as similar patterns of morphology and motor pattern. Both taxa A and B are encircled on all three planes, indicating that they are not significantly different from each other at any level. On the other hand, taxa A and C have similar patterns of muscle activation but divergent morphology and, as a result, show significantly different behaviours.

An analysis of this kind provides an overview of the trajectories through biomechanical space that lead to a given behavioural output, and it allows testing of explicit hypotheses about the mechanical causes of interspecific differences in behaviour. For example, taxa that differ in behaviour but show similar morphology would be predicted to show significant differentiation at other levels. Only by differentiation at the neuronal, physiological, or mechanical levels (*Table 1*) could taxa differ in behaviour but have similar morphology.

Salamander case study. To provide an example of such an analysis, we undertook a comparative investigation of feeding behaviour, head morphology, and jaw muscle function in four clades of salamanders (Lauder and Reilly, 1996). This analysis was built on earlier research on the biomechanics of the feeding mechanism in ambystomatid salamanders (Lauder and Reilly, 1988, 1990; Lauder and Shaffer, 1988; Reilly and Lauder, 1990a,b; Shaffer and Lauder, 1988) and then was extended to include three other clades. The final analysis was composed of a data set consisting of kinematic (behavioural data), morphometric measurements of the head, and muscle activity patterns from species representing four salamander families: Ambystomatidae, *Ambystoma*; Cryptobranchidae, *Cryptobranchus*; Sirenidae, *Siren*; Proteidae, *Necturus*.

High-speed videos (200 frames s^{-1}) of prey capture were obtained for a number of individuals ($n = 2$–10) in each of these species. Patterns of head and jaw movement were digitized frame by frame from these videos, and from those data files statistical variables were constructed. These variables consisted of maximal angular excursions of head elements and the relative timing of head bone movements. This provided the data set for the behavioural level of analysis. Morphometric data were obtained by measuring linear distances on the head to describe head shape. Additional variables relevant to the hydrodynamics of prey capture included variables such as the number of lateral gill openings. Muscle activity patterns were quantified using electromyography. Five homologous muscles were chosen that could be reliably identified across taxa (Lauder and Reilly, 1996) and these muscles represent those that are expected on biomechanical grounds to generate the movements quantified for the behavioural level. An explicit attempt was made to link the muscles studied electromyographically to the kinematic output observed during feeding. Statistical variables extracted from the electromyographic data include the duration of activity, relative onset times, and the total rectified area of electrical activity in each muscle.

Principal components analyses followed by MANOVA were conducted separately on each data set to locate taxa in multivariate space for each set of characters and to identify taxa that are significantly different from each other. The results of this analysis of the feeding mechanism of salamanders are presented diagrammatically in *Figure 5*. As in *Figure 4*, taxa are labeled by a single letter that reflects the mean position of all individuals studied in that species, and taxa that are *not* significantly different from each other are enclosed by an ellipse. The x-axis in each plot represents PC1 scores and the y-axis PC2. The only result of concern in the analysis is which taxa are significantly different from each other in each of the three planes. As PC scores are standardized around means of zero, the crossed dotted lines within each plane indicate values of zero on the x- and y-axes, and axis labels are omitted for clarity.

A number of interesting conclusions about the relationship of biomechanical traits and behaviour in these taxa can be drawn from this analysis. First, it is apparent that there is considerable complexity in the trajectories that taxa take with respect to each of the three hierarchical levels. No simple pattern emerges that provides an overall explanation of how behaviour differs interspecifically and how biomechanical traits have caused behavioural differentiation. Second, taxa such as *Ambystoma* and *Siren* show similar morphology (as they are grouped together and are thus not significantly different from each other in the multivariate space defined by the measured structural variables; *Figure 5*: A, S), but divergent behaviours. This could only result from significant differences in the motor patterns that generate behaviour or in tissue biomechanics. Indeed, these two taxa show the expected significant differences at the motor pattern level.

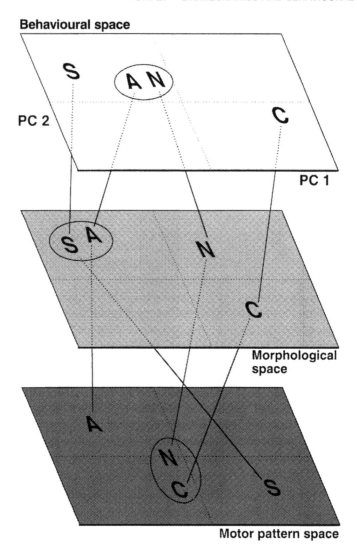

Figure 5. *A comparative analysis of feeding behaviour (jaw kinematics), cranial morphology, and jaw muscle activity patterns in salamanders reveals patterns of biomechanical and behavioural evolution. It should be noted that taxa such as A and S may share similar cranial morphology but possess different motor patterns and hence display divergent behaviours. However, taxa A and N possess similar behaviours but display different morphology and motor pattern. A, Ambystoma; C, Cryptobranchus; N, Necturus; S, Siren. (See text for further discussion.) Modified from Lauder and Reilly (1996).*

Finally, and most interestingly, two of the taxa, *Ambystoma* and *Necturus*, have a common feeding behaviour but achieve this similarity via different mechanistic routes. These two taxa are significantly different from each other in both morphology and motor pattern, and yet do not show significant differences in behaviour. This can occur only if traits at the morphological and motor pattern levels interact mechanistically in such a way as to generate a common behavioural output.

The significance of this result is that it demonstrates clearly that taxa can take different mechanistic trajectories though the biomechanical hierarchy and that such trajectories indicate that the causal underpinnings of phenotypically similar behaviours may be rather different. Mechanistic biology is concerned with the causal explanations, and instances where phenomenological descriptions of behavioural similarity mask underlying mechanistic differences are of particular interest. This result demonstrates the value of biomechanical analysis for the study of behaviour, and also the interplay between behavioural analysis and its causal mechanical basis.

4. Discussion and prospectus

The discipline of animal behaviour has not traditionally involved much of a contribution from the study of organismal biomechanics. Examination of texts such as those by Alcock (1998), Roe and Simpson (1958), and Slater and Halliday (1994) shows that the field of biomechanics has had relatively little impact on the issues seen as central by ethologists. On the other hand, the area of neuroethology is an exciting enterprise integral to the study of ethology. Behavioural biologists have increasingly incorporated data from comparative and mechanistic neuroethology into their texts (Alcock, 1998), and the number of books and papers addressing significant issues in the neural mechanisms underlying behaviour continues to grow (Bradbury and Vehrencamp, 1998; Camhi, 1984; Cohen and Strumwasser, 1985; Cohen et al., 1988; Eaton, 1984; Harris-Warrick and Marder, 1991; Roeder, 1998). Neuroethologists have also embraced comparative and phylogenetic approaches to behaviour (e.g., Bass, 1998; Hoy et al., 1988; Paul, 1981,1991) as evolutionary biologists have themselves rediscovered the value of phylogenetic analyses of behaviour (Brooks and McLennan, 1991; Burghardt and Gittleman, 1990; De Queiroz and Wimberger, 1993; Greene and Burghardt, 1978; Lauder, 1986; Martins, 1996; McLennan, 1994; Wenzel, 1992) after a period of quiescence.

Given the wealth of behavioural analyses in the literature and the continuing influence of neuroethology and phylogenetics on the study of behaviour, it is perhaps surprising that the discipline of biomechanics has not had a greater influence. Many neuroethological studies involve biomechanical concepts, and as behaviour results from the interaction of central neural processing with the structure and function of peripheral tissues, it would seem that the mechanics of those peripheral tissues would be a prime candidate for inclusion in behavioural studies.

The importance of biomechanical studies for the proximate analysis of behaviour is even more clear when detailed studies of the generation of behaviour within individuals are extended into the comparative and phylogenetic arena. As discussed above, explaining how and why species differ in behaviour necessitates understanding the comparative biomechanics of tissues. What are the properties of skeletal, muscular, and connective tissue elements, and how do these properties interact with patterns of neuronal activation to generate behaviour? Patterns of evolution at the different causal levels underlying behaviour are complex and yet critical for understanding the evolution of behaviour, and yet have received far less attention than studies in behavioural ecology, for example (e.g. Krebs and Davies, 1997).

In addition to generating improved understanding of proximate hypotheses for the generation of behaviour, ultimate explanations of both intraspecific and interspecific behavioural differences also benefit from a biomechanical perspective (see Chapter 1).

The increased integration of biomechanical analyses into the comparative study of behaviour promises to assist in all forms of causal explanation, and hence to enhance greatly our understanding of the evolution of behaviour and its physiological basis.

Acknowledgments

I thank Paolo Domenici, Bob Blake, Bob Josephson, Eliot Drucker, Steve Reilly, Alice Gibb, Lara Ferry-Graham, Cheryl Wilga, Gary Gillis, and an anonymous referee for general discussions on the topic of the symposium or comments on the manuscript. Support was provided by NSF grant IBN 98-07012.

References

Alberch, P. (1980) Ontogenesis and morphological diversification. *Am. Zool.* **20**: 653–667.

Alcock, J. (1998) *Animal Behavior: An Evolutionary Approach.* Sinauer, Sunderland, MA.

Arratia, G. (1996) The Jurassic and the early history of teleosts. In: *Mesozoic Fishes— Systematics and Paleoecology* (eds G. Arratia and G. Viohl). Dr. F. Pfeil, Munich, pp. 243–259.

Bass, A.H. (1998) Behavioral and evolutionary neurobiology: a pluralistic approach. *Am. Zool.* **38**: 97–107.

Bradbury, J.W. and Vehrencamp, S.L. (1998) *Principles of Animal Communication.* Sinauer, Sunderland, MA.

Brooks, D.R. and McLennan, D.A. (1991) *Phylogeny, Ecology, and Behavior: A Research Program in Comparative Biology.* University of Chicago Press, Chicago, IL.

Burghardt, G.M. and Gittleman, J.L. (1990) Comparative behavior and phylogenetic analysis. In: *Interpretation and Explanation in the Study of Behavior* (eds M. Bekoff and D. Jamieson). Westview Press, Boulder, CO.

Camhi, J.M. (1984) *Neuroethology: Nerve Cells and the Natural Behavior of Animals.* Sinauer, Sunderland, MA.

Chatfield, C. and Collins, A.J. (1980) *Introduction to Multivariate Statistics.* Chapman and Hall, London.

Cohen, M.J. and Strumwasser, F. (eds) (1985) *Comparative Neurobiology: Modes of Communication in the Nervous System.* John Wiley, New York.

Cohen, A.H., Rossignol, S. and Grillner, S. (eds) (1988) *Neural Control of Rhythmic Movements in Vertebrates.* John Wiley, New York.

Denny, M.W. (1993) *Air and Water. The Biology and Physics of Life's Media.* Princeton University Press, Princeton, NJ.

De Queiroz, A. and Wimberger, P.H. (1993) The usefulness of behavior for phylogeny estimation — levels of homoplasy in behavioral and morphological characters. *Evolution,* **47**: 46–60.

Domenici, P. and Blake, R.W. (1997) The kinematics and performance of fish fast-start swimming. *J. Exp. Biol.* **200**: 1165–1178.

Eaton, R.C. (1984) *Neural Mechanisms of Startle Behavior.* Plenum, New York.

Eaton, R.C., DiDomenico, R. and Nissanov, J. (1988) Flexible body dynamics of the goldfish C-start: implications for reticulospinal command mechanisms. *J. Neurosci.* **8**: 2758–2768.

Fink, W.L. (1988) Phylogenetic analysis and the detection of ontogenetic patterns. In: *Heterochrony in Evolution* (ed. M.L. McKinney). Plenum, New York, pp. 71–91.

Full, R.J. and Weinstein, R.B. (1992) Integrating the physiology, mechanics, and behavior of rapid running ghost crabs: slow and steady doesn't always win the race. *Am. Zool.* **32**: 382–395.

Greene, H. and Burghardt, G.M. (1978) Behavior and phylogeny: constriction in ancient and modern snakes. *Science* **200**: 74–77.

Harper, D.G. and Blake, R.W. (1990) Fast-start performance of rainbow trout *Salmo gairdneri* and northern pike *Esox lucius. J. Exp. Biol.* **150**: 321–342.

Harris, R.J. (1975) *A Primer of Multivariate Statistics*. Academic Press, New York.

Harris-Warrick, R.M. and Marder, E. (1991) Modulation of neural networks for behavior. *Annu. Rev. Neurosci.* **14**: 39–57.

Hoy, R.R., Hoikkala, A. and Kaneshiro, K. (1988) Hawaiian courtship songs: evolutionary innovation in communication signals of *Drosophila. Science* **240**: 217–219.

Krebs, J.R. and Davies, N.B. (1997) *Behavioural Ecology : An Evolutionary Approach*. Blackwell Scientific, Cambridge, MA.

Lauder, G.V. (1983) Functional design and evolution of the pharyngeal jaw apparatus in euteleostean fishes. *Zool. J. Linnean Soc.* **77**: 1–38.

Lauder, G.V. (1986) Homology, analogy, and the evolution of behavior. In: *The Evolution of Behavior* (eds M. Nitecki and J. Kitchell). Oxford University Press, Oxford, pp. 9–40.

Lauder, G.V. (1991) Biomechanics and evolution: integrating physical and historical biology in the study of complex systems. In: *Biomechanics in Evolution* (eds J.M.V. Rayner and R.J. Wootton). Cambridge University Press, Cambridge, pp. 1–19.

Lauder, G.V. and Liem, K.F. (1983) The evolution and interrelationships of the actinopterygian fishes. *Bull. Mus. Comp. Zool.* **150**: 95–197.

Lauder, G.V. and Reilly, S.M. (1988) Functional design of the feeding mechanism in salamanders: causal bases of ontogenetic changes in function. *J. Exp. Biol.* **134**: 219–233.

Lauder, G.V. and Reilly, S.M. (1990) Metamorphosis of the feeding mechanism in tiger salamanders (*Ambystoma tigrinum*): the ontogeny of cranial muscle mass. *J. Zool.* **222**: 59–74.

Lauder, G.V. and Reilly, S.M. (1996) The mechanistic bases of behavioral evolution: a multivariate analysis of musculoskeletal function. In: *Phylogenies and the Comparative Method in Animal Behavior* (ed. E.P. Martins). Oxford University Press, New York, pp. 104–137.

Lauder, G.V. and Shaffer, H.B. (1988) The ontogeny of functional design in tiger salamanders (*Ambystoma tigrinum*): are motor patterns conserved during major morphological transformations? *J. Morphol.* **197**: 249–268.

Liem, K.F. (1970) Comparative functional anatomy of the Nandidae (Pisces: Teleostei). *Fieldiana Zool.* **56**: 1–166.

Liem, K.F. (1973) Evolutionary strategies and morphological innovations: cichlid pharyngeal jaws. *Syst. Zool.* **22**: 425–441.

Martins, E.P. (ed.) (1996) *Phylogenies and the Comparative Method in Animal Behavior*. Oxford University Press, New York.

McKinney, M.L. (1988) *Heterochrony in Evolution*. Plenum, New York.

McLennan, D.A. (1994) A phylogenetic approach to the evolution of fish behavior. *Rev. Fish Biol. Fish.* **4**: 430–460.

Paul, D.H. (1981) Homologies between neuromuscular systems serving different functions in two decapods of different families. *J. Exp. Biol.* **94**: 169–187.

Paul, D.H. (1991) Pedigrees of neurobehavioral circuits: tracing the evolution of novel behaviors by comparing motor patterns, muscles and neurons in members of related taxa. *Brain Behav. Evol.* **38**: 226–239.

Reilly, S.M. and Lauder, G.V. (1990a) The evolution of tetrapod prey transport behavior: kinematic homologies in feeding function. *Evolution* **44**: 1542–1557.

Reilly, S.M. and Lauder, G.V. (1990b) The strike of the tiger salamander: quantitative electromyography and muscle function during prey capture. *J. Comp. Physiol.* **167**: 827–839.

Reilly, S.M. and Lauder, G.V. (1992) Morphology, behavior, and evolution: comparative kinematics of aquatic feeding in salamanders. *Brain Behav. Evol.* **40**: 182–196.

Roe, A. and Simpson, G.G. (eds) (1958) *Behavior and Evolution*. Yale University Press, New Haven, CT.

Roeder, K.D. (1998) *Nerve Cells and Insect Behavior*. Harvard University Press, Cambridge, MA.

Rome, L.C., Swank, D. and Corda, D. (1993) How fish power swimming. *Science* **261**: 340–343.

Sanford, C.P.J. and Lauder, G.V. (1989) Functional morphology of the 'tongue-bite' in the Osteoglossomorph fish *Notopterus*. *J. Morphol.* **202**: 379–408.

Sanford, C.P.J. and Lauder, G.V. (1990) Kinematics of the tongue-bite apparatus in osteoglossomorph fishes. *J. Exp. Biol.* **154**: 137–162.

Shaffer, H.B. and Lauder, G.V. (1988) The ontogeny of functional design: metamorphosis of feeding behavior in the tiger salamander (*Ambystoma tigrinum*). *J. Zool.* **216**: 437–454.

Slater, P.J.B. and Halliday, T.R. (eds) (1994) *Behavior and Evolution*. Cambridge University Press, Cambridge.

Wenzel, J.W. (1992) Behavioral homology and phylogeny. *Annu. Rev. Ecol. Syst.* **23**: 361–381.

3

Intermittent locomotor behaviour alters total work

Randi B. Weinstein and Robert J. Full

1. Introduction

The limits within which behaviour operates can be defined best by integrating biomechanics with physiology. In particular, the mechanistic bases for locomotion allow predictions about the work capacity available for migration, mating, foraging and fleeing. In this chapter, we use the terrestrial locomotion of ghost crabs as an example. The wealth of biomechanical and physiological laboratory data allows for testable predictions of field behaviour involving locomotion. First, we make predictions from biomechanics and energetics based on laboratory studies of steady-state locomotor behaviour. Second, we provide predictions based on laboratory studies of non-steady-state locomotor behaviour. Non-steady-state locomotor behaviour includes intermittent locomotion consisting of repeated bouts of exercise interspersed with pause periods. Third, we test laboratory-based predictions from biomechanics and energetics by reporting measured locomotor behaviour in the field. Data reveal that the total amount of work performed during a behavior depends on the rate of work, the duration of work and the extent to which the work is conducted intermittently. Finally, we examine the generality of the consequences of intermittent locomotor behaviour on total work. We compare species with diverse metabolic responses to locomotion.

2. Predictions from biomechanics and energetics based on laboratory studies of steady-state locomotor behaviour

Laboratory studies of ghost crabs (*Ocypode quadrata*) have examined the relationships between morphology, physiology and performance during steady-state, continuous locomotion. Studies of the biomechanics (Blickhan and Full, 1987), energetics and endurance (Full, 1987) of continuous locomotion by ghost crabs have proved to be important in the search for general principles of locomotion (see reviews by Full and Weinstein (1992) and Herreid and Full (1988).

Biomechanics in Animal Behaviour, edited by P. Domenici and R.W. Blake.

2.1 *Biomechanics*

The steady-state biomechanics and kinematics of terrestrial locomotion by ghost crabs have been characterized on miniature force platforms and motorized treadmills. At slow speeds, less than 0.4 m s^{-1}, 30 g ghost crabs use a walking gait (Blickhan and Full, 1987). In a walking gait, the potential energy of the center of mass is out of phase with its kinetic energy and is analogous to an inverted pendulum. The maximum energy exchange (transfer between kinetic and potential energy) is achieved at 0.2 m s^{-1}, a slow walking speed (*Figure 1(a)*). At speeds greater than 0.4 m s^{-1}, ghost crabs trot or run slowly (Blickhan and Full, 1987). In a trot, the potential and kinetic energy of the center of mass are in phase and the energy recovery from pendulum-like exchange is reduced. To maximize energy exchange, ghost crabs should move continuously at a speed of approximately 0.2 m s^{-1}.

Stride frequency increases linearly with exercise speed in walking and trotting gaits and it becomes independent of speed at exercise speeds greater than 0.8 m s^{-1}. At these fast speeds, ghost crabs increase stride length and have an aerial phase (Blickhan and Full, 1987) and the gait is called a gallop. Relative peak strain in the exoskeleton during a gallop increases significantly over that observed in a walk or trot (*Figure 1(b)*; Blickhan et al., 1993). To minimize the relative peak strain in the exoskeleton, ghost crabs should move continuously at speeds less than 0.6 m s^{-1}.

2.2 *Energetics*

During continuous terrestrial locomotion, steady-state oxygen consumption increases linearly with speed until a maximal rate of oxygen consumption is attained (*Figure 2(a)*). The slowest speed that elicits the maximal rate of oxygen consumption is the maximum aerobic speed (MAS; John-Alder and Bennett, 1981). At speeds below the MAS (i.e. < 0.2 m s^{-1} for 30 g ghost crabs), the energy required for sustained, constant speed locomotion is supplied by aerobic ATP production. Ghost crabs can increase their rates of oxygen consumption as much as 12-fold greater than resting rates (Full and Herreid, 1983). Their aerobic factorial scope is comparable with that found in exercising mammals (five- to 15-fold; Taylor et al., 1980). Accelerated glycolysis in ghost crabs contributes little to energy production during steady-state, submaximal exercise, even at speeds that elicit 70–90% of maximal rates of oxygen consumption (Full, 1987). At speeds above the MAS (i.e. > 0.2 m s^{-1}), ghost crabs rely heavily on nonaerobic sources (*Figure 2(a)*). For these experiments, crabs were exercised on a treadmill and then whole animals were frozen in liquid nitrogen. Lactate and arginine phosphate concentrations in walking legs were assayed spectrophotometrically. Crabs exercising at the walk–trot transition (i.e. 0.4 m s^{-1}) show large increases in muscle lactate and a significant depletion of the high-energy phosphate, arginine phosphate

Figure 1. *Biomechanics of continuous locomotion by the ghost crab. (a) Energy recovery as a function of speed and gait (Blickhan and Full, 1987). (b) Peak strain in the exoskeleton as a function of speed and gait (Blickhan et al., 1993). It should be noted that the gait transitions are shifted to lower speeds because the individuals in this study were of smaller mass. (c) Preferred speed of free locomotion by ghost crabs obtained on a track (Blickhan and Full, 1987). The preferred speed distribution attained a maximum in the middle of the walking gait where mechanical energy exchange is maximal and strain is minimized. It should be noted that speeds above the maximum aerobic speed (>0.2 m s^{-1}) were frequent.*

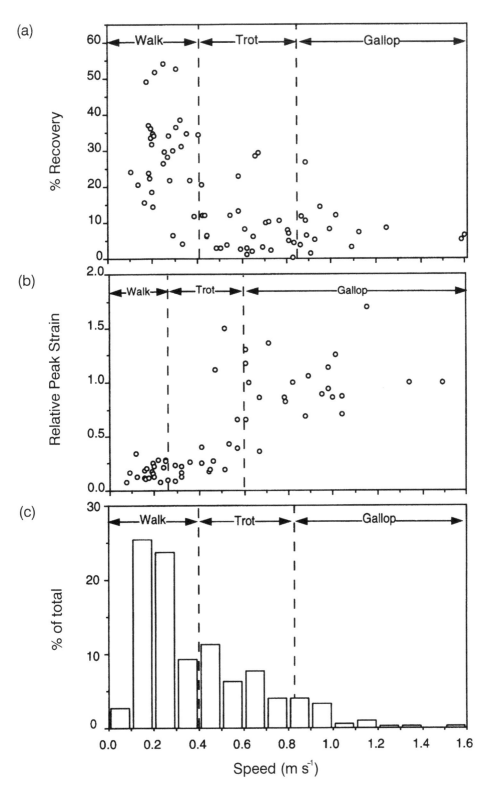

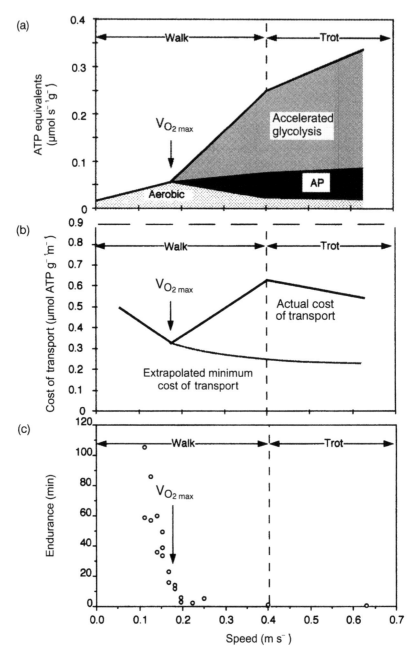

Figure 2. *Energetics and endurance of continuous locomotion by the ghost crab. (a) Aerobic, high-energy phosphate (AP) and glycolytic contributions to walking and trotting in ghost crabs (Full and Prestwich, 1986). Aerobic contributions decline at speeds above the maximal oxygen consumption (V_{O_2}) because oxygen uptake does not attain a steady state before fatigue. (b) Total cost of locomotion as a function of speed. The total cost includes high-energy phosphate (AP) and glycolytic contributions. The total cost of locomotion is minimized at the maximum aerobic speed. (c) Endurance or time to fatigue as a function of speed (Full, 1987). Endurance decreases to low values at speeds above the maximum aerobic speed. From Full and Weinstein (1992).*

(Full and Prestwich, 1986). The rates of lactate accumulation and arginine phosphate depletion are even faster at speeds that fall in the middle of the trot (i.e. 0.6 m s^{-1}). The contribution of ATP from aerobic metabolism may actually decrease at these high speeds because the crabs fatigue before attaining their maximal rate of oxygen consumption. The total energy utilization rate appears to increase curvilinearly as speed is increased (Full and Prestwich, 1986). To maximize the reliance on aerobic metabolism and minimize the contribution of nonaerobic energy sources, ghost crabs should move continuously at speeds of 0.2 m s^{-1} (i.e. the MAS).

The metabolic cost to travel a given distance (i.e. the cost of transport) can be expressed as an aerobic or total cost of transport. The aerobic cost of transport, which is calculated from the amount of oxygen consumed per distance, decreases and approaches the minimum cost of locomotion (i.e. C_{min}) as exercise speed increases to approach the MAS. The C_{min} is typically restricted to aerobically supported speed ranges (Heglund et al., 1982; Taylor et al., 1970). This speed range is narrow for ghost crabs and most other ectotherms relative to endotherms of the same body mass. If the total cost of transport, which is calculated from the total ATP consumed per distance and includes both aerobic and nonaerobic energy sources, is considered for ghost crabs, then an energetic minimum appears to be attained in the middle of the walking gait (i.e. at 0.2 m s^{-1}; *Figure 2(b)*). To minimize the aerobic cost of transport, ghost crabs should move continuously at a speed equal to or greater than 0.2 m s^{-1}, provided that the exercise speed is aerobically sustainable. To minimize the total cost of transport, ghost crabs should move continuously at a speed of 0.2 m s^{-1}.

2.3 *Endurance*

Exercise at speeds below the MAS has been termed 'sustainable' because it can be maintained by aerobic metabolism. Exercise at speeds above the MAS requires supplemental energy provided by nonaerobic sources and is classified as 'nonsustainable', as it rapidly leads to exhaustion. Indeed, ghost crab endurance declines significantly at speeds above the MAS (*Figure 2(c)*; Full, 1987). The MAS determined in the laboratory has been used to predict the speed of animal locomotion in the field, leading to the general prediction that under natural conditions, animals travel at or below the MAS (Gleeson, 1979; Hertz et al., 1988; John-Alder et al., 1983). Using this criterion, ghost crabs should move continuously at speeds equal to or less than their maximum aerobic speed (0.2 m s^{-1}) to maximize endurance.

3. Predictions based on laboratory studies of non-steady-state locomotor behaviour

Although continuous, steady-state exercise has played an essential role in defining the limits of locomotor performance, steady-state conditions represent an artificial situation for many animals and the systems that support their activity. Few animals move continuously. Instead, most animals move intermittently – starting and stopping frequently. Energy demand, oxygen transport, and biomechanics of musculoskeletal systems are well documented during steady-state, constant speed locomotion. However, physiological and biomechanical systems must function under non-steady-state, transient conditions during intermittent activity (*Figure 3*). Therefore, we must re-evaluate the capacity of biomechanical and physiological systems with respect to

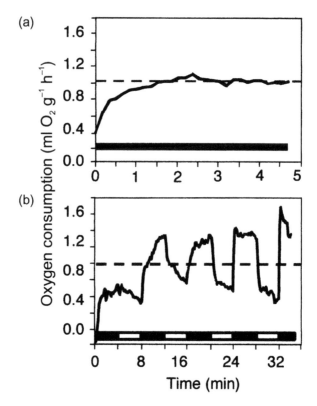

Figure 3. *Oxygen uptake kinetics of ghost crabs as a function of time for continuous versus intermittent locomotion. (a) Continuous locomotion at a constant speed of 0.15 m s⁻¹. (b) Intermittent locomotion at a constant average speed of 0.15 m s⁻¹. Exercise periods were conducted at speeds of 0.30 m s⁻¹ and interspersed with pause periods of zero speed. Protocols are shown with filled bars representing exercise and open bars pause periods. Modified from Weinstein and Full (1992). The average rate of oxygen consumption is shown by the dashed line.*

repeated transitions to advance our understanding of systems design, function, and control. Furthermore, we should consider intermittent locomotor behaviour when we formulate predictions for performance in the field.

Intermittent exercise is defined by three variables: work duration, work rate, and pause duration. Locomotor speed is an indicator of the amount of mechanical work done by animals such as ghost crabs moving on land. The rate of mechanical energy required to move the center of mass increases linearly with speed (Blickhan and Full, 1987). Therefore, as a ghost crab moves faster it does more mechanical work. When a ghost crab travels at a constant work rate (e.g. at a constant speed), the distance it travels, calculated by integrating the speed function, increases linearly (*Figure 4(a)*). However, if the crab moves intermittently, alternating a brief work period with brief pause periods, the instantaneous work rate varies over time and the amount of work done (e.g. distance traveled) increases in a step-like manner (*Figure 4(b)*). The average work rate of a complete work–pause cycle is calculated from the absolute work rate during the work period and the duty cycle (per cent of complete work–pause cycle the animal spends moving). Thus, at the end of a complete work–pause cycle, the total

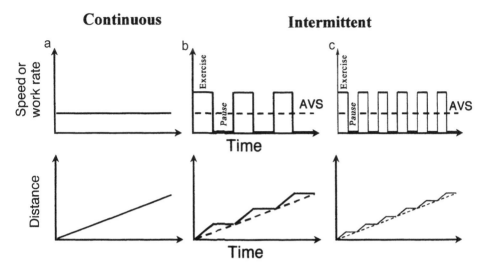

Figure 4. *Comparison of continuous versus intermittent work. (a) Instantaneous work rate (or speed) and work (or distance) versus time for continuous locomotion. (b) Instantaneous work rate (or speed) and work (or distance) versus time for intermittent locomotion. The dashed line represents the same average continuous work rate or speed as shown in (a). (c) Instantaneous speed and distance versus time for intermittent locomotion at a shorter exercise and pause duration than in (b). Exercise and pause duration of intermittent locomotion can be varied while maintaining the same average speed or work rate. AVS, average speed.*

work done during intermittent locomotion is the same as during continuous loco-motion at the same average work rate. Intermittent protocols that differ in exercise duration, pause duration or instantaneous work rate can still lead to the same average work rate (*Figure 4(c)*).

3.1 *Distance capacity*

In cases where ghost crabs perform the work at the same average rate, we might predict similar endurance capacity. We find distance capacity, defined as the total distance traveled before fatigue (Weinstein and Full, 1992), to be a more useful measure of endurance capacity for intermittent exercise, as endurance usually implies continuous activity. Despite the same average work rates for continuous and intermittent movement, the frequent dynamic adjustments that characterize intermittent loco-motion can alter distance capacity. When ghost crabs move intermittently, by alter-nating brief periods of exercise at a speed greater than the MAS with brief pauses, they can travel as much as two to five times farther before they fatigue than if they move continuously at the same submaximal average speed (*Figure 5*; Weinstein and Full, 1992, 1998). Alternatively, under different exercise and pause intervals, the distance capacity for intermittent locomotion can be reduced to a tenth that of continuous locomotion at the same average speed (Weinstein and Full, 1992). Ghost crabs moving at slow, 'sustainable' speeds (i.e. below the MAS) should be able to travel long distances before fatigue whether they move continuously or intermittently. Faster moving crabs should move intermittently to maximize distance capacity although not all movement–pause intervals will result in improved performance.

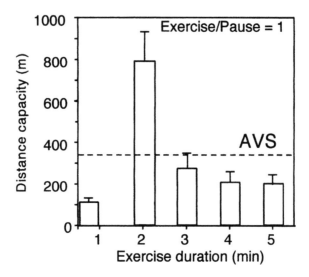

Figure 5. *Distance capacity as a function of exercise and pause duration for the ghost crab. Distance capacity is defined as the distance traveled before fatigue. The ratio of exercise to pause duration was kept at one. Values are means ± SE. All protocols yielded distance capacities that were significantly different from those for continuous locomotion at the same average speed (AVS) except for the 3 min exercise duration experiment. From Weinstein and Full (1992).*

3.2 *Energetics*

Predictions regarding the supply of ATP to working muscle during steady-state exercise do not predict the supply of ATP during intermittent exercise, which is characterized by frequent transitions from rest to exercise and vice versa (*Figure 3*). Whereas ATP for steady-state submaximal continuous exercise is derived primarily from oxidative metabolism, a significant non-negligible fraction of the total ATP must be derived from anaerobic metabolism during both submaximal and supramaximal intermittent exercise. If the exercise duration is shorter than the time to reach and sustain a steady-state rate of oxygen consumption, then ATP must be generated by nonoxidative pathways to meet the increased ATP demand. However, as short exercise bouts are alternated with brief pauses, there is a potential for at least partial recovery from metabolic fatigue.

The average aerobic cost of intermittent exercise is determined by integrating segments of the intermittent exercise records that contain at least one complete exercise–pause cycle in which the sum of the increase and decrease in oxygen consumption are within a constant percentage of the average oxygen consumption of the animal (Weinstein and Full, 1992). Whereas the average aerobic cost of intermittent locomotion is similar to or higher than the cost of continuous exercise at the same average speed (Weinstein and Full, 1992, 1998), the total metabolic cost of intermittent locomotion has yet to be addressed. As ghost crab muscles have elevated lactate and depleted high-energy phosphate stores at the end of a bout of intermittent exercise (Weinstein and Full 1992, 1998), the total metabolic cost of intermittent exercise should include both aerobic and anaerobic energy production. If anaerobic costs are added to aerobic costs, the total metabolic cost of intermittent exercise is likely to be even greater than for continuous locomotion at the same average speed. Elevated muscle lactate in intermittently exercising Christmas Island red crabs (*Gecarcoidea natalis*) also supports an increased total cost for intermittent locomotion (Adamczewska and Morris, 1998). To minimize the total metabolic cost of inter-mittent locomotion per unit time, ghost crabs should move at submaximal speeds during the exercise intervals. Pauses should be adequate for at least partial recovery from fatigue incurred during the previous exercise period. In addition, the movement durations of fast-moving stressed crabs should be shorter than those of crabs moving at slow speeds to minimize fatigue associated with movement at rapid speeds.

In the laboratory, changes in ghost crab leg muscle metabolites are correlated with changes in whole-animal intermittent locomotor performance (*Figure 6*). Alternating 120 s of exercise periods with 120 s pause periods increases distance capacity by two-fold compared with continuous locomotion at the same average speed (Weinstein and Full, 1992). During a 120 s pause period, there is net clearance of lactate from leg

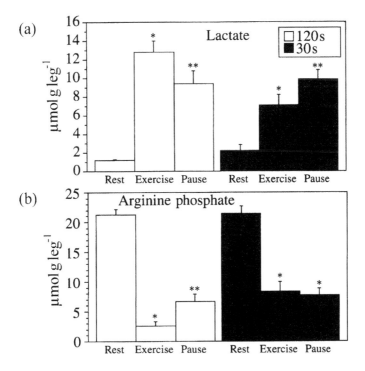

Figure 6. *Metabolite changes in the ghost crab leg in response to intermittent locomotion. (a) Lactate for two intermittent exercise protocols during rest, at the end of exercise and at the end of a pause period. (b) Arginine phosphate for two intermittent exercise protocols. Open bars represent a 30 s exercise, 30 s pause protocol, which results in a decrease in distance capacity compared with continuous locomotion at the same average speed. Closed bars represent a 120 s exercise, 120 s pause protocol, which results in an increase in distance capacity compared with continuous locomotion at the same average speed. A single asterisk indicates a difference from rest. A double asterisk indicates a difference from rest and from the exercise group. The exercise speed was 160% MAS during exercise (0.3 m s⁻¹) for both protocols. The muscle samples were frozen at the end of the fourth exercise or pause period. From Weinstein and Full (1992; the 120 s data) and Full and Weinstein (1992; the 30 s data).*

muscle and net resynthesis of arginine phosphate (Weinstein and Full, 1992). In contrast, intermittent exercise with 30 s exercise bouts alternated with 30 s pauses resulted in a decrease in distance capacity, no net clearance of lactate from leg muscle and no resynthesis of arginine phosphate during the pause period (Full and Weinstein, 1992).

4. Testing laboratory-based predictions from mechanics and energetics by measuring locomotor behaviour in the field

Laboratory studies of ghost crab biomechanics and energetics defined performance limits for continuous and intermittent locomotion and generated specific predictions for locomotor behaviour in the field. To determine where ghost crabs naturally operate within these performance limits, an IR videotaping system and focal animal sampling were used to observe ghost crabs in the field (Weinstein, 1995).

As predicted by laboratory studies of continuous locomotor performance limits, ghost crabs generally moved at speeds below their MAS. During routine foraging behaviour, voluntarily active crabs move at a mean instantaneous velocity of 0.08 m s^{-1}, as measured by three-dimensional motion analysis (Weinstein, 1995). The mean stride frequency of voluntarily active focal animals (2.1 strides s^{-1}) is equivalent to a velocity of about 0.1 m s^{-1} (Blickhan and Full, 1987). Therefore, most of the time, voluntarily active ghost crabs move at speeds below their MAS of approximately 0.2 m s^{-1} (Full, 1987). Mechanically, these voluntarily active ghost crabs primarily use a walking gait, as the mean stride frequency is less than the stride frequency at the walk–run transition (about 4 strides s^{-1}; Blickhan and Full, 1987). The metabolic energy for these slow speeds is probably supplied by aerobic metabolism and although lactate concentrations were not measured in the field, they are likely to be close to resting levels (Full, 1987). A ghost crab moving continuously at 0.08 m s^{-1}, a speed that is 46% of its MAS, will have an endurance capacity of more than 2 h (body temperature 24°C; Full, 1987).

Intermittent locomotion is not likely to alter the voluntarily active ghost crab's distance capacity at the slow speeds measured in the field. The mean movement duration is 11 s and the mean pause duration is 23 s (Weinstein, 1995). If the ghost crab moves intermittently, alternating 11 s of movement at 0.08 m s^{-1} with 23 s pause periods, the average speed is only 0.03 m s^{-1} (15% MAS). The endurance capacity for continuous exercise at 0.03 cm s^{-1} is greater than 2 h (Full, 1987). Starting and stopping frequently may increase the metabolic cost of movement at these slow speeds, but is not likely to constrain the ghost crab's distance capacity.

Stressed crabs, such as those escaping from predators or incoming waves and engaging in intra-specific aggression, move at speeds greater than their MAS. Stressed crabs move with a mean instantaneous velocity of 0.83 m s^{-1}, as determined by three-dimensional motion analysis (Weinstein, 1995). Mechanically, stressed crabs move at speeds that correspond to a running gait, as the stride frequency estimated from the mean instantaneous velocity is greater than the stride frequency at the walk–run transition (Blickhan and Full, 1987). The rapid movements of the stressed crabs cannot be supported aerobically and require additional energy from nonoxidative energy sources. Lactate levels of these stressed crabs are likely to be elevated above resting levels (Full, 1987). Endurance capacity for continuous exercise at 0.83 m s^{-1} is less than 10 s, corresponding to a distance of less than 2 m (Full, 1987; Full and Prestwich, 1986). Despite the low endurance capacity for continuous exercise at the mean instantaneous velocities measured in the field, the stressed crabs moved over distances greater than 2 m, yet they did not appear to be completely fatigued (i.e. subsequent movements were observed following brief pauses; Weinstein, 1995). Stressed ghost crabs behave differently from voluntarily active crabs. The stressed crabs alternate 1–2 s movement periods with 8 s pause periods, resulting in an average speed of only 0.13 m s^{-1} (65% MAS). Endurance capacity for continuous locomotion at 0.13 m s^{-1} is approximately 55 min, corresponding to a total distance of 410 m. The greater distance capacity observed in the field, compared with the predicted distance capacity based on continuous locomotion, suggests that the performance limits of the stressed crabs are increased as a result of intermittent movement. The long pause period duration, relative to the short movement duration, may allow the stressed crabs to partially recover from their brief, high-intensity movements and may explain their increased performance capacity.

5. Generality of the consequences of intermittent locomotor behaviour in other ectotherms

Because oxygen transport kinetics appears to play an important role in determining the effect of intermittent locomotion, we selected species that differ in their oxygen uptake kinetics (*Table 1*). We chose the American cockroach, *Periplaneta americana*, because it possesses very rapid onset kinetics and can increase oxygen consumption 20-fold over resting levels (Herreid and Full, 1984). Furthermore, *P. americana* does not accumulate significant anaerobic end-products such as lactate or α-glycerophosphate at speeds exceeding the MAS (Full and Min, 1990). We hypothesized that intermittent locomotion should have little or no effect on distance capacity for highly aerobic species with rapid kinetics such as *P. americana*. To ensure that our findings are not specific to insects because of some unique character such as tracheae, we also examined several other species with differing oxygen uptake kinetics, including the hissing cockroach, *Gromphadorhina portentosa*, which does accumulate a significant amount of lactate during exercise at speeds greater than the MAS (Lee *et al.*, 1993). We hypothesized that intermittent locomotion in *G. portentosa* should alter distance capacity under particular exercise–pause regimes just as it does in other invertebrate species with moderate oxygen uptake kinetics, such as the ghost crab, *O. quadrata*, and in some lower vertebrates such as the frog-eyed gecko, *Teratoscincus przewalskii* (Autumn *et al.*, 1994; Weinstein and Full, 1999). We also examined the effect of intermittent locomotion in the highly anaerobic fiddler crab, *Uca pugilator*, which relies heavily on anaerobic metabolism to move at the slowest walking speeds (Full and Herreid, 1984). We hypothesized that intermittent locomotion should have the greatest effect on *U. pugilator*, as it could benefit most by recovering during frequent pauses.

5.1 *Fast oxygen uptake kinetics (American cockroaches)*

As we predicted, intermittent locomotion did not increase distance capacity in *P. americana* (*Table 1*; Lee *et al.*, 1992). The distance capacity for *P. americana* exercising intermittently did not differ significantly from that for continuous exercise at the same average speed when the exercise speed was 150% MAS, the exercise–pause ratio was 1.0 (duty cycle 50%), and the exercise duration was 15 or 30 s (*Table 2*). However, intermittent exercise resulted in a 25% decrease in distance capacity compared with continuous exercise at the same average speed when the exercise speed was 143%

Table 1. Comparison of continuous and intermittent locomotor performance in species with different oxygen uptake kinetics

Species	$t_{1/2on}$ (s)	Factorial aerobic scope (V_{O2max}/V_{O2rest})	Maximum % change in distance capacity relative to continuous exercise at same average speed
American cockroach (*P. americana*[a])	<30	29	−25
Hissing cockroach (*G. portentosa*[b])	60	16	+92
Ghost crab (*O. quadrata*[c])	29	10	+119
Gecko (*T. przewalskii*[d])	83	5	+67
Fiddler crab (*U. pugilator*[e])	120	5	−65

$t_{1/2on}$, time to reach 50% of the steady-state rate of oxygen consumption; V_{O2max}, maximal rate of oxygen consumption, V_{O2rest}, resting rate of oxygen consumption. [a]Lee *et al.* (1992); [b]Lee *et al.* (1993); [c]Weinstein and Full (1992); [d]Weinstein and Full (1999); [e]Van Laarhoven *et al.* (1993).

Table 2. Comparison of distance capacity for continuous and intermittent exercise at the same average speed in several ectothermic species

Species	Exercise duration (s)	Pause duration (s)	Average speed (% MAS)	Absolute speed (% MAS)	Maximum % change in distance capacity relative to continuous exercise at same average speed
American cockroach (*P. americana*[a])	15	15	75	150	NS
	30	30	75	150	NS
	10	5	95	143	−25
Hissing cockroach (*G. portentosa*[b])	5	5	90	180	+92
	10	5	90	135	+72
	30	15	90	135	+59
	15	15	90	180	−15
Ghost crab (*O. quadrata*[c])	120	120	83	166	+119
	30	30	83	166	−71
Gecko (*T. przewalskii*[d])	15	30	90	270	+67
	30	30	90	180	NS
	120	120	90	180	NS
Fiddler crab (*U. pugilator*[e])	9	3	85	113	NS
	10	10	85	170	−65

MAS, maximum aerobic speed; NS, not significantly different from continuous exercise at the same average speed. [a]Lee *et al.* (1992); [b]Lee *et al.* (1993); [c]Weinstein and Full (1992); [d]Weinstein and Full (1999); [e]Van Laarhoven *et al.* (1993).

MAS, the exercise duration was 10 s and the pause duration was 5 s (duty cycle 66%). The average aerobic cost of intermittent exercise was not significantly different from the corresponding values for continuous exercise at the same average speed for any of the intermittent exercise protocols.

5.2 *Moderate oxygen uptake kinetics (hissing cockroaches, ghost crabs, and frog-eyed geckos)*

Moving intermittently increased distance capacity in species with moderate oxygen uptake kinetics, including the ghost crab (*Table 1*; Weinstein and Full, 1992). For *G. portentosa*, three intermittent exercise protocols increased distance capacity and one decreased distance capacity compared with continuous exercise at the same average speed (90% MAS; Lee *et al.*, 1993). When the exercise speed was 180% MAS, the exercise–pause ratio was 1.0 (duty cycle 50%), and the exercise duration was 5 s; intermittent exercise increased distance capacity by 92% compared with continuous loco-motion at the same average speed (*Table 2*). Alternating 30 s of exercise at 135% MAS with 15 s pauses, or 10 s of exercise at 135% MAS with 5 s pauses (duty cycle 66%), increased distance capacity by 59–72% compared with continuous locomotion at the same average speed. Intermittent exercise decreased distance capacity by 15% compared with continuous exercise at the same average speed when the exercise speed was 180% MAS, the exercise duration was 15 s, and the pause duration was 15 s (duty cycle 50%). The average aerobic cost of intermittent exercise was not significantly different from the corresponding values for continuous exercise at the same average speed for any of the intermittent exercise protocols.

At an absolute exercise speed of 270% MAS, *T. przewalskii* exercising intermittently with a 15 s exercise duration and a 30 s pause duration (duty cycle 33%) exhibited an 11-fold increase in distance capacity compared with lizards exercised continuously at the same absolute speed (270% MAS) and a 67% increase in distance capacity compared with geckos exercised continuously at the same average speed (90% MAS; *Table 1*; Weinstein and Full, 1999). At an exercise speed of 180% MAS, geckos alter-nating 30 s exercise periods with 30 s pause periods or 120 s exercise periods with 120 s pause periods (duty cycle 50%) had similar distance capacities compared with geckos exercising continuously at the same average speed (90% MAS; *Table 2*). The average aerobic cost of intermittent exercise for *T. przewalskii* was not significantly different from the maximal rate of oxygen consumption.

5.3 *Slow oxygen uptake kinetics (fiddler crabs)*

Contrary to our prediction, intermittent locomotion reduced or did not alter distance capacity in *U. pugilator* (*Table 1*; Van Laarhoven *et al.*, 1993). When fiddler crabs alter-nated 10 s of exercise at 170% MAS with 10 s pauses (duty cycle 50%), distance capacity was reduced by 65% compared with continuous locomotion at the same average speed (85% MAS; *Table 2*). In the field, the average movement duration of droving *U. pugilator* was 9 s and the average pause duration was 3 s (duty cycle 66%; Weinstein, 1993). When these intervals were tested in the laboratory, the distance capacity for intermittent locomotion was not significantly different from that for continuous locomotion at the same average speed (85% MAS; Van Laarhoven *et al.*, 1993). The average aerobic cost of intermittent exercise was not significantly different from the corresponding values for continuous exercise at the same average speed for any of the intermittent exercise protocols.

6. Conclusion

The total amount of work performed during a behaviour depends on the rate at which the work is done, the duration of the work, and the extent to which the behaviour is conducted intermittently. Low work rate behaviours performed continuously can be sustained for long periods of time, resulting in substantial amounts of total work. High work rate behaviours performed continuously can be sustained only for short periods of time, allowing only small amounts of total work to be accomplished. Surprisingly, if a high work rate behaviour is performed intermittently, the total work accomplished can far exceed the work done during the same behaviour performed continuously at the same average work rate. Intermittent locomotor behaviour appears to have its greatest effect on species that rely on both aerobic and nonaerobic energy sources. Highly aerobic or nonaerobic species are affected less by the intermittent nature of the behaviour. Clearly, the biomechanics of behaviour must be considered in the context of the animal's physiological capacity.

Many behaviours involve intermittent activity. Intermittent locomotion, also called stop-and-go or saltatory search behaviour by ethologists, can improve prey detection by providing predators with more time to scan a given visual field (Andersson, 1981; Avery, 1993; Avery et al., 1987; Gendron and Staddon, 1983), reduce the rate of attack by predators (Martel and Dill, 1995), and improve detection of predators (McAdam and Kramer, 1998). No matter what the advantage of an intermittent behaviour may be, the ability of an animal to sustain the performance of the behaviour will be determined by its physiological capacity and biomechanical limitations.

Finally, we contend that non-steady-state behaviour involving transitions must be afforded the same degree of attention as given to the steady-state behaviour. Perhaps the design of musculoskeletal, circulatory and respiratory systems is determined primarily by non-steady-state behaviour. These further studies of intermittent locomotion could lead to improved predictions of field behaviour. At the same time, a knowledge of field behaviour can set the biomechanical and physiological protocols by which we test future hypotheses of function.

References

Adamczewska, A.M. and Morris, S. (1998) Strategies for migration in the terrestrial Christmas Island red crab *Gecarcoidea natalis*: intermittent versus continuous locomotion. *J. Exp. Biol.* 201: 3221–3231.

Andersson, M. (1981) On optimal predator search. *Theor. Pop. Biol.* 19: 58–86.

Autumn, K.A., Weinstein, R.B. and Full, R.J. (1994) Low cost of locomotion increases performance at low temperature in a nocturnal lizard. *Physiol. Zool.* 67(1): 238–262.

Avery, R.A. (1993) Experimental analysis of lizards pause–travel movement: pauses increase probability of prey capture. *Amphib. Rep.* 14: 423–427.

Avery, R.A., Mueller, C.F., Smith, J.A. and Bond D.J. (1987) The movement patterns of lacertid lizards: speed, gait and pauses in *Lacerta vivipara*. *J. Zool.* 211: 47–63.

Blickhan, R. and Full, R.J. (1987) Locomotion energetics of the ghost crab. II. Mechanics of the center of mass. *J. Exp. Biol.* 130: 155–174.

Blickhan, R., Full, R.J. and Ting L. (1993) Exoskeletal strain: evidence for a trot–gallop transition in rapidly running ghost crabs. *J. Exp. Biol.* **179**: 301–321.

Full, R.J. (1987) Locomotion energetics of the ghost crab. I. Metabolic cost and endurance. *J. Exp. Biol.* **130**: 137–153.

Full, R.J. and Herreid, C.F. (1983) The aerobic response to exercise of the fastest land crab. *Am. J. Physiol.* **244**: R530–R536.

Full, R.J. and Herreid, C.F. (1984) Fiddler crab exercise: The energetic cost of running sideways. *J. Exp. Biol.* **109**: 141–161.

Full, R.J. and Min, C. (1990) Do insects have a maximal oxygen consumption? *Physiologist* **33**(4): A89.

Full, R.J. and Prestwich. K.N. (1986) Anaerobic metabolism of walking and bouncing gaits in ghost crabs. *Am. Zool.* **26**(4): 88A.

Full, R.J. and Weinstein, R.B. (1992) Integrating the physiology, mechanics and behavior of rapid running ghost crabs: slow and steady doesn't always win the race. *Amer. Zool.* **32**: 382–395.

Gendron, R.P. and Staddon, J.E.R. (1983) Searching for cryptic prey: the effect of search rate. *Am. Nat.* **121**: 172–186.

Gleeson, T.T. (1979) Foraging and transport costs in the Galapagos marine iguana, *Amblyrhynchus cristatus. Physiol. Zool.* **52**: 549–557.

Heglund, N.C., Fedak, M.A., Taylor, C.R. and Cavagna, G.A. (1982) Energetics and mechanics of terrestrial locomotion. IV. Total mechanical energy changes as a function of speed and body size in birds and mammals. *J. Exp. Biol.* **97**: 57–66.

Herreid, C.F. and Full, R.J. (1984) Cockroaches on a treadmill: Aerobic running. *J. Insect Physiol.* **30**: 395–403.

Herreid, C.F. and Full, R.J. (1988) Energetics and locomotion. In W. Burygen and B.R. McMahon (eds). *Biology of the Land Crabs*, pp. 333–377. Cambridge University Press, New York.

Hertz, P.E., Huey, R.B. and Garland, T. (1988) Time budgets, thermoregulation, and maximal locomotor performance: are reptiles olympians or boy scouts? *Am. Zool.* **28**: 927–938.

John-Alder, H.B. and Bennett, A.F. (1981) Thermal dependence of endurance and locomotory energetics in a lizard. *Am. J. Physiol.* **241**: R342–R349.

John-Alder, H.B., Lowe, C.H. and Bennett, A.F. (1983). Thermal dependence of locomotory energetics and aerobic capacity of the Gila monster (*Heloderma suspectum*). *J. Comp. Physiol.* **151**: 119–126.

Lee, C.S., Full, R.J. and Weinstein, R.B. (1992) Intermittent locomotion in insects. *Am. Zool.* **32**(5): 39A.

Lee, C.S., Full, R.J. and Weinstein. R.B. (1993) Exercising intermittently increases endurance in an insect. *Am. Zool.* **33**(5): 29A.

Martel, G. and Dill, L.M. (1995) Influence of movement by Coho salmon (*Oncorhynchus kistuch*) parr on their detection by common mergansers (*Mergus merganser*). *Ethology* **99**: 139–149.

McAdam, A.G. and Kramer, D.L. (1998) Vigilance as a benefit of intermittent locomotion in small mammals. *Anim. Behav.* **55**: 109–117.

Taylor, C.R., Schmidt-Nielsen, K. and Raab, J.L. (1970) Scaling of energetic cost to body size in mammals. *Am. J. Physiol.* **210**: 1104–1107.

Taylor, C.R., Maloly, G.M.O., Weibel, E.R., Langman, V.A., Kamau, J.M.Z., Seeherman, H.J. and Heglund, N.C. (1980) Design of the mammalian respiratory system. III. Scaling maximum aerobic capacity to body mass: wild and domestic mammals. *Respir. Physiol.* **44**: 25–37.

Van Laarhoven, M., Weinstein, R.B. and Full, R.J. (1993) Intermittent locomotion does not increase performance in anaerobic fiddler crabs. *Am. Zool.* **33**(5): 139A.

Weinstein, R.B. (1993) Crabs maintain a constant relative metabolic workload in nature in response to changes in body temperature. *Am. Zool.* **33**(5): 139A.

Weinstein, R.B. (1995) Locomotor behavior of nocturnal ghost crabs on the beach: focal animal sampling and instantaneous velocity from three-dimensional motion analysis. *J. Exp. Biol.* **198**: 989–999.

Weinstein, R.B. and Full, R.J. (1992) Intermittent exercise alters endurance in an eight-legged ectotherm. *Am. J. Physiol.* **262**: R852-R859.

Weinstein, R.B. and Full, R.J. (1998) Performance of low-temperature, continuous locomotion is exceeded when locomotion is intermittent in the ghost crab. *Physiol. Zool.* **71**: 274–284.

Weinstein, R.B. and Full, R.J. (1999) Intermittent locomotion increases endurance in a gecko. *Physiol. Biochem. Zool.* **72**: 732–739.

4

Walking and running strategies for humans and other mammals

R. McNeill Alexander

1. Introduction

This chapter asks questions about travelling on legs. How fast should an animal go, using which gait? What route should it take? And in what circumstances should it be prepared to take risks? Only mammals will be discussed, because more relevant information is available for them than for other walking and running animals. However, a discussion of the intermittent running strategies of crabs, cockroaches and lizards will be found in Chapter 3.

2. Gaits and speeds

Humans walk to go slowly and run to go fast, adults making the change at a speed of about 2 m s^{-1}. Horses walk at low speeds, trot at intermediate speeds and gallop at high speeds. Kangaroos travelling at low speeds shuffle along on all four feet and the tail, but they hop to go faster. In all these cases it has been shown, by measurements of oxygen consumption, that the preferred gait at each speed is the one that minimizes metabolic energy costs (Dawson and Taylor, 1973; Hoyt and Taylor, 1981; Margaria, 1976). The same is presumably true of other mammals, and birds, which make similar changes of gait.

Let us look in more detail at the case of the horse. Hoyt and Taylor (1981) trained ponies to walk, trot or gallop on command, so that walking at speeds at which they would usually have trotted, and vice versa, could be induced. They measured the rates of oxygen consumption of these ponies as they ran on a moving belt, matching their speed to that of the belt, and obtained the data shown in *Figure 1(a)*. The crossing curves show that at speeds up to 1.7 m s^{-1} walking was the most economical gait, between 1.7 and 4.6 m s^{-1} trotting was more economical than walking or galloping, and above 4.6 m s^{-1} galloping was the most economical gait. The bars on the speed axis show the ranges of speed at which each gait was used, when the ponies were moving spontaneously in their paddock; each gait was preferred in the speed range in which it was the most economical.

Biomechanics in Animal Behaviour, edited by P. Domenici and R.W. Blake.
© 2000 Taylor & Francis, Oxford.

The energy needed to travel unit distance is the metabolic rate divided by the speed; thus it is the slope of the line joining the appropriate point on the graph to the origin. The maximum range speed is the speed for which the ratio of energy cost to distance is least, enabling the animal to travel furthest on a given food store. It can be found by drawing tangents through the origin on graphs such as *Figure 1(a)*; the tangent drawn on this graph shows that the maximum range speed for walking was 1.2 m s⁻¹. Similar tangents to the trotting and galloping curves would show that the maximum range speeds for these gaits are 3.2 and about 7 m s⁻¹. The slopes of all three tangents would be about the same, showing that energy/distance was 300–320 J m⁻¹ for all three gaits, at their maximum range speeds. At other speeds the ratio is higher; for example, about 400 J m⁻¹ both for walking and for trotting at 1.7 m s⁻¹. A pony wishing to travel at 1.7 m s⁻¹ could save energy by walking part of the way at 1.2 m s⁻¹ and trotting the rest at 3.2 m s⁻¹, instead of keeping its speed constant. *Figure 1(a)* shows that in spontaneous movement the ponies chose speeds near the maximum range speeds for walking and trotting, avoiding the costly intermediate speeds.

The maximum range speed should be preferred only if the aim is to minimize the energy cost of the journey, irrespective of the time spent. In many circumstances it will be better to travel faster and arrive sooner. In particular, if the time spent travelling could otherwise have been spent feeding, account should be taken of this. Let the metabolic rate at speed v be $R(v)$, and let the net rate of energy gain while feeding (assimilation minus metabolism) be G. Then the time required to travel unit distance is $1/v$, and the energy cost per unit distance is $[R(v) + G]/v$. The speed that minimizes this cost can be found as shown in *Figure 1(b)*, by drawing the line of least slope from the point $(0, -G)$ to the graph of $R(v)$ against v. Unless G is very small, the optimum speed predicted in this way will be the highest galloping speed that can be sustained over the distance.

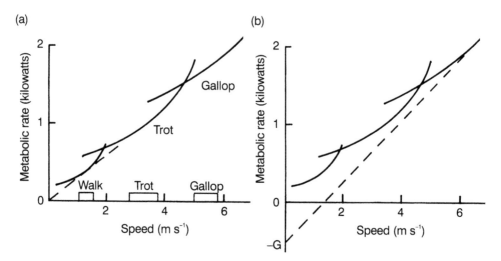

Figure 1. *Graphs of metabolic rate against speed for a 140 kg pony walking, trotting and galloping, from the data of Hoyt and Taylor (1981). Dashed lines show how optimum speeds are predicted, as explained in the text. Boxes on the speed axis of (a) show the ranges of speed in which each gait was used, when the pony was moving spontaneously.*

This suggests that migrating mammals should travel at a gallop, but migrating zebra generally walk, as also do gazelles; and migrating wildebeest sometimes walk at about 1 m s⁻¹ and sometimes canter at 4–5 m s⁻¹ (Pennycuick, 1975; a canter is a slow gallop). People travelling on foot, as distinct from taking exercise, also walk. This suggests that there may be other costs, in addition to energy costs, that influence choice of speed. Fatigue may be an important consideration (see the discussion of intermittent loco-motion in Chapter 3).

3. Routes

The shortest distance between two points is a straight line, but a straight route may cross hills or difficult terrain that could be avoided by a detour, which may save energy although it increases the length of the journey. We will discuss the possible advantages of detours, first on level ground that has soft or rough patches, and then on hills.

The energy cost of walking on soft or rough ground may be very much higher than that of walking the same distance on firm, smooth surfaces. For example, for walking humans, the metabolic energy cost per unit distance has been found to be about 2.5 times as high on dry sand (75 mm deep) as on concrete (Lejeune et al., 1998), and up to five times as high in deep snow as on a treadmill (Pandolf et al., 1976). Smaller increases have been recorded on a sandy beach compared with firm ground (Zamparo et al., 1992) and on dry tundra compared with a road (White and Yousef, 1978). Similar data have been collected for various ungulate mammals on tundra (White and Yousef, 1978), in snow (Dailey and Hobbs, 1989) and on ploughed land (Dijkman and Lawrence, 1997).

Figure 2(a) represents an expanse of firm ground interrupted by a triangular marshy area BKE. An animal wishes to walk from A to F, choosing the route that incurs the least energy cost. By taking the route AGJF, it shortens the distance travelled on the arduous marshy ground, although increasing the total distance. This involves deviating by an angle θ from the straight path, while on the firm ground. Is there a value of θ that minimizes the energy cost of the journey?

Let the energy cost of walking unit distance be C_{firm} on the firm ground and C_{marsh} on the marshy ground. The energy cost of the journey is given by

$$E = 2(AG\,C_{firm} + GH\,C_{marsh}). \tag{1}$$

Differentiating this with respect to θ,

$$dE/d\theta = 2[C_{firm}d(AG)/d\theta + C_{marsh}d(GH)/d\theta]. \tag{2}$$

When θ has its optimum value, which minimizes the cost of the journey, $dE/d\theta$ is zero. We will find this optimum.

Angle ABG = $(90° + \phi)$, where ϕ is the angle BKD; hence sin ABG = cos ϕ. Angle AGB = $(90° - \theta - \phi)$, whence its sine is $\cos(\theta + \phi)$. Hence, by the Law of Sines,

$$AG = AB \cos \phi \sec(\theta + \phi)$$
$$d(AG)/d\theta = AB \cos \phi \sec(\theta + \phi) \tan(\theta + \phi). \tag{3}$$

Also,

$$BG = AB \sin \theta \sec(\theta + \phi)$$

whence

$$BC = BG \sin \theta = AB \sin \theta \sin \phi \sec(\theta + \phi)$$

and

$$d(GH)/d\theta = -d(BC)/d\theta = -AB \sin \phi \sec(\theta + \phi)[\cos \theta + \sin \theta \tan(\theta + \phi)]. \tag{4}$$

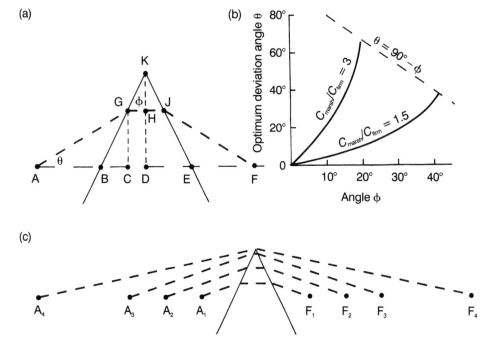

Figure 2. (*a*) *A plan of a schematic landscape, including a triangular patch of marshy ground, illustrating a discussion of the optimum path between the points A and F. (b) Graphs of the optimum angle* θ, *by which a walker should be prepared to deviate from the straight route, against the angle* φ *at the apex of the triangle of marshy ground, for two values of the ratio of energy costs,* C_{marsh}/C_{firm}. (*c*) *Examples of optimum paths for* $C_{marsh}/C_{firm} = 1.5$.

From Equations (2), (3) and (4), $dE/d\theta$ is zero when

$$C_{firm}\cos\phi\tan(\theta + \phi) = C_{marsh}\sin\phi\,[\cos\theta + \sin\theta\tan(\theta + \phi)]$$

$$\cos\theta\cot(\theta + \phi) + \sin\theta = (C_{firm}/C_{marsh})\cot\phi. \qquad (5)$$

Optimum angles of deviation θ, calculated by means of Equation (5), are shown in *Figure 2(b)*. They have been plotted against the angle φ of the edge of the marshy patch, for two values of the ratio of energy costs, C_{marsh}/C_{firm}. Comparison of the two curves shows (as expected) that the greater this ratio of energy costs, the greater the angle by which an animal should be prepared to deviate from a direct path. Each curve shows that the greater the angle φ, the greater the optimal angle of deviation. This is because the greater φ is, the more does a given angle of deviation reduce the distance to be walked over the marshy ground.

There is no point in deviating from the direct path by more than is necessary to avoid the marshy patch completely. The animal should deviate either by the angle θ given by Equation (5) and plotted in *Figure 2(b)*, or by the minimum angle required to avoid the marshy area, whichever is the less. *Figure 2(c)* illustrates this point. It represents a marshy area for which φ = 30°, $C_{marsh}/C_{firm} = 1.5$, and shows optimum paths from A_1 to F_1, from A_2 to F_2, and so on. *Figure 2(b)* tells us that in this case the optimum value of θ is 18°, irrespective of the distance AB. *Figure 2(c)* shows that this is indeed the case, for the journeys starting at A_1, A_2 and A_3. However, for an animal starting at A_4, deviation by 18° would take it unnecessarily far clear of the marshy patch, and it is more economical to deviate by only 12°, as shown. It can never be advantageous for θ to exceed $(90° - \phi)$, represented by the broken line in *Figure 2(b)*, because no matter how close the starting point is to the marshy area, the angle θ needed to avoid the marsh cannot exceed $(90° - \phi)$.

Humans or animals travelling over uniform, level ground can save energy by taking the shortest route, which is a straight line. However, initially uniform ground may become non-uniform as a result of trampling. The ground may be compacted and vegetation destroyed along paths that are frequently used. Helbing *et al.* (1997) studied path systems formed by human trampling in urban green spaces such as parks. They showed that there is a tendency for initially straight paths between different destinations to deviate from straight lines and merge, in places where they run close together. Their computer simulations showed how this could result from a preference for well-trampled paths.

Among hills, as on marshy ground, a detour is often more economical of energy than a straight route. Metabolic rates of humans walking up- and downhill on various slopes have been determined by measuring their oxygen consumption while they walked on a sloping treadmill. For each slope, a maximum range speed can be found, in the same way as for level walking. *Figure 3* shows the energy cost of walking unit

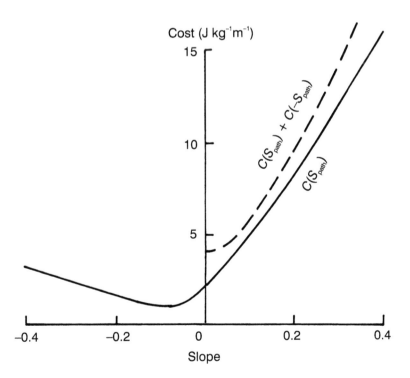

Figure 3. *Graphs showing the energy costs of human walking on slopes, plotted against the slope. The continuous line shows the energy costs per unit distance of walking up or down slopes at the maximum range speed for each slope. The dashed line shows the total cost of walking unit distance up the slope and unit distance down. The data are from Margaria (1976).*

distance at the maximum range speed, plotted against the slope. The energy cost of walking by other mammals is similarly increased on slopes (see, e.g. Blaxter (1989) and Taylor *et al.* (1972), and references in Full and Tullis (1990)).

Minetti (1995) used the data shown in *Figure 3* to calculate the energy cost of human walking up a hillside either on a straight path perpendicular to the contour lines, or on zig-zag paths. He concluded that for slopes up to 0.25 the straight path required least energy. On steeper paths, the most economical path was a zig-zag made up of straight segments, each of which ascended with a slope of 0.25. This conclusion applies only at moderate altitudes. At very high altitudes, at which atmospheric pressure is severely reduced, walkers would be unable to sustain the maximum range speed for a 0.25 slope, and the optimum path would have a shallower gradient. Minetti (1995) compared these predictions with the slopes of footpaths measured from contour maps of the Alps and Himalayas, and found that many paths conform to these predictions; but that, on steep slopes, paths are commonly much steeper or less steep than predicted.

Another problem raised by hills is whether it is more economical to walk in a straight line over the hill, or to skirt around it. *Figure 4(a)* is a contour map of the shoulder of a stylized hill. A walker wishes to travel from A to C. The shortest route would be the straight line ABC, which would involve climbing and then descending. An alternative route, longer but level, is along the contour line by way of E. There is a range of other possible routes, between these extremes, of which ADC is an example. In this simple landscape, the optimum path must consist of two straight segments, as shown, because the environment of any point on the hillside can be made identical to that of any other by a uniform change of scale. Wherever the walker is on the surface, the optimum direction of walking must be the same.

The path shown in *Figure 4(a)* consists of an uphill and a downhill segment, each of length (projected on a horizontal surface) equal to AB sec θ. The slopes of these path segments are $S \cos(\theta + \phi)$, where S is the slope of steepest ascent and ϕ is the angle

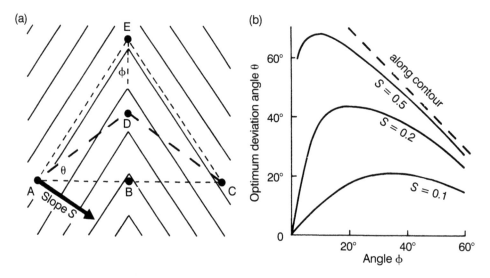

Figure 4. *(a) A contour map of the shoulder of a pyramidal hill, illustrating a discussion of the optimal path from A to C. (b) Graphs of the angle θ for the optimal path, against the angle φ, for hills of three slopes.*

shown in the figure. Let the energy cost per unit distance of walking up a path of slope S_{path} be $C(S_{path})$. Then the energy cost of walking the path shown in *Figure 4(a)* is

$$E = AB \sec \theta \{C[S \cos(\theta + \phi)] + C[-S \cos(\theta + \phi)]\}. \tag{6}$$

The empirical data of *Figure 3(a)* are well represented by the equation

$$C(S_{path}) + C(-S_{path}) = 46S_{path} + 4 \exp(-12S_{path}) \tag{7}$$

(units are J kg^{-1} m^{-1}). These two equations allow us to find the optimum value of θ, for given values of S and ϕ.

Results are shown in *Figure 4(b)*. The broken line ($\theta = 90° - \phi$) shows the value of θ that would take the walker along the contour, at constant elevation. The curves showing the optimum values of θ, which would minimize energy costs, all lie below this line, showing that the optimum path always involves a climb followed by a descent. Comparison of the curves for different slopes shows that the steeper the slope, the more should the walker deviate from the direct path over the hill and the more nearly should he or she follow the contours. On a hillside of slope 0.5, the optimum path is always within 5° of the direction of the contours, provided the angle ϕ is at least 25°. If the angle ϕ is small, a path that follows the contours is a very long way round.

If the slope S is negative, *Figure 4(a)* represents a valley instead of the shoulder of a hill. The optimum angle θ for crossing a valley is the same as for the shoulder of a hill, if the slopes have the same absolute value.

We have discussed optimum paths for stylized landscapes with very simple geometries designed for mathematical convenience (*Figures 2(a)* and *4(a)*). Paths for other geometries could be found by the calculus of variations, a development that would make the theory more amenable to testing by comparison of real paths with theoretical optimum paths. Comparisons with ancient paths that were used for travel on business seem more likely to be useful than comparisons with modern paths designed for leisure use and constrained by property boundaries.

For the present, only qualitative comparisons seem possible. Old paths through mountain country (e.g. the Brandy Pad through the Mourne Mountains, Northern Ireland) deviate widely from straight lines in many cases, and tend to follow the contours closely wherever the ground slopes steeply.

4. Taking risks

Howland (1974) showed that, in a chase, a prey animal may be able to escape a predator faster than itself, if it is able to swerve more sharply than the predator. Films of cheetahs hunting gazelles show that the gazelles attempt to escape by swerving. Let the prey run at speed v and be capable of swerving with radius of curvature r, whereas the predator runs at speed V and can swerve with radius R. The prey can escape if its transverse acceleration in a swerve (v^2/r) is greater than the predator's transverse acceleration (V^2/R). However, if the prey's advantage is small, it must delay swerving until the predator is very close behind it. The simple theory predicts that the swerve should be initiated only when the predator is infinitesimally close, but in real life allowance must be made for error. The optimum time for swerving will depend on the prey's judgement both of the probability distribution of error in its own decision, and of the magnitude of its advantage over the predator.

The faster the prey is running, the less sharply is it likely to be able to swerve. Let us suppose first that swerving radius is limited by the coefficient of friction with the ground; the prey will skid if it attempts to swerve too sharply. The vertical component of the vertical force on the ground, averaged over a stride, must equal its body weight mg (m is body mass and g is the gravitational acceleration). The horizontal component required to give the animal its transverse acceleration is mv^2/r. Thus the coefficient of friction required is v^2/gr, and the same transverse acceleration v^2/r should be possible on all radii small enough for speed to be limited by the danger of skidding. This suggests that it may not be to the prey's advantage to run at maximum speed. However, the faster the prey runs, the more slowly the distance between it and the predator will fall, and the better it may be able to judge the optimum moment for swerving.

The coefficient of friction with the ground may vary from place to place in a way that the prey animal cannot predict. If it skids and falls, it will probably be caught. For this reason it may swerve less sharply than it judges to be possible, allowing a margin of safety.

That argument depended on the assumption that swerving is limited by the coefficient of friction with the ground, which may often not be true. Greene (1985) performed a series of experiments in which people were asked to run as fast as possible round circles of different radii, on grass (wearing spiked running shoes) and on concrete. His results agreed reasonably well with a theory based on the assumption that speed is limited on all radii by the resultant force that the feet are required to exert on the ground.

Let us consider now the strategy of predators. A hunt is more likely to succeed if predators hunt in a group than if only one predator is involved. If the predators run in line abreast, those to either side of the prey's path will be well placed to intercept it if it swerves. If they run in single file, the prey must swerve before the leading predator overtakes it, which will be before the optimum time to escape other members of the predator group, some of whom may be able to intercept the prey's swerve. The advantage of group hunting, of improved chances of success, must be offset against the penalty of having to share the prey. Lions and African hunting dogs hunt in groups; tigers and cheetahs hunt alone (Kingdon, 1997).

Some predators such as the hunting dog pursue prey over long distances, but others such as the cheetah depend on a sprint of which the length is limited by their aerobic capacity. In the early stages of a chase, it may be wise for a sprinting predator to moderate its speed, because the probability of accidents such as stumbling is likely to increase with speed. As the chase continues, the remaining aerobic capacity will fall, and the predator should be prepared to take greater risks. It can be expected to accelerate to its maximum speed just before its aerobic capacity is exhausted, at which stage it must abandon the chase (Alexander, 1988).

The tactics that have been discussed in this section have parallels in human sports. Swerving is important both in rugby football and in the children's game of tig (also known as tag), and players recognize the importance of timing. Runners accelerate to maximum speed only at the end of races, aiming to exhaust their aerobic capacity just after crossing the finishing line. This does not apply to sprinting, in which even the highest speeds do not exhaust the athlete's aerobic capacity (Reilly et al., 1990).

References

Alexander, R.McN. (1988) The risks of the chase. Behav. Brain Sci. 11: 130.

Blaxter, K. (1989) Energy Metabolism in Animals and Man. Cambridge University Press, Cambridge.

Dailey, T.V. and Hobbs, N.T. (1989) Travel in alpine terrain: energy expenditures for locomotion by mountain goats and bighorn sheep. *Can. J. Zool.* **67**: 2368–2375.

Dawson, T.J. and Taylor, C.R. (1973) Energetic cost of locomotion in kangaroos. *Nature* **246**: 313–314.

Dijkman, J.T. and Lawrence, P.R. (1997) The energy expenditure of cattle and buffaloes walking and working in different soil conditions. *J. Agric. Sci.* **128**: 95–103.

Full, R.J. and Tullis, A. (1990) Energetics of ascent: insects on inclines. *J. Exp. Biol.* **149**: 307–317.

Greene, P.R. (1985) Running on flat turns: experiments, theory, and applications. *J. Biomech. Eng.* **107**: 96–103.

Helbing, D., Keltsch, J. and Molnár, P. (1997) Modelling the evolution of human trail systems. *Nature* **388**: 47–50.

Howland, H.C. (1974) Optimal strategies for predator avoidance: the relative importance of speed and manoeuvrability. *J. Theor. Biol.* **47**: 333–350.

Hoyt, D.F. and Taylor, C.R. (1981) Gait and the energetics of locomotion in horses. *Nature* **292**: 239–240.

Kingdon, J. (1997) *The Kingdon Field Guide to African Mammals*. Academic Press, San Diego, CA.

Lejeune, T.M., Willems, P.A. and Heglund, N.C. (1998) Mechanics and energetics of human locomotion on sand. *J. Exp. Biol.* **201**: 2071–2080.

Margaria, R. (1976) *Biomechanics and Mechanics of Muscular Exercise*. Clarendon Press, Oxford.

Minetti, A.E. (1995) Optimum gradient of mountain paths. *J. Appl. Physiol.* **79**: 1698–1703.

Pandolf, K.B., Haisman, M.F. and Goldman, R.F. (1976) Metabolic energy expenditure and terrain coefficients for walking on snow. *Ergonomics* **19**: 683–690.

Pennycuick, C.J. (1975) On the running of the gnu (*Connochaetes taurinus*) and other animals. *J. Exp. Biol.* **63**: 775–799.

Reilly, T., Secher, N., Snell, P. and Williams, C. (eds) (1990) *Physiology of Sports*. Spon, London.

Taylor, C.R., Caldwell, S.L. and Rowntree, V.J. (1972) Running up and down hills: some consequences of size. *Science* **178**: 1096–1097.

White, R.G. and Yousef, M.K. (1978) Energy expenditure in reindeer walking on roads and on tundra. *Can. J. Zool.* **56**: 215–223.

Zamparo, P., Perini, R., Orizio, C., Sacher, M. and Ferretti, G. (1992) The energy cost of walking or running on sand. *Eur. J. Appl. Physiol.* **65**: 183–187.

5

Fish swimming behaviour: predictions from physical principles

Paul W. Webb and Cynthia L. Gerstner

1. Introduction

Behaviour has been defined as the motor responses to all internal and external stimuli (Bond, 1996). To explore the role of biomechanics in swimming behaviour of fishes, we identify three categories in this otherwise all-encompassing definition (*Table 1*). The first is basic swimming behaviours, focusing on ways individual fishes utilize propulsors in gaits. A second level also applies to individuals but considers simple behaviours that expand the locomotor repertoire by choosing how, when and where to swim. The third category considers behaviours among individuals in pairs or groups, when swimming plays a role in predator–prey and social interactions. In each category, biomechanical principles are shown to influence behaviour. Sometimes, biomechanical analysis of propulsion systems reveals limits to locomotor behavioural options and the necessity for complementary systems.

Our emphasis is on the propulsors and the metabolic properties of the muscles that power them. Recent studies elucidating details of muscle mechanics and body deformation are not included. We also do not emphasize scaling, including larvae, which are discussed in Chapter 14.

2. Gaits as basic swimming behaviours

2.1 *Performance range fractionation*

The most basic swimming behaviours are those by which a fish propels the body. Fish have numerous muscle–propulsor systems from which to choose, largely because the high density of water makes supporting the body weight a smaller problem for fishes compared with terrestrial animals. Using various propulsion systems, fish can swim at speeds from zero to more than 20 body lengths s^{-1}, accelerate at rates in excess of 10 g, and turn in circles with radii as small as 3–10% of the body length (Beamish, 1978; Blake, 1983; Domenici and Blake, 1997; Gerstner, 1999; Videler, 1993; Webb, 1994a).

Biomechanics in Animal Behaviour, edited by P. Domenici and R.W. Blake.
© 2000 Taylor & Francis, Oxford.

Table 1. Summary of behaviours of fish in which swimming plays a major role, and the major measures of performance for those behaviours

		Complex behaviour	
Basic swimming behaviours (gaits)	Simple swimming behaviours	Predator–prey interactions	Social behaviour
• station holding • hovering • median and paired fin slow swimming • body and caudal fin propulsors powered by SO muscle in BCF–SO cruising • sprinting • fast start	Choosing *how* to swim: • periodic swimming • minimum cost of transport • tilting Choosing *when* to swim: • selective tidal stream transport • braking Choosing *where* to swim: • ground effect and wall effect • flow refuging or entrainment	Predation • feeding migration • search • stalk • strike • chase • subduing prey Anti-predator behaviour • avoidance • escape	Agonistic display Reproduction: • reproductive migration • courtship • nest-building • parental care Schooling

Primary Performance Measures

Speed, $0 \geq u \geq 0$ (forward and backwards swimming)
Linear acceleration rate, $0 > a > 0$ (acceleration and deceleration)
Rotations, $u = 0, r = 0;$.   turns, $u > 0, \infty > r > 0$

The power required to swim over these performance ranges spans several orders of magnitude. However, no single propulsion system is mechanically capable of efficiently supplying power over such a range (Alexander, 1989). Instead, the total loco-motor performance range is fractionated, a solution that is common to all systems with a large dynamic range. Fish fractionate the locomotor performance range with different muscle and body–fin combinations, each working over a limited range of speeds. Each muscle–propulsor combination is called a gait (*Figure 1*), defined in terms of discrete changes in the use of propulsion system components as speed increases (Alexander, 1989). Because gaits represent choices in the way body and appendages are used to vary performance level, we view them as basic swimming behaviours, necessi-tated by the mechanical requirements of range fractionation.

2.2 *Gait recruitment*

Different gaits are used as speed increases (Alexander, 1989; Jayne and Lauder, 1996; Webb, 1994a). An idealized recruitment pattern starts at zero ground speed; fish use the station-holding gait to remain stationary on the ground. The hovering gait is used at zero water speed to remain stationary relative to the water. Both gaits use median and paired fins (MPF) as propulsors. These behaviours can be sustained for long periods and hence are presumably powered by slow oxidative (SO) muscle (*Figure 1*).

In the idealized recruitment pattern, fish swim at low speeds using MPF propulsors powered largely by SO muscle, perhaps supplemented by fast glycolytic (FG) fibres towards the speed limits of the gait. The MPF–SO slow-swimming gait is succeeded by body and caudal fin (BCF) propulsion powered by myotomal SO muscle. This BCF–SO cruising gait is succeeded in turn by sprinting, the BCF gait powered by the large mass of myotomal FG muscle.

Fish can also achieve very high rates of acceleration and turning (Domenici and Blake, 1997). These high levels of unsteady motions are seen in a distinct fast-start gait, characterized by large-amplitude transient BCF movements powered by FG muscle.

2.3 *Gait expression*

By definition, each basic behaviour requires the necessary anatomical and morphological features for a gait to be expressed. In the absence of such traits, the behavioural repertoire must be constrained. For example, esocids have almost no myotomal SO, essentially eliminating the BCF–SO gait. Ocean sunfish (*Mola*) lack tails and hence the ability to express any BCF gait.

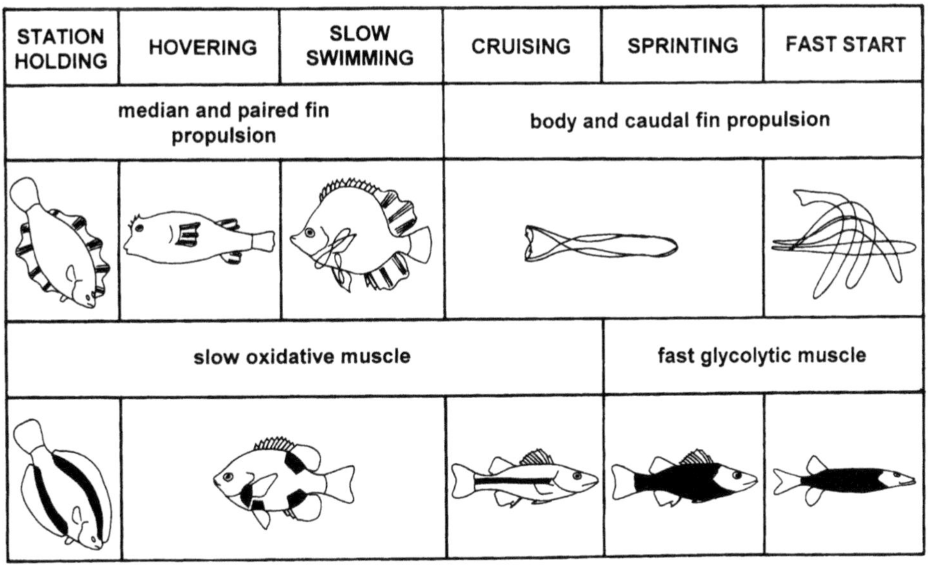

Figure 1. Fish use their body and fin systems to swim in a range of gaits that fractionates the total swimming performance range. The primary determinants of gaits are the use of median and paired fin and body and caudal fin propulsors (shown by figure-of-eight symbols and parallel shading), powered by slow oxidative or fast glycolytic muscle (dark shading, bottom row). In the station-holding gait (ground speed = 0, water speed > 0), fish remain stationary on the ground. Hovering is distinguishable from low swimming by zero water speed. The fast-start gait is distinguished by the transient large-amplitude motions used to achieve high rates of acceleration.

The inability to express a gait because of the absence of the necessary morphological traits is especially important in the locomotor repertoire during ontogeny. Thus early larvae typically lack differentiated fins and SO muscle, so gaits are largely restricted to BCF–FG gaits (Webb, 1994b). In this situation, the absence of the requisite anatomy and morphology and the resulting reduction in gait expression have a physical origin in the physical properties of water. Because larvae are small, oxygen can readily diffuse to supply the metabolic needs of the muscle so that the need for a separate oxidative muscle is reduced. In addition, these small larvae swim at low speeds, so that viscous forces are large. Metamorphosis to an elongate body form at first feeding begins to exploit inertial forces increasing thrust and increasing behavioural choices of how and when to swim (Weihs, 1980). Similarly, fins differentiate when it is hydrodynamically possible to develop propulsive forces (Webb and Weihs, 1986).

2.4 Gaits, speed, acceleration and turning

The original definition of gaits in terms of discrete kinematic changes with increasing speed reflects the use of treadmills, wind tunnels, and water tunnels to study terrestrial, aerial, and aquatic rectilinear locomotion, respectively (Alexander, 1983; Beamish, 1978). In contrast, the trajectory of a freely swimming fish is the vector sum of initial velocity and subsequent linear and angular accelerations. Therefore each gait fractionates not only the speed range, but also ranges of linear acceleration and turning (Table 1). Of these, speed changes and turns dominate routine swimming and greatly increase swimming costs (Boisclair and Tang, 1993; Krohn and Boisclair, 1994). Therefore, it is not known if range fractionation using gaits evolved because of the mechanical need to cover a large range of speeds, accelerations, or manoeuvring rates. Data to resolve such design questions are lacking, and may be impossible to obtain. Therefore, it is practical to continue to define gaits in terms of speed at this time, although recognizing that each gait also spans some range of acceleration and turning performance (Table 1).

In addition, it should be noted that speed and acceleration rate may be positive or negative (Table 1). Studies of swimming have focused on forward swimming, but many fish can swim backwards. For example, surgeonfish (Acanthurus bahianus) combine backwards swimming and rotation before moving forward in an overall manoeuvre resulting in a turn in a very small space (Gerstner, 1999). In general, however, data on swimming backwards are lacking, although backward swimming kinematics have recently been described in detail for eels (D'Août and Aerts, 1999). Furthermore, deceleration becomes especially important for braking (see Section 3.2).

2.5 Performance-limited behaviour

Gaits describe how fish propel themselves but not how fast they will swim or how rapidly they can change speed and direction. These measures of performance depend on the physical properties of muscle, linkage and propulsor elements, and body and fin form (Long and Nipper, 1996; Wainwright, 1983). Together, design features determine thrust and resistance, and hence performance, both within a gait and for the whole animal when summed across all gaits expressed. Designs resulting in high performance in a gait facilitate some behaviours. Thus external morphologies maximizing total body thrust production are typical of predators that lunge at prey from ambush sites, for example, esocids and cottids. Similarly, streamlined bodies driven by high aspect ratio caudal fins linked to the body by a narrow peduncle are typically associated with high-speed cruising and thus support pelagic life styles.

3. Simple swimming behaviours

Gaits describe the ways fish use their propulsors to propel the body at various rates. Gaits also provide basic ingredients that can be assembled into derived behaviours that expand individual performance capabilities. We define these as simple behaviours (*Table 1*) whereby individual fishes choose how, when, and where to swim.

3.1 *Choosing how to swim*

Fish may choose how to swim by alternating power twitches or bursts of propulsion with coasting (periodic swimming), by selecting the speed of swimming and by selecting swimming posture (tilting). The physical basis for all these behaviours is believed to be energy economy as discussed in Chapter 6. In general, periodic swimming behaviour is common near gait transitions (Rome *et al.*, 1990; Videler and Weihs, 1982). Twitch-and-coast behaviour typically occurs at the transition from MPF gaits to BCF–SO cruising, when the myotomal SO muscle supporting the BCF propulsor would function at low strain rates. Burst-and-coast behaviour is recruited around the BCF–SO cruising to BCF–FG sprint transition where capacity-limited FG muscle is recruited, also initially working at low strain rates.

Previously, twitch-and-coast and burst-and-coast behaviours were considered to be separate gaits (Webb, 1993a). Our current analysis shows these behaviours are actually assembled from the normal components of swimming within a gait and hence should not be viewed as discrete gaits (Gerstner, 1999).

Speed selection can affect energy expenditure. Here the physical basis for speed selection derives from the U-shaped relationship between the total energy cost of transport (COT) and swimming speed in a gait. As a result, there is a speed at which COT is least and fish swimming at this speed can maximize the distance travelled for a given fuel consumption (Weihs, 1973a). This behaviour is important for fish making long migrations. For example, adult salmon migrate towards their natal breeding streams at the speed minimizing COT (Brett, 1995).

Swimming posture can minimize transportation costs by reducing energy expenditure to control stability, especially when swimming slowly (He and Wardle, 1986; Webb, 1993b). The problem has its physical origin in the magnitude of correction forces at low speeds that are poorly matched with the inertial resistance necessary to rectify a disturbance (Machaj, 1988). To reduce the potential costs of 'erratic' or unstable motions, many fish tilt at low speeds, swimming with the body at an angle to the swimming trajectory. This helps provide larger thrust forces for stability control (He and Wardle, 1986; Webb, 1993b). Other fish increase the angle of attack between the paired fins and the incident water flow and still others increase the paired fin area (Bone *et al.*, 1995). Such behaviours increase energy expenditure compared with stable rectilinear swimming, but the energy costs associated with unstable trajectories would be very much higher.

3.2 *Choosing when to swim*

Many fish do not swim continuously. Many fish can minimize energy costs by choosing when to swim. For example, many fish live in tidal habitats and choose when to swim on the ebbs and floods, thereby minimizing transport costs by as much as 50% (Weihs, 1978). In this behaviour, called selective tidal-stream transport (STST), fish ride the tide in a desired direction and avoid swimming by station holding on the opposite tide (Harden-Jones *et al.*, 1978). This STST behaviour is used by plaice, *Pleuronectes platessa*, migrating in the North Sea (Metcalfe *et al.*, 1990).

In choosing when to swim, stopping can be as important as translocation. Braking is a swimming behaviour that is poorly known, but functionally important in adjusting to the motions of prey, to water currents near submerged objects, and in avoiding collisions. The physical basis of braking is a large increase in resistance, and this ability varies with body and fin form. Braking at low rates can be achieved by reducing the ratio of propulsive wave speed to forward speed or converting forward momentum into angular momentum in a turn. These may be the primary methods in aquatic vertebrates with reduced or rigid appendages (e.g. selachians, eels, snakes, and cetaceans), but data are lacking. Some fish can rotate the caudal fin forwards to brake (Aleyev, 1977; Breder, 1926; Geerlink, 1987). The most common mechanism used in braking behaviour is the forward rotation of flexible, low aspect ratio median and paired fins characteristic of actinopterygians (Breder, 1926; Geerlink, 1987; Harris, 1937; Videler, 1993; Webb, 1984b).

3.3 *Choosing where to swim*

As with many other simple behaviours, choosing where to swim can affect energy use. Structures such as the substratum, and weeds, corals, and stream banks present surfaces with which fish can interact. The interaction occurs between the wake generated by propulsors and these rigid surfaces, called ground effect, reducing the energy costs of swimming (Lighthill, 1979). The physical basis of behaviours using ground effect is the interference of the structure with the normal development and expansion of the wake leading to increased net thrust and reduced rates of working (Blake, 1983; Lighthill, 1979; Webb, 1981). Negatively buoyant mandarin fish (*Synchropus picturatus*) reduce power needed to hover by 30–60% (Blake, 1983), whereas BCF swimmers near walls can achieve 50% savings in total power, but only at low speeds (Webb, 1993c).

Although ground effect results from interactions between the wake created by a fish and a surface, wakes are also caused by flow around habitat protuberances such as rootwads, submerged branches, coral heads, and substratum protuberances such as ripples. These structures can be used by fish to avoid swimming or to reduce costs of swimming, termed flow refuging or entrainment (Arnold and Weihs, 1978; Gerstner, 1998; Gerstner and Webb, 1998; Webb, 1998; Webb *et al.*, 1996a). Fin-form and self-correcting fin distributions contribute to the physical basis for the behaviour. Use of wakes appears to reduce energy costs for upstream migrants, and the ability of fish to exploit wakes is a major feature in fish ladder design and stream improvement practices.

4. Complex behaviours

The discussion so far has been concerned with behaviours at an individual level. However, much behaviour involves interactions with other organisms. We define these as complex behaviours (*Table 1*), encompassing predator–prey relationships and social behaviour. As with simple behaviours, complex behaviours are assembled from options provided by basic behaviours and as such share the same physical basis for expression. However, many complex behaviours are well within the dynamic range of propulsion systems. As such, physical attributes may be less important because behaviours may be permitted within the scope of a propulsion system but not limited by its physical characteristics. For example, fish typically forage at low BCF–SO speeds, which may be less than 20% of the critical swimming speed (O'Brien et al., 1986). Thus, this behaviour is permitted by that gait. In contrast, fast starts are involved in some intense agonistic interactions, and higher performance by one of the participants might increase its success (Fernald, 1975). In this situation, behaviour is limited by the physical traits that determine performance in that gait (Webb, 1984b, 1986a,b). Overall, the properties of propulsion systems do appear to influence complex behaviour (e.g. Gerstner, 1999), but this is an area where further research is needed.

4.1 Predator–prey interactions

Foraging involves a series of phases, each of which makes different performance demands on a fish. The phases that can occur can be captured in a generalized predation cycle composed of a search, stalk, strike, chase, and subdue. The importance of each phase depends on food characteristics, including whether the fish is or is not a carnivore. For example, fish may search for plant material but the other phases will be diminished, whereas the search phase will be less important for an ambush predator waiting for its prey.

The goal of a search is to contact enough suitable prey to meet ration needs. Ideally, fish should maximize the space or volume searched for a given cost. Therefore, there is an optimal search speed, the physical basis for which is the same as choosing a speed minimizing COT. Indeed, continuous swimmers maximize net food gain by searching at an intermediate speed only slightly greater than that minimizing COT (Ware, 1975; Weihs, 1975a). Fish that more strongly express MPF gaits and eat small particulate prey appear to make greater use of saltatory search, alternating swimming to a viewing sight with hovering to survey the water for prey (O'Brien et al., 1986) but the energetics of this foraging pattern have not been analysed.

The goal of a stalk is to approach closely to evasive prey without triggering a startle response. In general, this requires a speed that decreases as prey are approached (Dill, 1974; Grobecker and Pietsch, 1979; Popper and Carlson, 1998; Webb, 1986a). As noted above, low speed swimming can present stability problems, which might affect stalking behaviour. However, data are lacking to determine if physical factors affect this locomotor behaviour.

Strikes have received considerable attention because they often involve high rates of acceleration in fast starts. The physical principles that maximize such performance have been studied in some detail, and have been shown to explain much of the behaviour of piscivores that ambush their prey (Harper and Blake, 1990,1991; Webb, 1984b, 1986b; Webb and Skadsen, 1980; and Chapter 14).

A failed strike may become a chase. An important physical requirement in making this transition is braking to avoid overshooting the prey escape path. More derived acanthopterygian fishes with antero-lateral pectoral fins and antero-ventral pelvic fins more readily adduct these normal to the flow to stop quickly, turn, and start a chase with least overshoot. Their chases are more successful than those of less-derived soft-rayed fish with antero-ventral pectoral fins and postero-ventral pelvic fins. Indeed, these latter fish are more likely to end an interaction if prey initiate an escape (Webb, 1984a, 1986a).

Searching, stalking and chasing tend to merge for fish feeding on small prey items. The tracks of such predators when foraging often include numerous turns in open water and near obstacles. Physical features that promote turning therefore affect foraging behaviour. For example, damselfish (*Stegastes leucostictus*) and butterflyfish (*Chaetodon capsitratus*) with low aspect ratio, highly mobile pectoral fins inserted at a large angle to the horizontal body axis are superior at turning during routine foraging than wrasse (*Thalassoma bifasciatum*) and surgeonfish with high aspect ratio fins, inserted at small angles to the horizontal body axis (Gerstner, 1999).

Fish tend to be generalized carnivores (Bond, 1996) and captured prey must be subdued and often broken into smaller pieces. Swimming movements may be involved in tearing prey into bite-size pieces, for example, by sharks, but the physical principles involved in such behaviour have not been analysed. In contrast, physical requirements for consuming much non-evasive food, often in hard to reach locations, usually involve the ability to orient the body in many ways. Clearly, body and fin plans that impair slow swimming and hovering will be at a disadvantage. Surprisingly, variation in body form and fin patterns among generalized actinopterygians does not appear to affect manoeuvrability and orientation behavioural capabilities (Schrank and Webb, 1998; Schrank *et al.*, 1999). For example, the ostariophysean goldfish (*Carassius auratus*) and silver dollar (*Metynnis hypsauchen*) orient the body to feed on patches of tubifex worms at various locations in the water column just as well as angelfish (*Pterophyllum scalare*) (P.W. Webb and C.L. Gerstner, unpublished observations).

Predation cycles focus on factors affecting predator success, but prey behaviour must also be considered. The measure of success for prey is escape (see Chapter 14). Timing of the escape, speed, acceleration rate and manoeuvring all play a role in this success (Howland, 1974), and hence escape abilities are affected by the physical properties of propulsion systems. Of the physical attributes once thought to be critical to high manoeuvrability, large body depth is now known to generally play a different anti-predator role in reducing capture success for gape-limited predators, the major risk for many fishes (Brönmark and Miner, 1992; Nilsson *et al.*, 1995; Pettersson, 1999; Wahl and Stein, 1988). Overall, advertising awareness, anti-predator defences and group behaviour may be most important in evading capture (Edmunds, 1974).

4.2 *Social behaviour*

Social behaviour involves interactions among conspecifics and heterospecifics, but unlike predation, is not involved with consumption of a participant. Much of this behaviour appears to build on basic behaviours, sometimes using the physical factors that promote high levels of performance in new ways.

Agonistic behaviours are used to avoid damaging fights with potential competitors. These signals often involve behaviours used for other locomotor functions. For example, territorial defence behaviours include lateral displays with extended median fins, a component of braking (Breder, 1926) and fast starts (Webb, 1977), and frontal displays with flared operculae and adducted paired fins, components of braking and non-locomotor feeding behaviours (Keenleyside, 1979). Physical factors affecting swimming speed are probably important in other territorial defence behaviours such as charging, nipping and chasing intruders. Again, specific biomechanical studies are lacking.

Reproductive success is critically dependent on social interactions usually involving swimming. Migration may be involved in bringing males and females together at spawning sites distant from feeding grounds. Swimming behaviour also is important in courtship and nest building, for example, in the famous stickleback dance (Keenleyside, 1979). Locomotor-type movements are also involved in spawning during the expulsion of eggs and sperm, and swimming is used in aspects of parental care. Some of these reproductive components are uses of simple swimming behaviours, such as migration at the speed minimizing COT (Brett, 1995) and 'social' fast starts (Fernald, 1975). Others involve agonistic behaviour as described above. Still others make novel use of swimming motions, for example, using fin wakes to dig redds by salmonids (Clemens and Wilby, 1961) or to ventilate nests in sticklebacks (Keenleyside, 1979), and using the toroid vortex shed by the caudal fin to slow the dispersion of eggs and sperm (Okubo, 1988). Again, studies on potential physical determinants of such behaviour are generally lacking.

Schooling is extremely common among fishes, and serves many functions in feeding, defence and spawning. Weihs (1973b, 1975b) postulated an additional physical advantage of schooling. He showed that certain spatial relationships among schooling fish could result in substantial energy savings. The necessary spatial relationships are not always found (Partridge and Pitcher, 1980), but must occur at least some of the time (Breder, 1976), and energy savings associated with changes in kinematics have been shown for schooling sea bass (Herskin and Steffensen, 1998).

5. From individual behaviours to whole organisms

Much of the diversity among the over 28 000 extant species of fishes, as well as the numerous extinct species, is associated with body or fin propulsor anatomy and morphology. We have shown how physical properties of propulsors affect swimming behaviour at all levels, from the way fish swim to their interactions with other individuals. These swimming behaviours are central to many essential activities such as food acquisition, predator evasion and breeding, but are not sufficient alone to meet these needs. Therefore, we now consider how individual components of the propulsion system are combined in whole organisms.

5.1 *Functional-design compromise*

A propulsion system is composed of many components of which we have focused on the form of the propulsor itself and the energetic properties of locomotor muscle. Hydrodynamic modelling and experimental studies have shown how these components affect performance, especially determining limits for basic behaviours.

These studies have also shown that design compromises often occur. In particular, biomechanical studies of propulsion system components have shown that trends towards specialization facilitating performance in one area or gait are often made at a cost of reduced swimming performance and hence behavioural options in other areas or gaits.

Such performance compromises occur within gaits and/or across gaits (*Table 2*) because various species generally show some morphological tendencies to improve some aspect(s) of performance within certain gaits. For example, the total volume for muscle is typically limited so that a large mass of SO muscle reduces the volume for

Table 2. Examples of convergence in locomotor function at various levels of organization.

	Level of swimming behaviour; whole-organism functional convergence				
	Within basic swimming behaviours (gaits)	Within basic swimming behaviours (gaits)	Among basic swimming behaviours (gaits)	Simple and complex swimming behaviour	Non-locomotor traits
Similarities and differences	• Similar locomotor performance • Similar morphology and/or physiology • Same gait • Historically distant clades	• Similar performance • Different morphology and/or physiology • Same gait	• Similar performance • Different morphology and/or physiology • Different gaits	• Similar performance • Different morphology and/or physiology • Different gaits • Different behaviour	• Different locomotor performance • Different morphology and/or physiology • Different non-locomotor adaptations
Examples of different groups or features leading to similarities in performance	Station-holding: • plaice-like fishes • ray-like fishes Thunniform cruisers: • tuna • lamnid sharks • ichthyosaur • cetaceans	Station holding: • plaice-like fishes • cottid-like fishes • torrential fishes • parr posture fishes Slow swimming: • rays • diverse ray-finned fishes Fast starts: • variable muscle mass • variable body or fin form • variable response latency	Cruising: • wrasse-like fishes • ocean sunfish-like fishes • trout-like fishes • sharks • rays	Predation: • Choosing non-evasive prey (e.g. large fishes) • Walking (e.g. frogfishes) group hunting (e.g. tuna) sideswiping and slashing (e.g. gars, sawfishes) Prey escape: • aerial flight (e.g. flying fishes) • vigilance • advertisement • refuges Social behaviour: • Schooling	Inertial suction feeding Defensive toxins and spines Internal fertilization

Various options in morphology result in convergence in gait expression, performance within gaits and performance in different gaits. Reduced performance occurs as a result of focus on specific basic behaviours resulting in a locomotor 'deficit'. Simple and complex swimming behaviour and non-locomotor adaptations are essential to compensate for such deficits.

FG muscle and vice versa. Esocids have the highest known peak acceleration rate among fishes, powered by a myotome composed almost entirely of FG muscle. BCF–SO cruising is, therefore, not an option. Thus specialists for acceleration sacrifice high performance in BCF–SO cruising behaviours (Domenici and Blake, 1997; Webb, 1982, 1994a). Similarly, lift, drag and acceleration reaction force components can be used to generate thrust, but thrust from each is maximized by different propulsor motions and shapes (Webb, 1988). As a result, high lift-based thrust production is not compatible with extensive development of resistance-based thrust mechanisms. Consequently, adaptations supporting high-speed BCF–SO (thunniform) cruising result in reduced acceleration rate, turning rate, hovering and station holding. Similarly, fin shape affects manoeuvrability in MPF gaits (Gerstner, 1998). Adoption of more pelagic lifestyles by acanthopterygian fish is usually associated with reduced fin flexibility and area (e.g. mackerel, *Scomber*), here trading-off specialization for higher speed in pelagic living with reduced performance in MPF gaits. Finally, we have noted that at extremes of specialization morphological structures necessary to express other gaits may be lost (e.g. ocean sunfish are specialized to cruise with their median fins and lack a tail).

Overall, the intra-specific tendencies towards specialization (even if small) of a component of the propulsion system for improved performance (u, a, r, etc., *Table 1*) in any area tend to reduce performance in other areas (*Table 2*). Thus, many important basic locomotor behaviours are mutually exclusive.

5.2 *Functional convergence*

In spite of the effects of design of propulsion system components on force generation and rates of working, smaller effects are often found at the whole animal inter-specific level. Thus, many fish show remarkable inter-specific functional convergence, as measured by locomotor performance.

In some situations, fish with very distant evolutionary histories have converged on similar external morphologies, which are used in a common basic behaviour (gait), thereby achieving similar levels of performance. In this situation, a common physical solution has evolved to meet a common physical need. For example, station holding is facilitated in both pleuronectiform fishes and rays by a convergent flattened, drag-minimizing form involving body shape in the flatfishes and pectoral fin shape in rays (Webb, 1989). Similarly, sharks and bony fishes have adopted a streamlined body with a lunate caudal fin, and probably endothermy, for their high-speed pelagic lifestyles, a solution mimicked by ichthyosaurs and cetaceans (Lighthill, 1971, 1975).

Other fishes differ in morphology but none the less achieve similar levels of performance within a given gait, regardless of evolutionary history. This is usually achieved by manipulating different components of the force balance. For example, the slip speed at which a station-holding fish is first displaced downstream is maximized if both lift and drag are simultaneously minimized. In practice, this cannot be achieved using form alone. Instead, flattened plaice-like fish minimize drag, but at the cost of high lift, whereas depressed but more fusiform cottid-like fish minimize lift but suffer high drag (Webb, 1989). Yet other fish have evolved suckers and rough surfaces to increase the friction force (Webb, 1994a; Webb *et al.*, 1996a).

Similarly, net fast-start performance is often similar among species (Webb, 1994a).

In some situations this results from compromises between positive effects of external body and fin shapes promoting thrust generation and negative effects of body shape on the volume of myotomal muscle (Law and Blake, 1996; Webb, 1978). In other situations, positive effects of external body shape are offset by kinematics. The large caudal area promoting thrust production in esociform fishes results in high peak acceleration rates. However, the shape cannot sustain high thrust between fast-start stages. As a result, peak deceleration rates are also high, offsetting much of the advantages of the initial high acceleration rate (see Webb, 1994a). Sometimes fish with forms that would be expected to produce little thrust use unexpected compensating mechanisms. For example, knifefish have elongate tapering tails with small depth, opposite to esocids. However, knifefish rotate the caudal portion of the body into the vertical plane, thereby greatly increasing the effective depth and hence thrust production in fast starts (Kasapi et al., 1993).

The examples above illustrate how different anatomical and morphological components (and resultant body motions) can be assembled to achieve similar levels of performance within a gait. In contrast, some fishes achieve similar levels of performance using different gaits. In particular, cruising speeds of some MPF swimmers overlap those of BCF–SO swimmers (Dorn et al., 1979; Drucker, 1996; Gordon et al., 1989).

An important area of convergence is now being recognized for manoeuvring in MPF gaits of generalized actinopterygian fishes. MPF gaits are most commonly used at low speeds, and are believed to be especially important in conferring high manoeuvrability. It has been postulated that a deep body promotes manoeuvring, but this is now known to be more important as a predator deterrent (Alexander, 1967; Gerstner, 1999; Pettersson, 1999). Similarly, large median fins, antero-lateral pectoral fins, and antero-ventral fins patterns common in many spiny-rayed fishes have been considered optimal for manoeuvring because of the prevalence of such fish in structurally complex habitats where slow-swimming manoeuvrability seems essential (Domenici and Blake, 1997; Webb, 1982, 1984a,c). However, Domenici and Blake (1997) have suggested that the fin plan of esocids also is specialized for manoeuvre because these fish hover and position themselves before striking prey. Esocids share a less derived 'soft-rayed' body and fin form of a slender body and antero-ventral pectoral and postero-ventral pelvic fins with numerous other species, few of which share esocid behaviours. Among soft-rayed fishes, cyprinids, for example, tend to be omnivorous and are found in marshes, in shallow exposed waters on beaches, and in pelagic habitats. Thus it now appears that fin-distribution pattern has little impact on manoeuvrability in MPF gaits (Rosen, 1982).

Recent experimental studies support this view. Very little difference in manoeuvrability during MPF swimming has been found for goldfish, silver dollar, and angelfish negotiating various obstacle courses in spite of their large differences in body and fin form (Schrank and Webb, 1998; Schrank et al., 1999; Webb et al., 1996b). Furthermore, the same three species show little difference in their abilities to swim backwards and to turn in MPF and low-speed BCF gaits (P.W. Webb and A. Gardiner, unpublished observations). Lastly, there is little support for a strong relationship between braking rate and fin pattern (Videler, 1993), although more derived fish with flexible fins appear to make more use of braking than do soft-rayed fishes (Webb, 1984b).

Manoeuvrability is undoubtedly important in fish swimming. Thus it appears that in spite of large variation in body and fin patterns, good manoeuvrability has been conserved in generalized bony fishes over a wide range of median and paired fin distribution patterns (Rosen, 1982). We suggest that the common denominator underlying such convergent functionality is high fin flexibility, which characterizes actinopterygians (Gerstner, 1999). Rather than primarily affecting manoeuvrability, fin patterns may relate more to stability and the nature of flow disturbances fish face (Webb, 1998).

5.3 *Mapping form and organism function*

The many interactions among propulsor-system design, swimming behaviour, and mode of life can be explored by gait-expression maps (Webb, 1993a, 1994b, 1997), an approach originating with Alexander (1989). These maps couple physical principles and behaviour. They begin with the physical necessities for range fractionation, achieved by using multiple gaits. An essential adaptation permitting swimming over a wide range of speeds is uncoupling BCF and MPF propulsion systems so that fish can have well-developed MPF slow-swimming gaits as well as high performance in BCF gaits (Domenici and Blake, 1997). It is interesting to note that the expansion of swimming behaviours to low speeds using MPF systems is evolutionarily more recent than BCF swimming and appears later during larval development (Webb, 1994b).

Physical attributes of the components that make up the propulsion systems of various species explain performance limits in each gait expressed. Gait-expression maps place each gait in organismic context, and physical principles provide the tools to understand the performance impacts of design trade-off within and among gaits.

Such analysis often shows that although there may be substantial convergence in performance among species, this is not complete. Typically, functional gains in some area are still associated with some losses in another area. As such, different body and fin plans can be thought of as having some residual locomotor 'deficits'.

5.4 *Overcoming locomotor 'deficits'*

A performance 'deficit' arises when mechanical properties associated with a physical design do not meet whole-organism performance needs. Simple and complex behaviours described above, together with several non-locomotor behaviours, are found to be important in expanding whole-organism performance options.

First, performance deficits can often be expanded by simple locomotor behaviours, for example, to increase slip speed in station holding. Here, low-drag–high-lift pleuronectiformes induce water flow beneath the body to reduce lift (Arnold and Weihs, 1978), whereas low-lift–high-drag cottids grasp the substratum to increase friction (Webb, 1989). Bentho-pelagic forms such as salmonids achieve the same result by using the pectoral fins to generate negative lift (Arnold *et al.*, 1991) and others shelter behind surface structures such as wave-formed ripples (Gerstner and Webb, 1998). Other simple behaviours, such as periodic swimming (see Chapter 6) and choosing where to swim expand performance range in other gaits (*Table 2*).

Second, locomotor 'deficits' can be ameliorated by complex behaviours. Examples include group hunting (Major, 1978), slashing schools of fish to cause injury by sawfishes (Bond, 1996), side-swiping by gar (Lauder and Liem, 1983), and shifting to non-evasive prey (Webb and de Buffrénil, 1990). Physical principles show that large aquatic vertebrates experience reduced acceleration and manoeuvrability, and many of these complex behaviours have their physical basis in this scaling phenomenon, which in turn leads to reduced strike capability (see Chapter 14).

Third, non-locomotor adaptations may make up for locomotor 'deficits'. For example, inertial suction feeding in actinopterygians (Lauder and Liem, 1983) replaces whole-body lunging for prey capture. This may be especially important in food limited deep-sea habitats (Hochachka and Somero, 1984).

5.5 *Ecological implications*

The diversity of propulsion systems and the wide range of behaviours they support results in substantial overlap by fish with very different body and fin forms in whole-organism functionality and hence habitats occupied (Law and Blake, 1996; Temple and Johnston, 1998). This is especially apparent in structurally complex, highly productive habitats such as reefs, which provide numerous niches. However, even the apparently uniform pelagic zone supports a structurally diverse fish fauna. For example, the open oceans are populated with fish as diverse as thunniform teleosts and elasmobranchs, ocean sunfish, manta rays and eels, as well as more common shapes such as those of salmon and herring.

Are, then, different locomotor body and fin plans ecologically equivalent (Rosen, 1982)? Body and fin plans affect locomotor behaviour and performance and are certainly a major determinant of where a fish may live (*Figure 2(a)*). The resultant fundamental niches clearly have substantial overlap. The actual habitat occupied, the realized niche, is modulated within the fundamental niche range by other factors. In temporally more stable situations, biotic interactions determine where fish are actually found. Then, marginal differences in the mechanical properties of propulsion (and other) systems affect the ability of various species to compete for common resources. This can lead to resource partitioning (*Figure 2(b)*) and segregation among sympatric species (Probst *et al.*, 1984; Schlosser, 1982; Werner, 1986). In less stable circumstances, there may be insufficient time for biotic factors to determine assemblage composition. Instead, chance events become important (Poff and Ward, 1989). Again, marginal differences in behaviours such as station holding and migration will be important for persistence and recolonization, resulting in segregation among species with different body and fin forms.

6. Conclusions and future studies

The relationships between physical characteristics and behaviour discussed above lean heavily on biomechanical studies of individuals, and often on experimental situations where behaviour has been constrained to focus on a particular activity. There are suggestions that more complex behaviours also reflect physical design properties, but studies are few, and further research is needed.

A biomechanical study of social behaviour in particular presents major challenges. The strong stimuli constraining behaviour that are the hallmark of most studies of swimming are not appropriate, necessitating different approaches (see also Chapter 14). First, a return to natural history is important (Gans *et al.*, 1997), especially for fish, because traditional collection methods do not make observations on the habitat scale experienced by the fishes. Scuba diving and snorkelling, streams with observation windows, and underwater cameras can allow observers to study fish behaviour *in situ*, and often with little

(a)

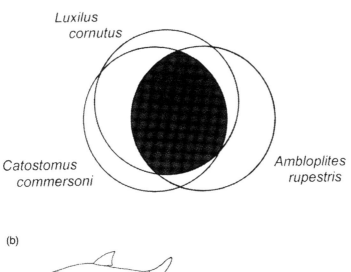

(b)

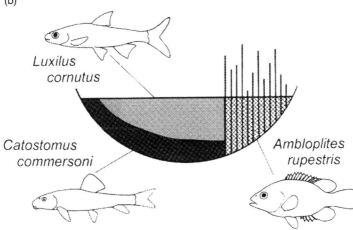

Figure 2. *In northern Michigan, minnows, such as the common shiner* (Luxilus cornutus), *white suckers* (Catostomus commersoni) *and rock bass* (Ambloplites rupestris) *are often found together in medium current-speed, warm-water pool–race–riffle streams. Experimental tests on these species shows their locomotor repertoires, represented by the circles in (a), overlap substantially (illustrated by the stippled area). As a result, preferred habitats of each species in allopatry also show substantial overlap. The marginal differences in the locomotor repertoire (non-overlapping regions of the locomotor-repertoire space) become important for habitat segregation in sympatry (b). Then more streamlined deep-tailed minnows, such as the common shiner, are more pelagic. White suckers, with a mouth for benthic feeding, combined with good station-holding abilities, tend to show bentho-pelagic habits. Rock bass, with a deep body and higher manoeuvrability, are found in slower water along stream banks and among macrophytes.*

disturbance to the subjects (Dolloff *et al.*, 1996). In addition, field and laboratory manipulations are required whereby natural behaviours are exploited, placing fish in situations where success is likely to depend on the ability to swim over a large performance range in various gaits. Situations holding most promise appear to be in the acquisition of resources, such as food, exposure to naturally occurring risks, and territoriality.

Acknowledgements

The preparation of this chapter was supported by the National Science Foundation (grants IBN 9507197 and 9973942), as was much of our research reported here. We thank Paolo Domenici and Robert Batty for their suggestions.

References

Alexander, R.McN. (1967) *Functional Design of Fishes*. Hutchinson, London.
Alexander, R.McN. (1983) The history of fish biomechanics. In: *Fish Biomechanics* (eds P.W. Webb and D. Weihs). Praeger, New York, pp. 1–35.
Alexander, R.McN. (1989) Optimization and gaits in the locomotion of vertebrates. *Physiol. Rev.* **69**: 1199–1227.
Aleyev, Y.G. (1977) *Nekton*. Junk, The Hague.
Arnold, G.P. and Weihs, D. (1978) The hydrodynamics of rheotaxis in the plaice (*Pleuronectes platessa*). *J. Exp. Biol.* **75**: 147–169.
Arnold, G.P., Webb, P.W. and Holford, B.H. (1991) The role of the pectoral fins in station-holding of Atlantic salmon (*Salmo salar* L). *J. Exp. Biol.* **156**: 625–629.
Beamish, F.W.H. (1978) Swimming capacity. In: *Fish Physiology*, Vol. 7, *Locomotion*, (eds W. S. Hoar and D.J. Randall). Academic Press, New York, pp. 101–187.
Blake, R.W. (1983) *Fish Locomotion*. Cambridge University Press, Cambridge.
Boisclair, D. and Tang, M. (1993) Empirical analysis of the swimming pattern on the net energetic cost of swimming in fishes. *J. Fish. Biol.* **42**: 169–183.
Bond, C.E. (1996) *Biology of Fishes*. Saunders, New York.
Bone, Q., Marshall, N.B. and Blaxter, J.H.S. (1995) *Biology of Fishes*. Chapman and Hall, London.
Breder, C.M. (1926) The locomotion of fishes. *Zoologica* **4**: 159–297.
Breder, C.M. (1976) Fish schools as operational structures. *Fish. Bull. US* **74**: 471–502.
Brett, J.R. (1995) Energetics. In: *Physiological-Ecology of Pacific Salmon* (eds J.R. Brett and C. Clark). Government of Canada, Department of Fisheries and Oceans, Ottawa, Ont., pp. 3–68.
Brönmark, C. and Miner, J.G. (1992) Predator-induced phenotypical change in body morphology in crucian carp. *Science* **258**: 1348–1350.
Clemens, W.A. and Wilby, G.V. (1961) *Fishes of the Pacific Coast of Canada. Bull. Fish. Res. Board Can.* **68**: 1–443.
D'Août, K. and Aerts, P. (1999) Kinematic comparison of forward and backward swimming in the eel *Anguilla anguilla*. *J. Exp. Biol.* **202**: 1511–1521.
Dill, L.M. (1974) The escape response of the zebra danio (*Brachydanio rerio*). I. The stimulus for escape. *Anim. Behav.* **22**: 711–722.
Dolloff, A., Kershner, J. and Thurow, R. (1996) Underwater observation. In: *Fisheries Techniques* (eds. B.R. Murphy and D.W. Willis). American Fisheries Society, Bethesda, MD, pp. 533–554.
Domenici, P. and Blake, R.W. (1997) The kinematics and performance of fish fast-start swimming. *J. Exp. Biol.* **200**: 1165–1178.
Dorn, P., Johnson, L. and Darby, C. (1979) The swimming performance of nine species of common California inshore fishes. *Trans. Am. Fish. Soc.* **108**: 366–372.
Drucker, E.G. (1996) The use of gait transition speed in comparative studies of fish locomotion. *Am. Zool.* **36**: 555–566.
Edmunds, M. (1974) *Defense in Animals*. Longman, New York.
Fernald, R.D. (1975) Fast body turns in a cichlid fish. *Nature* **258**: 228–229.
Gans, C., Gaunt, A.S. and Webb, P.W. (1997) Vertebrate locomotion. In: *Handbook of Physiology* (ed. W.H. Dantzler). American Physiological Society, Oxford University Press, Oxford, pp. 55–213.

Geerlink, P.J. (1987) The role of the pectoral fins in braking of mackerel, cod and saithe. *Neth. J. Zool.* **37**: 81–104.

Gerstner, C.L. (1998) Use of substratum ripples for flow refuging by Atlantic cod, *Gadus morhua*. *Environ. Biol. Fish.* **55**: 455–460.

Gerstner, C.L. (1999) Maneuverability of four species of coral-reef fish that differ in body and pectoral-fin morphology. *Can. J. Zool.* **77**: 1–9.

Gerstner, C.L. and Webb, P.W. (1998) The station-holding by plaice, *Pleuronectes platessa*, on artificial substratum ripples. *Can. J. Zool.* **76**: 260–268.

Gordon, M.S., Chin, H.G. and Vojkovich, M. (1989) Energetics of swimming in fishes using different modes of locomotion: I. Labriform swimmers. *Fish. Physiol. Biochem.* **6**: 341–352.

Grobecker, D.B. and Pietsch, T.W. (1979) High-speed cinematographic evidence for ultrafast feeding in antennariid anglerfishes. *Science* **205**: 1161–1162.

Harden-Jones, F.R., Arnold, G.P., Greer-Walker, M. and Scholes, P. (1978) Selective tidal stream transport and the migration of plaice (*Pleuronectes platessa*) in the southern North Sea. *J. Cons. Int. Explor. Mer* **38**: 331–337.

Harper, D.G. and Blake, R.W. (1990) Fast-start performance of rainbow trout *Salmo gairdneri* and northern pike *Esox lucius*. *J. Exp. Biol.* **150**: 321–342.

Harper, D.G. and Blake, R.W. (1991). Prey capture and the fast-start performance of northern pike *Esox lucius*. *J. Exp. Biol.* **155**: 175–192.

Harris, J.E. (1937) The mechanical significance of the position and movements of the paired fins in the Teleostei. *Papers Tortugas Lab., Carnegie Inst.* **31**: 173–189.

He, P. and Wardle, C.S. (1986) Tilting behaviour of the Atlantic mackerel, *Scomer scombrus*, at low swimming speeds. *J. Fish. Biol.* **29** (Suppl. A): 223–232.

Herskin, J. and Steffensen, J.F. (1998) Energy savings in sea bass swimming in a school: measurements of tail beat frequency and oxygen consumption at different swimming speeds. *J. Fish Biol.* **53**: 366–376.

Hochachka, P.W. and Somero, G.N. (1984) *Biochemical Adaptation*. Princeton University Press, Princeton, NJ.

Howland, H. (1974) Optimal strategies for predator avoidance: the relative importance of speed and manoeuverability. *J. Theor. Biol.* **47**: 333–350.

Jayne, B.C. and Lauder, G.V. (1996) New data on axial locomotion in fishes: how speed affects diversity of kinematics and motor patterns. *Am. Zool.* **36**: 642–655.

Kasapi, M.A., Domenici, P., Blake, R.W. and Harper, D.G. (1993) The kinematics and performance of knifefish *Xenomystus nigri* escape responses. *Can. J. Zool.* **71**:189–195.

Keenleyside, M.H.A. (1979) *Diversity and Adaptation in Fish Behaviour*. Springer, Berlin.

Krohn, M.M. and Boisclair, D. (1994) Use of a stereo-video system to estimate the energy expenditure of free-swimming fish. *Can. J. Fish. Aquat. Sci.* **51**: 1119–1127.

Lauder, G.V. and Liem, K.F. (1983) The evolution and interrelationships of the actinopterygian fishes. *Bull. Mus. Comp. Zool. Harvard Univ.* **150**: 95–197.

Law, T.C. and Blake, R.W. (1996) Comparison of the fast-start performances of closely related, morphologically distinct threespine sticklebacks (*Gasterosteus* spp.). *J. Exp. Biol.* **199**: 2595–2604.

Lighthill, J. (1971) Large-amplitude elongated-body theory for fish locomotion. *Proc. R. Soc. Lond., Ser. B* **179**:125–138.

Lighthill, J. (1975) *Mathematical Biofluiddynamics*. Society for Industrial and Applied Mathematics, Philadelphia, PA.

Lighthill, J. (1979) A simple fluid-flow model of ground effect on hovering. *J. Fluid Mech.* **93**: 781–797.

Long, J.H. and Nipper, K.S. (1996) The importance of body stiffness in undulatory propulsion. *Am. Zool.* **36**: 678–694.

Major, P.F. (1978) Predator–prey interactions in two schooling fishes, *Caranx ignobilis* and *Stolephorus purpureus*. *Anim. Behav.* **26**: 760–777.

Marchaj, C.A. (1988) *Aero-Hydrodynamics of Sailing.* International Marine Publishing, Camden, ME.

Metcalfe, J.D., Arnold, G.P. and Webb, P.W. (1990) The energetics of migration by selective tidal stream transport: an analysis for plaice tracked in the southern North Sea. *J. Mar. Biol. Assoc. UK* **70**: 149–162.

Nilsson, P.A., Brönmark, C. and Pettersson, L.B. (1995) Benefits of a predator induced morphology in crucian carp. *Oecologia* **104**: 291–296.

O'Brien, W.J., Evans, B.I. and Howick, G.L. (1986) A new view of the predation cycle of a planktivorous fish, white crappie (*Pomoxis annularis*). *Can. J. Fish. Aquat. Sci.* **43**: 1894–1899.

Okubo, A. (1988) Biological vortex rings: fertilization and dispersal of fish eggs. In: *Mathematical Ecology* (eds T.G. Hallam, L.J. Gross and S.A. Levin). World Scientific, Singapore, pp. 270–283.

Partridge, B.L. and Pitcher, T.J. (1980) Evidence against a hydrodynamic function for fish schools. *Nature* **279**: 418–419.

Pettersson, L.B. (1999) Phenotypic plasticity and the evolution of an inducible morphological defence in crucian carp. Ph.D. Dissertation, Lund University, Sweden.

Poff, N.L. and Ward, J.V. (1989) Implications of streamflow variability and predictability for lotic community structure: a regional analysis of streamflow patterns. *Can. J. Fish. Aquat. Sci.* **46**: 1805–1818.

Popper, A.N. and Carlson, T.J. (1998) Application of sound and other stimuli to control fish behaviour. *Trans. Am. Fish. Soc.* **127**: 673–707.

Probst, W.E., Rabeni, C.F., Covington, W.G. and Marteney, R.E. (1984) Resource use by stream-dwelling rock bass and smallmouth bass. *Trans. Am. Fish. Soc.* **113**: 283–294.

Rome, L.C., Funke, R.P. and Alexander, R. McN. (1990) The influence of temperature on muscle velocity and sustained swimming performance in swimming carp. *J. Exp. Biol.* **154**: 163–178.

Rosen, D.E. (1982) Teleostean interrelationships, morphological function and evolutionary inference. *Am. Zool.* **22**: 261–273.

Schlosser, I.J. (1982) Fish community structure and function along two habitat gradients in a headwater stream. *Ecol. Monogr.* **52**: 395–414.

Schrank, A.J. and Webb, P.W. (1998) Do body and fin form affect the abilities of fish to stabilize swimming during maneuvers through vertical and horizontal tubes? *Environ. Biol. Fishes.* **53**: 365–371.

Schrank, A.J., Webb, P.W. and Mayberry, S. (1999) How do body and paired-fin positions affect the ability of three teleost fishes to maneuver around bends? *Can. J. Zool.* **77**: 203–210.

Temple, G.K. and Johnston, I.A. (1998) Testing hypotheses concerning the phenotypic plasticity of escape performance in fish of the family Cottidae. *J. Exp. Biol.* **201**: 317–331.

Videler, J.J. (1993) *Fish Swimming.* Chapman and Hall, New York.

Videler, J.J. and Weihs, D. (1982) Energetic advantages of burst-and-coast swimming of fish at high speed. *J. Exp. Biol.* **97**: 169–178.

Wahl, D.H. and Stein, R.A. (1988) Selective predation by three esocids: the role of prey behaviour and morphology. *Trans. Am. Fish. Soc.* **117**: 142–151.

Wainwright, S.A. (1983) To bend a fish. In: *Fish Biomechanics* (eds P.W. Webb and D. Weihs). Praeger, New York, pp. 68–91.

Ware, D.M. (1975) Growth, metabolism, and optimal swimming speed of a pelagic fish. *J. Fish. Res. Board Can.* **32**: 33–41.

Webb, P.W. (1977) Effects of median-fin amputation on fast-start performance of rainbow trout (*Salmo gairdneri*). *J. Exp. Biol.* **68**: 123–125.

Webb, P.W. (1978) Fast-start performance and body form in seven species of teleost fish. *J. Exp. Biol.* **74**: 211–226.

Webb, P.W. (1981) The effect of the bottom on the fast start of a flatfish, *Citharichthys stigmaeus*. *Fish. Bull. US* **79**: 271–276.

Webb, P.W. (1982) Locomotor patterns in the evolution of actinopterygian fishes. *Am. Zool.* **22**: 329–342.

Webb, P.W. (1984a) Body form, locomotion and foraging in aquatic vertebrates. *Am. Zool.* 24: 107–120.

Webb, P.W. (1984b) Body and fin form and strike tactics of four teleost predators attacking fathead minnow prey. *Can. J. Fish. Aquat. Sci.* 41: 157–165.

Webb, P.W. (1984c) Form and function in fish swimming. *Sci. Am.* 251: 72–82.

Webb, P.W. (1986a) Effect of body form and response threshold on the vulnerability of four species of teleost prey attacked by largemouth bass. *Can. J. Fish. Aquat. Sci.* 43: 763–771.

Webb, P.W. (1986b) Locomotion and predator–prey relationships. In: *Predator–Prey Relationships* (eds M.E. Feder and G.V. Lauder). Chicago University Press, Chicago, IL, pp. 24–41.

Webb, P.W. (1988) Simple physical principles and vertebrate aquatic locomotion. *Am. Zool.* 28:709–725.

Webb, P.W. (1989) Station holding by three species of benthic fishes. *J. Exp. Biol.* 145: 303–320.

Webb, P.W. (1993a) Swimming. In: *The Physiology of Fishes* (ed. D.D. Evans). CRC Press, Boca Raton, FL, pp. 47–73.

Webb, P.W. (1993b) Is tilting at low swimming speeds unique to negatively buoyant fish? Observations on steelhead trout, *Oncorhynchus mykiss*, and bluegill, *Lepomis macrochirus*. *J. Fish. Biol.* 43: 687–694.

Webb, P.W. (1993c) The effect of solid and porous channel walls on steady swimming of steelhead trout *Oncorhynchus mykiss*. *J. Exp. Biol.* 178: 97–108.

Webb, P.W. (1994a) Exercise performance of fish. In: *Advances in Veterinary Science and Comparative Medicine*, 38B (ed. J.H. Jones). Academic Press, Orlando, FL, pp. 1–49.

Webb, P.W. (1994b) The biology of fish swimming. In: *Mechanics and Physiology of Animal Swimming* (eds L. Maddock, Q. Bone and J.M.V. Rayner). Cambridge University Press, Cambridge, pp. 45–62.

Webb, P.W. (1997) Swimming. In: *The Physiology of Fishes*, 2nd Edn, (ed. D.D. Evans). CRC Press, Boca Raton, FL, pp. 3–24.

Webb, P.W. (1998) Entrainment by river chub, *Nocomis micropogon*, and smallmouth bass, *Micropterus dolomieu*, on cylinders. *J. Exp. Biol.* 201: 2403–2412.

Webb, P.W. and de Buffrénil, V. (1990) Locomotion in the biology of large aquatic vertebrates. *Trans. Am. Fish. Soc.* 119: 629–641.

Webb, P.W. and Skadsen, J.M. (1980) Strike tactics of *Esox*. *Can. J. Zool.* 58: 1462–1469.

Webb, P.W. and Weihs, D. (1986) Functional morphology of early life-history stages of fishes. *Trans. Am. Fish. Soc.* 115: 115–127.

Webb, P.W., Gerstner, C.L., and Minton, S.T. (1996a) Station holding by the mottled sculpin, *Cottus bairdi* (Teleostei: Cottidae), and other fishes. *Copeia* 1996: 488–493.

Webb, P.W., LaLiberte, G.D. and Schrank, A.J. (1996b) Does body and fin form affect the maneuverability of fish traversing vertical and horizontal slits? *Environ. Biol. Fish.* 46: 7–14.

Weihs, D. (1973a) Optimal fish cruising speed. *Nature* 245: 48–50.

Weihs, D. (1973b) Hydromechanics of fish schooling. *Nature* 241: 290–291.

Weihs, D. (1975a) An optimum swimming speed of fish based on feeding efficiency. *Israel J. Technol.* 13: 163–167.

Weihs, D. (1975b) Some hydrodynamic effects of fish schooling. In: *Swimming and Flying in Nature* (eds T.Y. Wu, C.J. Brokaw and C. Brennan). Plenum, New York, pp. 703–718.

Weihs, D. (1978) Tidal stream transport as an efficient method of migration. *J. Cons. Int. Explor. Mer* 38: 92–99.

Weihs, D. (1980) Energetic significance of changes in swimming modes during growth of anchovy larvae, *Engraulis mordax*. *Fish. Bull. US* 77: 597–604.

Werner, E.E. (1986) Species interactions in freshwater fish communities. In: *Community Ecology* (eds J. Diamond and T.J. Case). Harper and Row, New York, pp. 344–357.

6

The biomechanics of intermittent swimming behaviours in aquatic vertebrates

Robert W. Blake

1. Introduction

For many years, physiologists have investigated the energetics of animal swimming through respirometry, employing flumes to set constant levels of activity. The animal is encouraged to move steadily against the flow for preset periods of time. Over the last 30 years or so, observations and the remote tracking (e.g. telemetry, underwater video systems, etc.) of the movements of animals in nature have made it clear that sustained continuous motion is the exception rather than the rule as far as the pattern of animal swimming is concerned. Left to their own devices, many aquatic vertebrates choose to move intermittently, alternating periods of force production with periods of passive motion. This chapter is about the energetic consequences of intermittent propulsion. Reviews of fish swimming in general have been given by Blake (1983a), Videler (1993) and in Chapter 5.

Burst-and-coast swimming involves cyclic bursts of movement followed by passive coasting. It is a common intermittent swimming mode in neutrally buoyant fish and certain marine mammals and birds. Burst-and-glide swimming occurs in negatively buoyant fish and involves changes in altitude. The kinematics, mechanics, and energetics of both swimming modes are compared with those of constant speed swimming at a given level, followed by a discussion of porpoising in cetaceans and aquatic birds and the unique locomotor behaviour of flying fish. The basic mechanics and energetics of porpoising are reviewed and extended by considering scaling issues, the influence of air resistance during the aerial phase, and the possible effects of sea conditions on leap length. Discussion of gliding flight in flying fish focuses on the performance of the cypselurids. Glide distance is predicted from a simple mathematical model of the process. Results are in good agreement with literature values of flight distance.

Biomechanics in Animal Behaviour, edited by P. Domenici and R.W. Blake.
© 2000 Taylor & Francis, Oxford.

2. Burst-and-coast swimming

Burst-and-coast swimming consists of cyclic bursts of swimming followed by a passive coasting phase in which the body is kept rigid. It is a common swimming behaviour of certain fish, marine mammals, and aquatic birds. Weihs (1974) gave a theoretical model showing the potential energetic advantages of burst-and-coast swimming relative to constant speed swimming at velocities lower than the maximum aerobic cruising speed. Videler and Weihs (1982) extended the analysis of Weihs (1974) to show the possible energetic advantages of burst-and-coast swimming at high average speeds where powering is anaerobic. Blake (1983c) developed a simple hydromechanical model to determine the influence of overall body form on burst-and-coast swimming performance. The hydromechanical analyses of the above-mentioned workers are not described in detail here. Only the basic results are outlined and discussed.

Weihs (1974) defined a nondimensional energy saving factor, R_1, that expresses the energy saving in burst-and-coast swimming relative to that in steady, constant speed swimming. The potential energy savings of burst-and-coast swimming over constant speed propulsion are predicated on the fact that the drag of an oscillating body exceeds that of the equivalent rigid body by a factor (κ) referred to as the drag augmentation factor (see Blake (1983a) for a review). Among other things, the extent of energy savings depends on the ratio of the drag power of the sum of the burst (rigorous oscillation) and coast (passive rigid body drag power) phases in burst-and-coast swimming relative to that of constant speed propulsion at the same average velocity. In addition, the non-dimensional energy saving factor depends upon the final velocity at the end of the burst phase, the average velocity of the burst-and-coast cycle, and the maximum sustainable aerobic cruising speed. These values are plotted in *Figure 1*, for $\kappa = 3$. *Figure 1* shows that when the normalized speed (ratio of the average cycle speed to the maximum sustainable cruising speed) equals one so does the energy saving factor and there is no advantage to burst-and-coast swimming over constant speed swimming. For values of the ratio of the average cycle speed to the maximum aerobic speed of less than one, burst-and-coast swimming is economical relative to constant speed swimming. However, the value of the ratio of the final burst phase speed to the maximum aerobic speed should not exceed the average cycle velocity by too much because as the ratio of the final burst speed to the maximum aerobic cruising speed tends to one so does the energy saving factor. *Figure 1* shows that the smaller the required average cycle velocity, the larger the benefit from burst-and-coast swimming.

It turns out that the theoretical minimum value of the energy saving factor ranges from one (for a ratio of the average cycle speed of one to $1/\kappa$, where κ is the drag augmentation factor) as the average cycle velocity tends to zero. *Figure 2* shows the influence of the drag augmentation factor on the energy saving factor in relation to the normalized speed. The graph is plotted for a ratio of the final burst speed to the maximum aerobic speed of 0.75. Small values of the drag augmentation factor correspond to small energy gains (i.e. to large values of the energy saving parameter). It is important to note that $\kappa > 1.0$ is required for any positive results.

Weihs (1974) also considered the influence of burst-and-coast swimming on the range available on a given energy store. In *Figure 3*, the range in burst-and-coast swimming relative to that in constant speed swimming is plotted against the ratio of the average burst-and-coast cycle speed to the maximum sustainable aerobic speed for salmon (*Onchorhynchus nerka*) and haddock (*Melanogrammus aeglefinus*). The graph

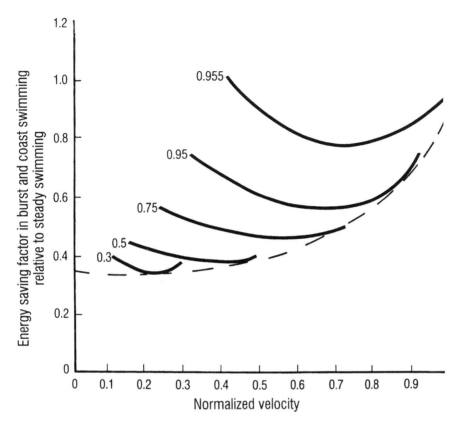

Figure 1. *The relative energy saving factor in burst-and-coast swimming relative to steady swimming is plotted against normalized swimming velocity (mean cycle speed/maximum sustainable aerobic speed) for a drag augmentation factor of three. Curves are for different values of the ratio of final coast phase speed to the maximum sustainable aerobic speed. The broken curve indicates the 'optimum' energy savings (R_1) for given values of the ratio of the final coast phase speed to the maximum sustainable aerobic speed in relation to the normalized speed (mean cycle speed/maximum sustainable aerobic speed) for $\kappa = 3$. Based on Weihs (1974).*

shows that range increases of the order of 50% and 80% are possible for salmon and haddock, respectively.

Blake (1981) pointed out that the Weihs model does not take account of differences in propulsive efficiency of the fish. During the burst phase of the burst-and-coast cycle it is less than that of steady swimming (e.g. McCutcheon, 1977). Taking these differences into account reduces the maximum energetic advantage of burst-and-coast swimming from about 50% to 30%.

Videler and Weihs (1982) considered the energetic advantage of high-speed burst-and-coast swimming. The model of Weihs (1974) is modified for this situation by considering the maximum anaerobic velocity and the associated maximum anaerobic thrust. The energy spent during the burst phase E_b, energy cost of constant speed swimming E_c, and the ratio of the two energy costs are given by

$$E_b = \frac{1}{\eta} \int_0^{t_1} TV \, dt \tag{1}$$

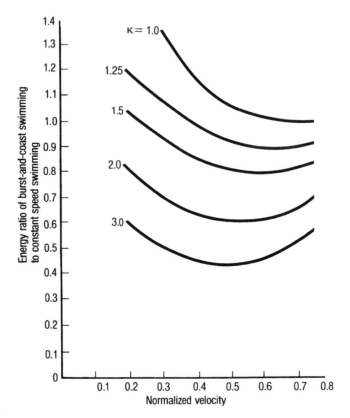

Figure 2. The energy cost ratio of burst-and-coast swimming relative to constant speed swimming is plotted against normalized velocity (mean cycle speed/maximum sustainable aerobic speed) for different values of the drag augmentation factor. Based on Weihs (1974).

$$E_c = \frac{1}{\eta} T_c V_c (t_1 + t_2) \tag{2}$$

$$R_1 = \frac{E_b}{E_c} = \frac{\int_0^{t_1} TV \, dt}{T_c V_c (t_1 + t_2)} \tag{3}$$

where η, t_1, t_2, T, T_c, and V_c are the efficiency (assumed to be constant), duration of the burst phase, duration of the coast phase, thrust, constant speed swimming thrust, and average velocity of the burst-and-coast cycle (= constant speed velocity), respectively. Assuming that the thrust during the burst phase is the maximum anaerobic thrust T_e, R_1 may be written as

$$R_1 = \frac{T_e \int_0^{t_1} V \, dt}{T_c V_c (t_1 + t_2)} \tag{4}$$

Equation (4) is not convenient for practical purposes. Noting that the integral term is equal to the distance covered during the burst phase, s_1, that U_c is taken as the average

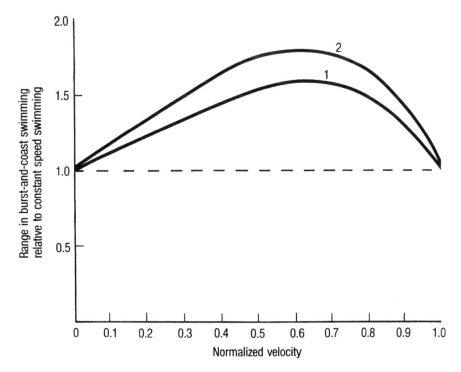

Figure 3. *The range in burst-and-coast swimming relative to constant speed swimming is plotted against normalized velocity (mean cycle speed/maximum sustainable aerobic speed) for salmon and haddock (curves 1 and 2, respectively). Based on Weihs (1974).*

cycle velocity, and that because thrust equals drag and equals the average thrust for two-phase swimming ($T = D = \frac{1}{2}\rho S_w \kappa C_D V_{avg}^2$, where T, D, ρ, S_w, C_D, and V_{avg}^2 are the thrust, drag, water density, wetted surface area, drag coefficient, and average swimming velocity, respectively.) Equation (4) can be written as

$$R_1 = \frac{1}{(V_c/V_e)^2} \frac{s_1}{s_1 + s_2}$$
$$\approx \frac{1}{(V_c/V_e)^2} \frac{1}{1 + s_2/s_1} \qquad (5)$$

where s_2 is the distance covered during the coasting phase of the burst-and-coast cycle and V_e is the maximum anaerobic speed. Videler and Weihs (1982) applied Equation (5) (for $\kappa = 3$) to cod (*Gadus morhua*) and saithe (*G. virens*) swimming rapidly in the burst-and-coast mode. With the correct choice of optimal combinations of initial and final velocities, the intermittent swimming mode can be of the order of 2.5 times cheaper than constant speed swimming.

Blake (1983c) modelled fish as prolate spheroids of a given fineness ratio ($f = L/d$, where d is the maximum diameter of the body measured perpendicular to body length L). Employing the equations of motion for burst-and-coast swimming (Videler and Weihs, 1982; Weihs, 1974), Blake (1983c) determined the influence of fineness ratio on the distance traversed over a burst-and-coast cycle of bodies of equal volume, mass, and initial and final velocities. The predicted ratio of the distance covered by prolate spheroids of various fineness ratios relative to a sphere ($f = 1$) of the same volume, mass, and initial and final velocities in the coast phase is shown in *Figure 4*. The maximum values of about 10 and 40 (for the case of a turbulent and laminar boundary layer, respectively) occur at about $f = 5$. Strictly speaking, the curves in *Figure 4* apply only to the coasting phase of the cycle. However, if it is assumed that the drag augmentation factor as a result of bodily oscillation and the maximum velocity possible are independent of fineness ratio, then the same curves also apply to the overall cycle. This is unlikely, because added mass (mass of water entrained by the fish during the 'burst' phase of the cycle) will decrease with increasing fineness ratio. In any event, the analysis suggests that active fish should be characterized by a fineness ratio of about five. Values of f close to five are found for frequent burst-and-coast swimmers such as cod (Gadidae, $f = 4.5$–5.0). Forms of low f ($f < 2$) (e.g. Diodontidae, Tetraodontidae, Ostraciidae) do not exhibit intermittent swimming.

It has been argued that the morphological requirements for good steady and unsteady (burst) swimming performance in fish are mutually exclusive (Webb, 1984). It has long been known that a fineness ratio of about five is optimal for maximizing steady swimming. Arguably, many pelagic and nektonic fish are of a form that allows for good steady and intermittent swimming performance.

Videler and Weihs (1982) have shown that a fish may choose many combinations of initial and final velocities that will result in intermittent propulsion being energetically less costly than steady swimming at the same average velocity. Experimental data from Videler and Weihs (1982) on cod and saithe are plotted in *Figure 5* to show the influence

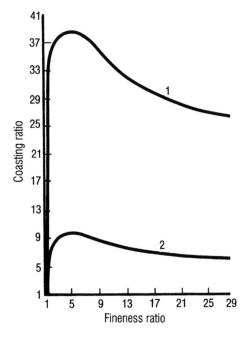

Figure 4. *The influence of fineness ratio on the coasting ratio (coasting distance of a prolate spheroid of given fineness ratio relative to that of a sphere) for the case of a laminar (curve 1) and turbulent (curve 2) boundary layer. Based on Blake (1983c).*

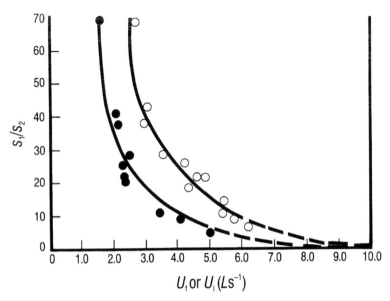

Figure 5. The ratio of coasting distance (s_2) to burst phase distance (s_1) is plotted against the final (U_f, ●) and initial (U_p, ○) coasting velocities for cod and saithe. Data from Videler and Weihs (1982).

of initial and final velocities of the burst-and-coast cycle on the ratio of s_2 to s_1. Lower values result in a proportionately longer coast phase relative to the burst phase. For a constant acceleration rate in the burst phase, the distance traversed between given values of the initial and final cycle velocity is independent of their absolute values. This is not so for the coasting phase, however, where deceleration is proportional to $-kV^2$ (where k is a constant). In the coasting phase higher values of s_2 occur for low absolute initial and final cycle velocities. However, more time is required to complete the cycle at lower velocities. A greater number of cycles will be completed at higher V_i and V_f, resulting in a high average cycle velocity and a greater absolute distance traversed.

Weihs (1980) has assessed the energetic significance of changes in swimming modes during the ontogeny of larval anchovy (*Engraulis mordax*). He observed that burst-and-coast swimming becomes economical for larvae of about 5 mm in length and beyond. At lower Reynolds numbers ($L < 5$ mm) 'bursts' of continuous swimming are energetically more efficient.

Seals and penguins can be observed burst-and-coast swimming in large aquaria. Both animals maintain an essentially rigid body while coasting. Seals and penguins are propelled by their flukes and wings, respectively, during the 'burst' phase of the cycle. Drag augmentation as a result of oscillation of these appendages is not known. However, energy savings should be realized for $\kappa > 1$.

3. Energy efficient swimming strategy in negatively buoyant fish

Some fish are negatively buoyant (e.g. certain scombrids, sharks, rays) and have the option of changing their altitude as they swim, as opposed to swimming at a constant speed and level. The pattern proposed by Weihs (1973) is shown diagrammatically in *Figure 6*. The fish glides down from position 1 to position 2 at a constant angle θ_1 with a

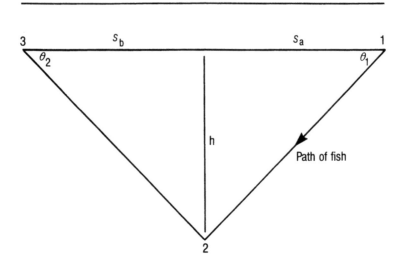

Figure 6. *Possible swimming paths of a negatively buoyant fish. This figure is explained in the text. Based on Weihs (1973).*

loss of altitude h. Once at position 2, the fish swims upward to point 3 at an angle θ_2. In constant speed swimming at a constant level the fish traverses the distance $(s_a + s_b)$ directly from point 1 to point 3. Weihs (1973) considered the relative energy cost of the two modes of swimming. The theoretical analysis of two-phase swimming was given in detail by Weihs (1973) and reviewed by Blake (1983a). Only the basic results of the model are given here. The discussion below focuses on field work on free-ranging tuna that gives support to the basic model and indications of the actual energy savings in nature.

Figure 7 is a plot of a non-dimensional relative energy saving parameter ($R_2 = (E_c - E_G)/E_c$, where E_c and E_G are the energy costs of constant speed swimming at a constant level and the two-phase mode, respectively) plotted against the ascent angle θ_2, for various values of the descent angle θ_1, for a drag augmentation factor as a result of bodily oscillations of $\kappa = 3$. *Figure 7* shows that the maximum value of the relative energy saving parameter decreases with increasing glide angle. When glide angles exceed about 30° for high values of the ascent angle ($\theta_2 > 60°$), negative values of R_2 occur. For glide values of the order of 10° energy savings of the order of more than 50% are theoretically possible.

In addition to considering potential energy savings from this form of two-phase swimming, Weihs (1973) also considered the range increases possible over steady constant speed and level swimming. A non-dimensional range increase factor (R_2) is plotted against θ_2 for various values of θ_1 in *Figure 8*. For $\theta_1 = 12°$, a range increase of over 80% is predicted for the two-phase mode over the constant speed and level option.

The non-dimensional energy saving factor R_2 can be written as

$$R_2 = \frac{\eta_m^{-1} \kappa D \, (s_a + s_b) - \eta_m^{-1} T \, (s_b^2 + h^2)^{1/2}}{\eta_m^{-1} \kappa D \, (s_a + s_b)} = \frac{E_c - E_G}{E_c} \tag{6}$$

where η_m and D refer to the muscle efficiency and drag of the fish, respectively.

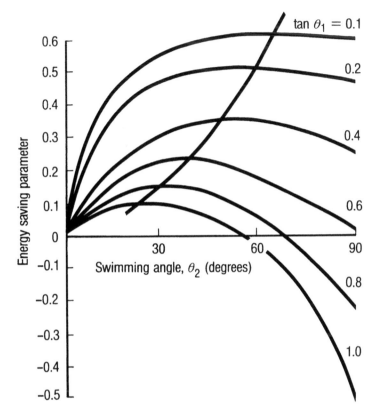

Figure 7. *A non-dimensional energy saving parameter for burst-and-glide relative to constant speed and level swimming is plotted against ascent angle (θ_2) for different values of the descent (glide) angle (θ_1) for $\kappa = 3$. Maximum savings are indicated by a line passing through the maxima of each curve. Based on Weihs (1973).*

Holland *et al.* (1990) have tracked the horizontal and vertical movements of yellowfin (*Thunnus albacares*) and bigeye (*T. obesus*) tuna off the coast of Hawaii using pressure-sensitive ultrasonic transmitters. Records of both species confirm the 'sawtooth' pattern of movement characteristic of the two-phase swimming mode. *Figure 9* shows a typical record for *T. obesus*. However, neither species showed the two-phase pattern of swimming all of the time. The overall significance of this locomotor strategy, as in all others, depends not only on the energy savings relative to other options, but also on the proportion of the active periods for which it is maintained.

Block *et al.* (1997) used acoustic telemetry to track the movement patterns of yellowfin tuna in the California Bight. They found that the fish move in an oscillatory motion consistent with the burst-and-glide pattern and that the descent speed was constant and ascent rates variable. This is expected if active swimming motions are confined to the ascent phase of the burst-and-glide cycle. Block *et al.* (1997) concluded that the oscillatory diving pattern of yellowfin tuna is consistent with the energy conserving strategy of burst-and-glide swimming.

More detailed analysis of field tracking data over long periods of time will be required before the overall importance of two-phase swimming to the daily energy

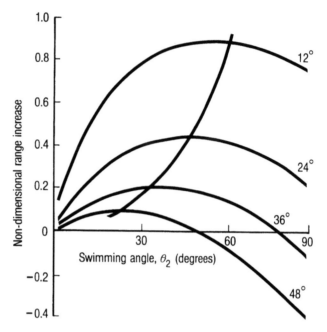

Figure 8. *A non-dimensional range increase parameter is plotted against ascent angle (θ_2) for various values of the descent angle. Maximum increases are indicated. Based on Weihs (1973).*

budget of tuna and other negatively buoyant fishes can be assessed. Two-phase swimming of the type discussed above is not limited to negatively buoyant fishes. Webber and O'Dor (1986) and O'Dor (1988) have recorded the behaviour in free-swimming squid. According to O'Dor (1988), two-phase swimming may reduce previous estimates for the gross cost of transport for squid under natural conditions

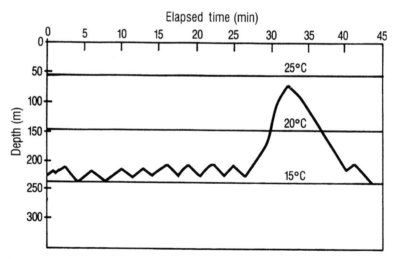

Figure 9. *Horizontal and vertical movements of a specimen of Thunnus obesus. (Brill, personal communication).*

by at least 35%. Records of diving grey seals (*Halichoerus grypus*) obtained from transmitted ultrasonic signals also show a 'sawtooth' pattern when dive depth is plotted
against time (Thompson and Fedak, 1993). The seals are probably close to neutrally
buoyant and so the energy cost of swimming horizontally is roughly the same as that
for swimming up or down. It is unlikely that seals gain any locomotor energetic
advantage from their up-and-down movement pattern.

4. Porpoising

Porpoising is a characteristic swimming behaviour of certain marine mammals (e.g.
small cetaceans, seals) and birds (e.g. penguins). Phases of active swimming alternate
with leaps from the water. The centre of mass of the animal follows an approximately
ballistic trajectory. The possible energetic significance of porpoising has been explored
using simple mathematical models by Au and Weihs (1980), Blake (1983a,b), Blake and
Smith (1988) and Gordon (1980). Hui (1987) considered the implications of
porpoising to respiratory ventilation. The discussion below centres on the energetic
analyses.

Au and Weihs (1980) described three modes of swimming in dolphins: slow
swimming close to the surface, cruising just beneath the surface, and 'fast running'. In
fast running, the dolphin leaps clear of the surface in a series of parabolic (porpoising)
leaps. Porpoising was modelled to determine if the energy cost of leaping is less than
that for continuous swimming close to the surface. The model was modified and
extended by Blake (1983b).

The energy required to traverse a unit distance in continuous subsurface swimming
is given by the product of the swimming drag and the distance travelled. The energy
cost of leaping can be found by multiplying the weight of the animal by the jump
height. The relative cost of leaping and subsurface swimming R_3 is

$$R_3 = \frac{1.2 \ (mg) \ (V_e^2/12 \ g)}{{}^1/_2 \rho \, S_w V_e^2 \, (V_e^2/g) \, \delta \, C_D} \tag{7}$$

Equation (7) assumes that the animal leaves the water at 45°, giving a maximum leap
distance V_e^2/g, where V_e is the emergence velocity. The height of the jump is $V_e^2/12 \ g$.
In Equation (7), the weight of the animal, mg, is multiplied by a factor of 1.2, which
accounts for the mass of the fluid (longitudinal added mass) that the animal takes with
it as it leaves the water.

The drag of an oscillating body moving close to the water's surface is greater than
that of a deeply submerged rigid form (as a result of gravitational wave drag) and this is
accounted for in Equation (7) by a drag augmentation factor δ. There are two effects
determining the value of δ, drag augmentation owing to wave drag, and drag augmentation owing to bodily oscillation. The latter effect has already been mentioned. For
streamlined bodies, the additional drag caused by the influence of the water's surface is
a function of the submersion depth index (depth of submersion/maximum body
depth, where submersion depth is measured from the surface to the centreline of the
body). Drag augmentation as a result of the surface effect is plotted against the
submersion index in *Figure 10*, which shows that drag may be increased by about five
times above the deeply submerged rigid body value when the animal is very close to the
surface (Hertel, 1966). Because such creatures as cetaceans and seals are not perfect

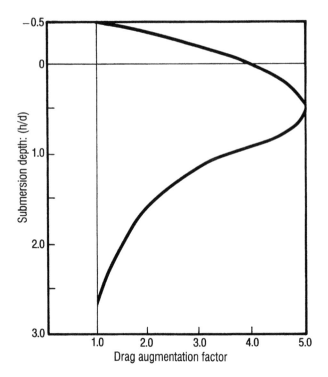

Figure 10. *Drag augmentation as a result of surface proximity is plotted against a nondimensional index of submersion depth (depth of submersion, h/maximum body depth, d). Based on Hertel (1966).*

bodies of revolution, Au and Weihs (1980) and Blake (1983a) assigned a maximum value of 4.5 to the magnitude of the surface drag effect. The combined influence of drag augmentation as a result of the surface and bodily oscillation may increase the drag on a streamlined body by a factor of about 20 relative to a rigid equivalent that is deeply submerged.

In *Figure 11*, the energy required to swim close to the surface, deeply submerged, and in porpoising, is plotted against velocity for a 'typical' dolphin. The graph shows that leaping is energetically more costly (i.e. $R_3 > 1$) than swimming close to the surface up to a certain speed (crossover speed, V_{cross}), after which it is more efficient (i.e. $R_3 < 1$). In addition, the curve for deeply submerged swimming also crosses that for leaping, showing that at higher speeds, leaping is also the most energy efficient locomotor strategy. Blake (1983a,b) developed a simple mathematical model that equates the power required for swimming with that available from the propulsive musculature and showed that dolphins ought to be able to attain predicted crossover speeds even in the presence of a turbulent boundary layer.

By substituting a value for δ and empirical equations for the relationships between wetted surface area and mass with respect to body length into Equation (7), and setting $R_3 = 1.0$, the crossover speed can be simply expressed as V_{cross} = (constant L)$^{1/2}$ Au and Weihs (1980) predicted crossover speeds that are about twice those predicted by Blake (1983a,b) because of the omission of a term to account for drag augmentation owing to body oscillation in their model. Au et al. (1988) reported field

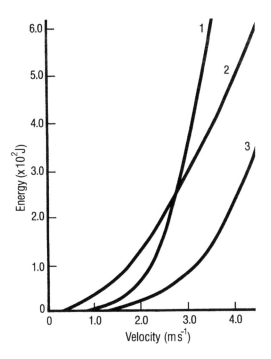

Figure 11. *The energy requirements for swimming close to the surface, leaping, and swimming deeply submerged (curves 1, 2 and 3, respectively) for a dolphin of about 2 m in length. From Blake (1983a).*

observations and porpoising speed measurements for *Stenella* spp. Their mean speed was 3.5 m s^{-1}. This value compares reasonably well with the value of V_{cross} (2.7 m s^{-1}) predicted by Blake (1983a,b). However, the crossover speed must be viewed as a lower bound for the velocity at which porpoising should occur, not as a prediction of porpoising speed *per se.* The predicted crossover speed from the model of Au and Weihs (1980) of about 5 m s^{-1} is significantly higher than the measured speeds. Williams and Kooyman (1985) predicted values of V_{cross} of 5.3 and 2.4 m s^{-1} for an 85 kg harbour seal (*Phoca vitulina*) based on the Au and Weihs (1980) and Blake (1983a,b) models, respectively. The animal was observed to begin porpoising at approximately 2.5–3.0 m s^{-1}. These results indicate that drag augmentation owing to oscillatory swimming is an important aspect of marine mammal locomotion in general and porpoising in particular.

Hui (1987) and Blake and Smith (1988) discussed porpoising in penguins. Hui (1987) found a mean emergence angle for captive porpoising humboldt (*Spheniscus humboldti*) and blackfoot (*S. demersus*) penguins, and free-ranging adélie penguins (*Pygoscelis adeliae*) of 32°, a value significantly different from the 45° assigned by Au and Weihs (1980) and Blake (1983a,b). For emergence angles, α_e, other than 45°, the energy cost ratio equation for leaping relative to swimming is

$$R_3 = \frac{mg\,(1 + m')\,(V_e^2 \sin^2 \alpha_e/2\,g)}{{}^1\!/{}_2\rho\,S_w\,\delta\,C_D\,(V_e^2/g)\,(\sin 2\alpha)\,V^2} \qquad (8)$$

where m' is a correction term for longitudinal added mass. Penguins are propelled underwater by their wings and the body is rigid. Blake and Smith (1988) assigned a value of 0.07 to m' for penguins (equivalent to the factor of 1.2 in Equation (7)), corresponding to that for a technical body of revolution of the same fineness ratio (Landweber, 1961).

On the basis of literature values, lower and upper bounds to C_D of 0.003 and 0.005 were assigned. Drag augmentation caused by oscillation of the flippers was assumed to be between 1.0 and 2.0 times the rigid value. A value of 4.5 was assigned to the drag augmentation factor as a result of surface proximity. Substituting these values and morphometric data into Equation (8), the energy costs and crossover speeds for small (humboldt) and large (emperor) penguins were calculated for different values of the oscillatory drag augmentation factor, drag coefficient, and emergence angle (32° and 45°). Predicted values of V_{cross} ranged from 0.81 m s⁻¹ (for δ = 9, C_D = 0.005, α_e = 32°) to 1.88 m s⁻¹ (for δ = 4.5, C_D = 0.003, α_e = 45°) for a 4.2 kg humboldt penguin. For a 30 kg emperor penguin, predicted values of V_{cross} ranged from 1.70 m s⁻¹ (for δ = 9, C_D = 0.005, α_e = 32°) to 3.93 m s⁻¹ (for δ = 4.5, C_D = 0.003, α_e = 45°). *Figure 12* shows the calculated hydromechanical energy cost of swimming for a humboldt penguin in porpoising and subsurface swimming. In *Figure 13*, crossover speeds for penguins (Blake and Smith, 1988, Equation (2)) are plotted against body length. Observed speeds are also shown. The model predicts that points falling below and above the curve should correspond to subsurface swimming and porpoising, respectively. Three of the four subsurface values fall below the curve and all of the porpoising values are above it. However, the porpoising birds did not do so continuously, and the overall number of measurements is too small to draw definite conclusions.

Small penguins are the smallest aquatic animals that porpoise. The largest are probably small whales (e.g. killer whales ~ 10 m in length). The dependence of crossover speed for streamlined forms of different size (expressed as body volume) on different values of drag augmentation caused by surface and body oscillation effects is shown in *Figure 14*. The graph shows that crossover speed increases with decreasing drag for an animal of any given size and that relative to their size, small animals have high crossover speeds. A small fish of about 20 cm in length (volume 0.0001 m³) swimming close to the surface has a predicted crossover speed of about 2 m s⁻¹, or

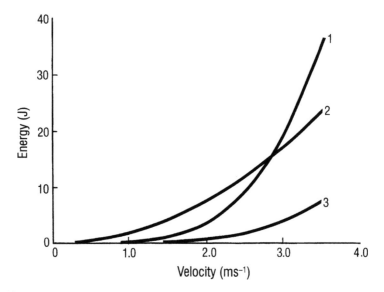

Figure 12. The energy requirements of a 30 kg emperor penguin swimming close to the surface, leaping, and swimming deeply submerged.

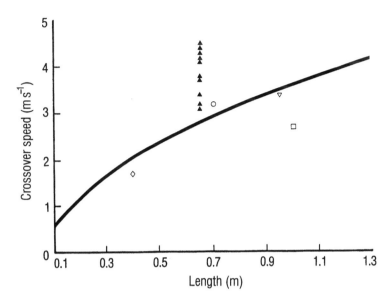

Figure 13. *Crossover speed is plotted against body length for penguins. The points above the curve (▲, humboldt penguins, from Hui, (1987); ○, blackfoot penguins, from Clark and Bemis (1979)) are for porpoising animals. Those below the curve (◊, ▽, □ for little blue, king, and emperor penguins, respectively) are from Clark and Bemis (1979).*

approximately $4\,L\,s^{-1}$. This is a relatively high performance level, which could probably not be sustained for a long period.

The scaling of crossover speed in relation to size for streamlined aquatic animals is more clearly shown in *Figure 15*. The graph shows that relative to their size, smaller forms must perform at a higher level to attain their crossover speed. Porpoising may be

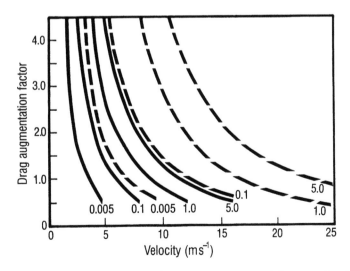

Figure 14. *Effect of the drag augmentation factor on crossover speed for streamlined animals of different body volume. The continuous and broken curves refer to bodily oscillation drag augmentation factors of four and one, respectively. From Blake (1983a).*

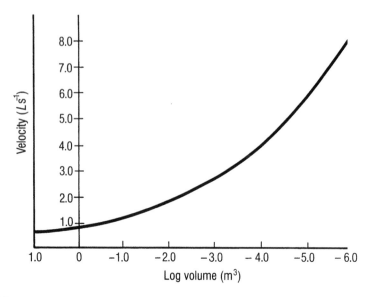

Figure 15. *Crossover speed (in lengths per second) is plotted against body volume for streamlined animals. The curve is based on a total drag augmentation factor (surface augmentation, plus that owing to bodily oscillation of 8.5). Data from Blake (1983a).*

viewed as an energy efficient locomotor strategy for maintaining a relatively high level of performance over a long period of time. As such it is probably an aerobic activity. It is likely that smaller forms do not porpoise because their crossover speeds are relatively high and can be attained only on the basis of anaerobic powering. In *Figure 16*, sustainable speed for fish (from Wu (1977, Figure 10), for 'active' fish) is divided by

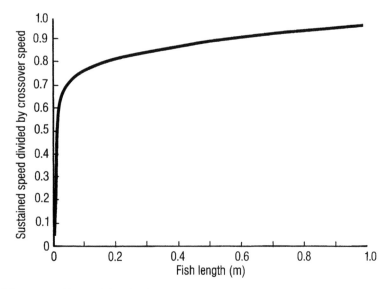

Figure 16. *Sustainable speed divided by crossover speed is plotted against body length for fish. Data from Wu(1977).*

the predicted crossover speed for fish up to 1.0 m in length. The graph shows that the crossover speed exceeds the sustainable speed over this size range. However, larger percomorph fish, comparable in size with dolphins, should be capable of attaining predicted crossover speeds aerobically but do not porpoise. Why not? In cetaceans the principal plane of flexure of the backbone is dorso-ventral, allowing the body to arch during the aerial phase of porpoising and a 'clean' head-first re-entry into the water. In fish, the principal plane of backbone flexure is lateral, precluding arching of the body. A porpoising leap in fish would therefore end in an energy wasting splash.

There are other factors that will influence the practical lower bound to porpoising speed. At small size, leap length and height may be too low to allow for efficient porpoising. For example, for a 4.2 kg humboldt penguin with $\delta = 9$, $C_D = 0.005$, and $\alpha_e = 32°$, $V_{cross} = 0.81$ m s^{-1}. Emerging at this speed, the maximum height of the centre of mass (height $= (V^2 \sin^2 \alpha_e)/2g$) would be 8.6 cm. On the basis of $L = 0.65$ m and a fineness ratio of 4.5, the maximum width of the body would be about 14 cm. Assuming that the centre of mass occurs midway along this point, then the leap height would have to exceed 7 cm for the animal to clear the water at its point of maximum elevation. Not doing so would result in energy wasting splashing. In addition, smaller forms are more likely to be constrained by sea conditions. The model assumes a level water surface, but this is unlikely to be so in nature.

The energetic models of porpoising outlined above are by no means complete. They ignore the effect of air resistance during the aerial phase and the influence of sea conditions, and do not take account of the exact behaviour of the animals during the submerged phase of the porpoising cycle.

Working from the basic equations of motion and force balance (see Lamb, 1929), it can be shown that the decrease in leap length as a result of the influence of air resistance is given by

$$L' = L_0 - 0.67 \left(\frac{\rho S_w C_D}{2m} \right) L_0^2 \tag{9}$$

where L' and L_0 are the decrease in leap length and leap length in a vacuum, respectively. It can be seen from Equation (9) that the effect of air resistance is insignificant, even for small forms of relatively high C_D. Results for a dolphin with a body length of 2 m are shown in *Figure 17*. For practical purposes, the curve for $C_D = 0.0025$ for leap lengths up to about 4 m is relevant.

A first approximation to the effects of sea conditions on the performance of a porpoising animal can be made by considering the influence that emerging on an inclined plane has on the leap length; that is to say, we suppose that the animal emerges on the windward side of a wave and re-enters the water on the same incline. It can be shown that leap length on a plane inclined at an angle φ relative to the horizontal is given by

$$L_i \approx \frac{2V_e^2 \cos \alpha_e \sin(\alpha_e - \varphi)}{g \cos^2 \varphi} \tag{10}$$

where L_i is the leap length on an inclined plane. For $\alpha_e = 45°$ (when leap length $= V^2/g$), leap length on an inclined plane is reduced by a factor I relative to a horizontal surface:

$$I = \frac{2\cos \alpha_e \sin(\alpha_e - \varphi)}{\cos^2 \varphi} \tag{11}$$

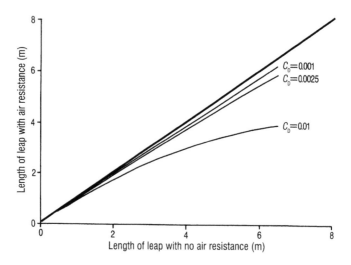

Figure 17. The effect of air resistance on the leap length of a dolphin.

For example, for $\alpha_e = 45°$ and $\varphi = 10°$, $I = 0.85$. An animal emerging on an incline of 10° at a given velocity will cover only 85% of the distance that would have been travelled over a flat surface. However, animals may compensate for this by increasing their emergence velocity, albeit at increased energy cost. Leap length $(L_0 = V^2 \sin 2\alpha_e/g)$ is much more sensitive to changes in V_e than to those in α_e.

The model does not take any account of the underwater phase of the porpoising cycle. It is known that dolphins re-enter the water with their pectoral fins extended (Au *et al.*, 1988). It is possible that the fins produce lift during the ascending stage of the underwater phase of the cycle. Modifications would be required for the force balance, taking account of the lift and its associated induced drag, if this is so. Changes in the force balance will change predicted values for the crossover speed.

5. Flying fish

Intermittent, gliding flight is reported to occur among three teleost families, the Exocoetidae, Pantodontidae, and Hemirhamphidae. The greatest degree of adaptation towards gliding is shown by the Exocoetidae. Specialized species belong to two genera, *Exocoetus* (two-winged forms, where only the pectoral fins function as wings) and *Cypselurus* (four-winged species, where the pectoral and pelvic fins act as lifting surfaces). In monoplane forms such as *Exocoetus volitans*, the momentum for flight is gained solely from rapid swimming close to the water's surface. Upon breaking through the surface, the pectoral fins are extended and the fish glides for about 10–20 m (Hubbs, 1933). Flight behaviour in *Cypselurus* is more elaborate, involving rapid subsurface swimming, emergence and a taxiing phase on the surface before take-off.

There is no absolute agreement on the exact flight performance of flying fish. Most information refers to cypselurids, and the account that follows focuses on these forms. An approximate description can be constructed from literature reports (e.g. Aleyev, 1977; Breder, 1930; Carter and Mander, 1935; Davenport, 1994; Edgerton and Breder, 1941; Gill, 1905; Hankin, 1920; Hertel, 1966; Hubbs, 1933, 1937; Klausewitz, 1960; Shoulejkin, 1929).

The fish accelerates in the water up to an emergence speed of about 10 m s⁻¹. Upon breaking through the surface, the body is inclined at a low angle (5–10°) to the horizontal, and the pectoral fins are extended. The force balance during this stage has been described by Hertel (1966). The lower lobe of the caudal fin remains in the water and beats rapidly (50–70 Hz), accelerating the fish in a taxiing run to take-off velocities of the order of 20 m s⁻¹. At this point the pelvic fins are extended. After a glide of about 10 m, the lower lobe of the caudal fin re-enters the water. A second taxiing phase may then propel the fish back up to take-off speed and a second flight. In this way, a total flight path of up to 400 m may be achieved in less than a minute.

Hertel (1966) considered speed in relation to possible values of lift coefficient for the fins in the influence of head winds of different strengths (*Figure 18*). *Figure 18* shows that an emergence speed of about 10 m s⁻¹ corresponds to a lift coefficient of about 1.3 in the absence of a head wind. Hertel (1966) also considered the trajectories of *Cypselurus californicus* for a constant value of the lift coefficient in relation to incident airflow ranging from 20 to 25 m s⁻¹ (*Figure 19*). The trajectory for an incident air flow (forward speed plus wind speed) of 20 m s⁻¹ corresponds to a total distance flown of 48 m. This corresponds well to the flight pattern described by Shoulejkin (1929) and my own observations of the performance of flying fish off the coast of Bermuda (unpublished observations). Most of the parabolic trajectories involving changes in height shown in *Figure 19* do not occur in nature. It is of interest then to consider the mechanics of low-level gliding in *Cypselurus*.

The total drag force acting on the fish during the aerial phase can be written as

$$F = K_1 V^2 + K_2 V^{-2} \qquad (12)$$

where K_1 and K_2 are constants; K_1 pertains to the body drag (parasite drag) and fin skin friction drag (profile drag) and K_2 to drag caused by lift (induced drag)

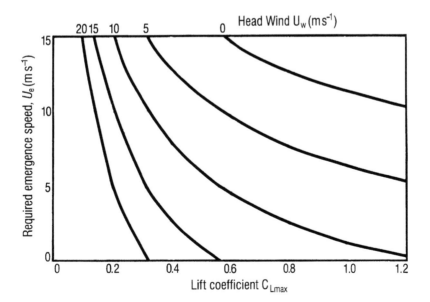

Figure 18. *Required emergence speed is plotted against the maximum lift coefficient for various values of head-wind speed for a 35 cm specimen of* Cypselurus. *Based on Hertel (1966).*

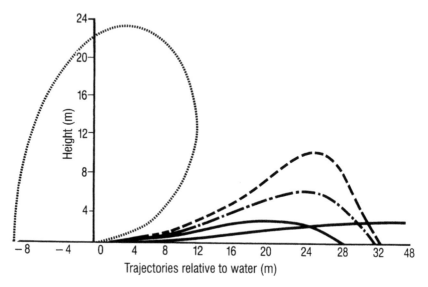

Figure 19. *Possible trajectories for a flying fish. This figure is explained in the text. Based on Hertel (1966).*

generated by the fins. From Newton's second law, the equation of motion of the gliding fish is

$$-F = mV \frac{dV}{ds} \tag{13}$$

where m is the fish mass and s is the glide distance. With some manipulation, Equation (13) can be written as

$$ds = -m \frac{dV}{K_1 V + K_2 V^{-3}} \tag{14}$$

Further manipulation gives the glide distance as

$$s = \frac{m}{4K_1} \left[\ln \left(V_0^4 + (K_2/K_1) \right) - \ln \left(V_f^4 + (K_2/K_1) \right) \right] \tag{15}$$

$$= \frac{m}{4K_1} \ln \left[\frac{V_0^4 + (K_2/K_1)}{[V_f^4 + (K_2/K_1)]} \right]$$

where V_0 and V_f are the initial and final glide velocities, respectively.

I have made estimates of m, K_1 and K_2 from a museum specimen of *Cypselurus californicus*. The fish was 0.35 m in length, with $m = 0.39$ kg ($mg \approx 3.9$ N), the planform area of the fins was 0.11 m², and the wing span was 0.63 m. Assuming an air density of 1.18 kg m⁻³ and a drag coefficient of 0.07 (Shoulejkin, 1929), values of 0.00454 kg m⁻¹ and 20.6 kg m³ s⁻⁴ are calculated for K_1 and K_2, respectively. For $V_0 = 20$ m s⁻¹ and $V_f = 10$m s⁻¹, $s = 52.2$ m, a predicted glide distance that compares well with the generally agreed performance estimate of about 50 m for this velocity interval (e.g. Aleyev, 1977; Hertel, 1966). Equation (15) shows that for given m, V_0 and V_f, glide distance increases with decreasing K_1 and K_2. However, changes in K_1 affect both terms on the right-hand side of the glide distance equation in the same direction and changes in K_2 influence only the second term. Therefore, errors in K_1 are more important than are those in K_2. A given per cent change in K_1 brings about a roughly corresponding change in s.

The adaptive significance of flight in flying fish is not clear. Stephens (1964) briefly discussed a variety of possibilities, ranging from escape from predators to finding prey and mates. Flying fish feed on zooplankton and occur in large schools and so there seems little reason to suggest flight as a requirement for finding food or mates. Rayner (1981) suggested that the cost of transport in gliding may be lower than that for continuous swimming over a given distance. In this sense, the behaviour may be viewed as an energy efficient transport strategy akin to porpoising. However, the high emergence speeds (about 10 m s^{-1}, 20–25 L s^{-1}) and high frequency of caudal fin oscillation during taxiing (50–70 Hz) imply that the process is anaerobically powered and cannot be sustained. In this connection, the results of counts of successive flights by *Cypselurus* are pertinent. Hubbs (1933) reported that about 55% of flights made by *Cypselurus* in the Celebes Sea were not continued and that 94% involved no more than three successive flights.

Escape from predators and finding prey are not mutually exclusive possible explanations. The principal predator of *Cypselurus* is the dolphin fish, *Coryphaena hippurus* (Stephens, 1964). The flying fish could escape from the visual field of the predator by leaving the water. The minimum distance required to escape detection upon re-entry is difficult to assess. It will depend, among other things, on the visual acuity of the predator, the contrast difference between the flying fish and its background, and the extent of light absorption and backscattering. Olson (1964) estimated a sight range for marine fish in clear water of about 10 m. It would seem that the flight performance of *Cypselurus* can exceed this estimate by a factor of about five.

Baylor (1967) observed that Atlantic flying fish *Cypselurus heterurus* off the south coast of Bermuda are often found in association with windrows of *Sargassum* weed and frequently move between patches of weed by flying. My observations made in the same area in August 1987 confirm this behaviour. It is possible that *C. heterurus* may utilize flying as a rapid way of moving between rich feeding sites, which also serve as refuges. Flying fish possess a high degree of visual acuity and a dual-purpose cornea, which permits effective vision in water and air (Baylor, 1967). This and other adaptations toward 'amphibious vision' are consistent with both the predator escape and prey-associated weed patch location hypothesis.

6. Conclusion

I have already emphasized the need for more field observations and assessments of the behavioural context and ecological significance of intermittent locomotor strategies in aquatic animals. Most of the models of intermittent propulsion discussed here make explicit predictions about the circumstances in which significant energetic benefits will accrue to animals engaged in particular intermittent movement styles. Field measurements of speed and movement pattern are needed to establish the extent of energetic gains. In particular, further research on the implications of burst-and-coast swimming to the energetics of free-ranging aquatic animals would clearly be rewarding. Also, more research is needed on forms other than fish. For example, I know of no work comparable with that on fish for burst-and-coast swimming in marine mammals and birds. In addition, a survey of the occurrence and pattern of intermittent swimming in fish would be most worthwhile. The overall significance of intermittent swimming to the foraging efficiency and daily energy budget of fish, marine mammals, and aquatic birds should be assessed. Similarly, depth-sensitive telemetry and side-scanning sonar could be employed to determine the likely realized energy savings associated with burst-and-glide swimming in tuna and other aquatic animals.

A comparative approach to evaluating the biological significance of intermittent propulsion is currently lacking and would be rewarding. For example, burst-and-coast swimming is common among fish and a comparative assessment of the relative usage, energy savings and contexts of burst-and-coast swimming in species of different form, individually and in schools and shoals is needed. Relatively 'fine-scale' observation of the kinematic similarities and differences between different forms has yet to be undertaken. It has been noted that some species collapse their median and caudal fins during the coasting phase of burst-and-coast swimming and some do not (Blake, personal observations).

It is also interesting to consider the extent to which members of given taxonomic groups engage in particular intermittent locomotor strategies. For example, many dolphins porpoise but some do not. Law and Blake (1994) studied the swimming behaviours and speeds of wild Dall's porpoises (*Phocoenoides dalli*). They showed that although *P. dalli* can attain and exceed the speeds at which porpoising is energetically efficient, they do not porpoise. Law and Blake (1994) argued that predation pressure from killer whales (*Orcinus orca*) may mean that swimming submerged at high speeds is an efficient means of avoiding predation in *P. dalli*.

The basic biomechanics of intermittent swimming strategies in aquatic vertebrates is currently fairly well understood in a behavioural context. However, as indicated above, there is scope for further research. Physiological measurements on free-ranging animals, documentation of 'fine-scale' kinematics and behaviour during intermittent propulsion (e.g. correlations of fin movements during the burst-and-coast swimming of fish), overall assessments of the impact of intermittent propulsion in important life history events (e.g. migrations), and the significance of the use of intermittent swimming strategies in conjunction with other locomotor strategies (e.g. use of currents and waves) seem promising areas.

7. Notation

Distances (m)

s	glide distance
s_1	distance covered during the burst phase of burst-and-coast swimming
s_2	distance covered during the coast phase of burst-and-coast swimming
d	maximum diameter of a body
L	body length
L'	leap length of a porpoising animal
L_0	leap length of a porpoising animal in a vacuum
L_i	leap length of a porpoising animal on an inclined plane (wave)

Area (m²)

S_w	characteristic reference (wetted) area of a swimming animal

Velocity (m s⁻¹)

V	velocity

V_c average velocity of the burst-and-coast cycle of a fish
V_{an} maximum anaerobic of a fish
V_i initial velocity of a burst-and-coast cycle in a swimming fish, initial glide velocity of a flying fish
V_f final velocity of a burst-and-coast cycle in a swimming fish, glide velocity of a flying fish
V_e emergence velocity of a porpoising animal
V_{cross} crossover speed, at which porpoising is energetically less expensive than other modes of swimming.

Mass (kg)

m mass
m' longitudinal added mass

Density (kg m⁻³)

ρ fluid density

Angles (degrees)

θ_1 glide angle of a negatively buoyant fish
θ_2 ascent angle of a swimming negatively buoyant fish
α_e emergence angle of an animal as it leaves the water
φ angle of an inclined plane (water surface) from which an animal leaps

Force (N)

F force
T thrust force of a swimming animal
T_c thrust of fish swimming at constant speed
T_{an} maximum anaerobic thrust of a fish
D hydrodynamic drag

Energy (J)

E_b energy expended in the burst phase of burst-and-coast swimming
E_G energy cost of burst-and-glide swimming

Factors, coefficients and ratios

R_1 an energy saving factor expressing the relative energy saving of burst-and-coast swimming to constant speed swimming
R_2 energy saving parameter for burst-and-glide swimming
R_3 relative energy cost of leaping (porpoising) to subsurface swimming
κ drag augmentation factor as a result of bodily oscillation
η hydrodynamic efficiency
η_m muscle efficiency

C_D drag coefficient
f fineness ratio
δ drag augmentation as a result of proximity to the water's surface of a swimming animal
K_1 a constant pertaining to the sum of the profile and parasite drag of an animal
K_2 a constant pertaining to the induced drag of an animal
I proportionate decrease in the leap length of a porpoising animal

References

Aleyev, Yu.G. (1977) *Nekton*. Junk, The Hague.

Au, D.W. and Weihs, D. (1980) At high speeds dolphins save energy by leaping. *Nature* **284**: 348–350.

Au, D.W., Scott, M.M. and Perryman, W.L. (1988) Leap–swim behaviour of 'porpoising' dolphins. *Cetus* **8**: 7–10.

Baylor, E.R. (1967). Air and water vision of the Atlantic flying fish, *Cypselurus heterurus*. *Nature* **284**: 348–350.

Blake, R.W. (1981) Mechanics of ostraciiform propulsion. *Can. J. Zool.* **59**: 1067–1071.

Blake, R.W. (1983a) *Fish Locomotion*. Cambridge University Press, Cambridge.

Blake, R.W. (1983b) Energetics of leaping in dolphins and other aquatic animals. *J. Mar. Biol. Assoc. UK* **63**: 61–71.

Blake, R.W. (1983c) Functional design and burst-and-coast swimming in fishes. *Can. J. Zool.* **61**: 2491–2494.

Blake, R.W. and Smith, M.D. (1988) On penguin porpoising. *Can. J. Zool.* **66**: 2093–2094.

Block, B.A., Keen, J.E., Castillo, B., Dewar, H., Freund, E.V., Marcinek, D.J., Brill R.W. and Farwell, C. (1997) Environmental preferences of yellowfin tuna (*Thunnus albacares*) at the northern extent of its range. *Mar. Biol.* **130**: 119–132.

Breder, C.M. (1930) On the structural specialization of flying fishes from the standpoint of aerodynamics. *Copeia* **4**: 114–121.

Carter, G.S. and Mander, J.A.H. (1935) The flight of the flying fish *Exocetus*. *Rep. Br. Assoc. Adv. Sci.* **105**: 383–384.

Clark, B. and Bemis, W. (1979). Kinematics of swimming of penguins at the Detroit Zoo. *J. Zool.* **188**: 411–428.

Davenport, J. (1994). How and why do flying fish fly? *Rev. Fish Biol. Fish.* **4**, 184–214.

Edgerton, H.E. and Breder, C.M. (1941) High speed photographs of flying fishes in flight. *Zoologica* **24**: 311–314.

Gill, T. (1905) Flying fish and their habits. *Annu. Rep. Smithsonian Inst.* **1904**: 495–515.

Gordon, C.N. (1980) Leaping dolphins. *Nature* **287**: 759.

Hankin, E.H. (1920) Observations on the flight of flying fishes. *Proc. Zool. Soc. Lond.* **2**: 467–474.

Hertel, H. (1966) *Structure, Form and Movement*. Reinhold, New York.

Holland, K.N., Brill, R.W. and Chang, R.K.C. (1990) Horizontal and vertical movements of tunas (*Thunnus* spp.) associated with fish aggregating devices. *Fish. Bull.* **88**: 493–507.

Hubbs, C.L. (1933) Observations on the flight of fishes. *Pap. Mich. Acad. Sci. Arts Letts* **17**: 575–611.

Hubbs, C.L. (1937) Further observations and statistics on the flight of fishes. *Pap. Mich. Acad. Sci. Arts Letts* **22**: 641–660.

Hui, C.A. (1987) The porpoising of penguins: an energy-conserving behavior for respiratory ventilation? *Can. J. Zool.* **65**: 209–211.

Klausewitz, W. (1960) Fliegende Tiere des Wassers. In: *Der Flug der Tiere* (ed. H. Schmidt). Waldemar Kramer, Frankfurt am Main.

Lamb, H. (1929) *Dynamics*. Cambridge University Press, Cambridge.

Landweber, L. (1961) Motion of immersed and floating bodies. In: *Handbook of Fluid Dynamics* (ed. V.L. Streeter). McGraw–Hill, New York, pp. 13.1–13.50.

Law, T. and Blake, R.W. (1994) Swimming behaviours and speeds of wild Dall's porpoises. *Mar. Mamm. Sci.* **10**: 208–213.

McCutcheon, C.W. (1977) Froude propulsive efficiency of a small fish measured by wake visualization. In: *Scale Effects in Animal Locomotion* (ed. T.J. Pedley). Academic Press, London, pp. 339–363.

O'Dor, R.K. (1988) The forces acting on swimming squid. *J. Exp. Biol.* **137**: 421–442.

Olson, F.C.W. (1964) The survival value of fish schooling. *J. Cons. Int. Explor. Mer* **29**: 115–116.

Rayner, J.M.V. (1981) Flight adaptations in vertebrates. In: *Vertebrate Locomotion* (ed. M.H. Day). *Symp. Zool. Soc. Lond.* **48**: 137–172.

Shoulejkin, W. (1929) Airdynamics of flying fish. *Int. Rev. Ges. Hydrobiol. Hydrogr.* **22**: 102–110.

Stephens, W.M. (1964) Flying fishes. *Sea Frontiers* **10**: 66–72.

Thompson, D. and Fedak, M.A. (1993) Cardiac responses of grey seals during diving at sea. *J. Exp. Biol.* **174**: 139–164.

Videler, J.J. (1993) *Fish Swimming*. Chapman and Hall, London.

Videler, J.J. and Weihs, D. (1982) Energetic advantages of burst and coast swimming of fish at high speeds. *J. Exp. Biol.* **97**: 169–178.

Webb, P.W. (1984) Body form, locomotion and foraging in aquatic vertebrates. *Am. Zool.* **24**: 107–120.

Webber, D.M. and O'Dor, R.K. (1986) Monitoring the metabolic rate and activity of free-swimming squid with telemetered jet pressure. *J. Exp. Biol.* **126**: 205–224.

Weihs, D. (1973) Mechanically efficient swimming techniques for fish with negative buoyancy. *J. Mar. Res.* **31**: 194–209.

Weihs, D. (1974) Energetic advantages of burst swimming of fish. *J. Theor. Biol.* **48**: 215–229.

Weihs, D. (1980) Energetic significance of changes in swimming modes during growth of larval anchovy, *Engraulis mordax*. *Fish. Bull.* **77**: 597–604.

Williams, T.M. and Kooyman, G.L. (1985) Swimming performance and hydrodynamic characteristics of harbor seals *Phoca vitulina*. *Physiol. Zool.* **58**: 576–589.

Wu, T.Y. (1977) Introduction to scaling of aquatic animal locomotion. In: *Scale Effects in Animal Locomotion* (ed. T.J. Pedley). Academic Press, London, pp. 203–232.

7

Bird migration performance on the basis of flight mechanics and trigonometry

Thomas Alerstam

1. Introduction

The process of bird migration is the sum of alternating periods of stopover, when energy (fuel) is accumulated and suitable travelling weather is awaited, and of flight, when fuel is consumed and distance is covered.

Mechanics of flight provides a fundamental basis for evaluating and understanding the limits and possibilities of bird migration performance. This was put forward three decades ago in a paper on 'The mechanics of bird migration' by Pennycuick (1969). However, in that paper focus was mainly on flight performance of migrating birds and not on the total process of migration, where the resulting performance must be integrated over a number of stopover periods and flight steps. Alexander (1998) took a more general approach, evaluating and comparing the total travelling time and energy costs for animals of different sizes that migrate shorter or longer distances by walking, swimming or flying.

The aim of this paper is to indicate the significance of a simple theoretical framework based on flight mechanics and aerodynamics in combination with trigonometry of flight vector summation, for analysing how the speed, range and precision of the birds' migratory journeys are affected by flight capacity, wind and orientation behaviour. After introducing in Section 2 some elementary relationships between power and speed based on flight mechanics, I will use these relationships to evaluate the expected overall migration speed by flapping and soaring flight (Section 3) as well as fuel loads and ranges for the flight steps of the migratory journey (Section 4). The impact of wind and orientation will be illustrated by simple trigonometrical considerations about how the birds' flight vector may be adapted to the wind vector (Section 5), how summation of flight vectors determines the angular and spatial precision of migration (Section 6) and, finally, to what degree travelling on the surface of the Earth will be favoured by great circle flight steps rather than rhumb vectors (Section 7).

Biomechanics in Animal Behaviour, edited by P. Domenici and R.W. Blake.

2. Gliding and flapping birds

In stable gliding flight the gravitational force (weight) is balanced by the aerodynamic forces lift (L) and drag (D), whereas in flapping flight the bird produces lift and thrust (T) to balance the weight and drag. The flight power P is related to speed according to

$$P = (D_{ind} + D_{par} + D_{pro})\, V \qquad (1)$$

where D_{ind} is induced drag, D_{par} parasite drag, D_{pro} profile drag and V is flight velocity. The induced drag is a consequence of the lift generation, whereas the parasite and profile drag are the pressure and friction resistance acting on the body and wings, respectively, moving through the air. For gliding, this power corresponds to the rate of loss of potential energy ($P = mgV_z$, where m is body mass, g acceleration due to gravity and V_z the vertical speed component), whereas in flapping flight power is produced by the flight muscles generating the wing strokes ($P = T\,V$).

The drag components are related in different ways to speed—D_{ind} becomes reduced whereas D_{par} and D_{pro} increase with increasing flight speed—giving as a result U-shaped power curves as illustrated in *Figure 1*. The power curve for gliding flight may be easily transformed to the well-known glide polar (sinking speed V_z in relation to gliding speed) by dividing power by the weight.

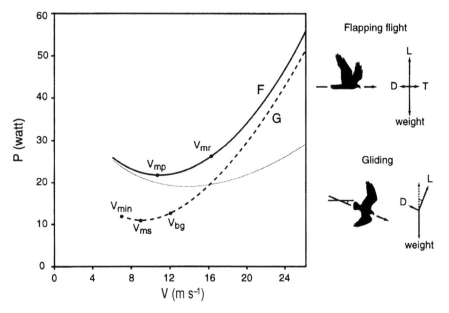

Figure 1. *Power in relation to speed for flapping (F) and gliding (G) flight as exemplified for the osprey, according to the theory of flight mechanics. Power refers to the mechanical output. The relationship predicted for flapping flight according to Pennycuick (1989; indicated by thin dotted line) has been modified by the addition of a component of profile power increasing with speed, because the curve for flapping flight should always be at a higher mechanical power level than the corresponding curve for gliding flight. To obtain the total chemical energy metabolized by a bird in flapping flight, the mechanical power output should be divided by the conversion efficiency. For gliding flight (G) the power corresponds to the rate of loss of potential energy. Characteristic gliding and flapping flight speeds are indicated as well as force diagrams for the two flight modes (see text).*

Both the power curve for flapping flight and the glide polar for birds have been extensively analysed within the framework of flight mechanical theory (e.g. Greenewalt, 1975; Pennycuick, 1969, 1975, 1989; Rayner, 1979; Thomas, 1996; Tucker, 1973, 1987, 1998). All analyses agree about the general U-shape of the power curve but the exact level and form of the curve differ depending on the assumptions and approximations suggested by the different researchers. Recent experiments designed to measure the chemical (metabolic) power for birds flying in wind tunnels have indicated both more and less pronounced U-shapes of the flapping flight power curves in comparison with the power curves predicted from flight mechanical theory (Kvist *et al.*, 1998; Rayner, 1999). This means that quantitative predictions from this theory, for example about absolute power and speed, are uncertain and provisional, whereas conclusions about qualitative differences in flight performance, for example between species or between flapping and soaring flight, are mostly unaffected by the uncertainties about the exact shape of the power curve. In addition, the power–speed relationship is not a unique curve for a given bird in either gliding or flapping flight but rather a family of curves depending on the wing and tail morphology adopted by the flying bird, and the bird is expected to adjust its wings and tail optimally at different speeds (Thomas, 1996; Tucker, 1987, 1998).

For the calculations in this paper I have used the procedure suggested by Pennycuick (1989), with modified assumptions about the profile power ratio in flapping flight (= 8.4/(aspect ratio); Pennycuick, 1995) and about the coefficient of body drag in flapping as well as gliding flight (= 0.1; Pennycuick *et al.*, 1996). However, comparing the estimated power (mechanical equivalents) in flapping and gliding flight shows that the calculation programmes for the two flight modes are incongruent, as the power level in flapping flight falls below that in gliding flight at moderate and high speeds. This cannot reasonably be the case, and to remedy this anomaly within the conceivable speed range for birds, I have provisionally added to the flapping flight power a profile power component increasing with speed (with a profile drag coefficient of 0.007, i.e. half the profile drag coefficient assumed for gliding flight). This means that the profile power in flapping flight becomes the sum of two components—one that is independent of speed as suggested by Pennycuick (1989) and one additional component increasing with speed. In the flapping flight power is also included the mechanical equivalents of metabolic power for basal maintenance and for extra costs of blood circulation and lung ventilation according to Pennycuick (1989). The calculations for gliding flight take into account the effect of a variable wing span.

Considering the significant uncertainties about the power components (not least the profile power) of bird flight, one must use quantitative predictions from the mechanical theory as rough and provisional estimates only, whereas qualitative predictions (about changes in performance related to size, wind, flight mode, etc.) are more robust.

In this paper I have calculated power–speed relations for a few selected species on the basis of the measurements given in *Table 1*. Power curves are illustrated for the osprey in *Figure 1*. V_{mp} and V_{mr} designate the flapping flight velocities that are associated with minimum power and with minimum energy cost per distance covered (mr is the maximum range on a given energy reserve), respectively. V_{min}, V_{ms} and V_{bg} are the stalling speed, the speed of minimum sink and the speed of best glide ratio, respectively, in gliding flight (see above references).

The osprey to a large degree migrates by thermal soaring flight and resorts to flapping flight mainly during sea crossings or in poor weather. Typical thermal soaring

Table 1. Measurements for species used as examples in calculations of migration performance on the basis of flight mechanical theory.

		Body mass (kg)	Lean body mass (kg)	Wing span (m)	Wing area (m²)	Aspect ratio
Dunlin	*Calidris alpina*	0.050	0.042	0.37	0.0165	8.3
Arctic tern	*Sterna paradisaea*	0.13	0.11	0.80	0.0571	11.2
Brent goose	*Branta bernicla*	1.35	1.25	1.2	0.146	9.9
Osprey	*Pandion haliaetus*	1.9	1.7	1.65	0.350	7.8
Whooper swan	*Cygnus cygnus*	10	9	2.4	0.670	8.6

Aspect ratio is a measure of the wing shape, reflecting the ratio of span to mean chord, and it is calculated as b^2/S where b is the wing span and S is the wing area.

flight is characterized by an alternation between circling flight in thermals, when the bird gains height in the rising air, and gliding flight in a rather straight line towards the bird's preferred direction. The bird will lose height as it glides in its preferred direction, but will resume circling and regain height when encountering a new thermal along its way. In typical thermal soaring flight the resulting cross-country speed V_x is

$$V_x = V \frac{V_c}{(V_z + V_c)} \tag{2}$$

where V is the horizontal interthermal gliding speed, V_z the interthermal sinking speed and V_c the climbing speed when circling in thermals. $V_c/(V_z + V_c)$ corresponds to the fraction of time that the bird glides between thermals, and the remaining time is spent circling in the thermals. Soaring theory predicts that the bird, to maximize V_x, should glide between thermals at a speed exceeding V_{bg}, and this optimal gliding speed increases with increasing V_c. For most birds the glide polar is such that the predicted fraction of time spent gliding between thermals is close to 50% (for most relevant climb rates) in typical thermal soaring flight.

By way of example, for the osprey the predicted interthermal gliding speed is 16 m s⁻¹ when thermals are weak (allowing $V_c = 1$ m s⁻¹) and 22 m s⁻¹ with stronger thermals ($V_c = 2$ m s⁻¹). The predicted fractions of gliding time for these two cases are 47 and 49%, and the resulting maximal cross-country speeds are 8 and 11 m s⁻¹, respectively. Even if this kind of thermal soaring flight is typical for many migrating raptors, storks and cranes, one should be aware that there is important variation in soaring techniques. In thermal streets or in slope lift along ridges birds may indulge in linear soaring, gliding during 100% of the time and achieving twice the resulting cross-country speed compared with ordinary thermal soaring when they have to stop to circle in thermals (Pennycuick, 1998).

3. Speed of migration

The total speed of migration (V_{migr}) taking into account fuel deposition as well as flight, is given by

$$V_{migr} = V_{flight} \frac{P_{dep}}{(P_{flight} + P_{dep})} \tag{3}$$

where P_{dep} and P_{flight} are the rates of energy deposition during stopover (or during the initial fuelling period at the breeding or wintering area, before the first flight of the migratory journey) and energy consumption during flight, respectively. The ratio $P_{dep}/(P_{flight} + P_{dep})$ corresponds to the fraction of total migration time that the bird is spending in actual flight.

Drawing a line connecting P_{dep} (plotted as a negative value on the ordinate) with the relevant flight power and speed according to the graphs in *Figure 2*, gives V_{migr} as the intercept of the speed axis (see Alerstam, 1991; Hedenström and Alerstam, 1998). P_{dep} will be limited for physiological reasons, and Lindström (1991) showed that 2.5 BMR (basal metabolic rate) is the expected maximum rate under ideal conditions. Energy deposition will be slower than this in many cases because of limited food availability and significant foraging costs, and a P_{dep} at 1.5 BMR has been chosen for the calculations in *Figure 2*.

Birds migrate by three main strategies: flapping flight, thermal soaring flight and soaring flight over the sea (often a combination of dynamic soaring and slope soaring; see Wilson, 1975). This classification is very schematic but useful for an evaluation of expected migration speeds.

3.1 *Flapping flight*

In flapping flight the bird produces the power required to fly by its flight muscles, and P_{flight} is given in relation to V_{flight} according to the power curve. Hence, to maximize V_{migr} a bird is expected to fly at a speed exceeding V_{mr}, and this optimal speed V_{mt} (speed giving the minimum total time of migration) increases with increasing P_{dep} (Alerstam and Lindström, 1990; Hedenström and Alerstam, 1995, 1998). Because P_{dep} is small in relation to P_{flight}, V_{mt} is only slightly faster than V_{mr}. Hence, with P_{dep} at

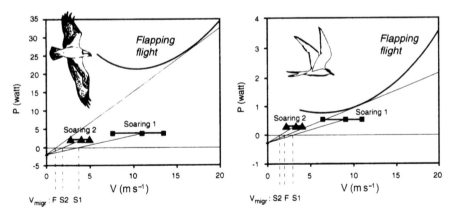

Figure 2. *Expected migration speed for osprey and arctic tern in flapping and thermal soaring flight. Migration speed is found as the intercept of the speed axis of a straight line connecting the rate of energy deposition at stopover (P_{dep} is assumed to be equal to 1.5 BMR) plotted on the power axis extended downwards, and the relevant flight power and speed (P_{flight} and V_{flight}). Cross-country speeds are calculated for three thermal climb rates ($V_c = 1, 2$ and $3\ m\ s^{-1}$) for unrestricted (soaring 1) and restricted (soaring 2) thermal soaring migration, with the gliding flight metabolism assumed to be 3 BMR (see text). Power is given in its mechanical equivalents (conversion efficiency 0.23). Resulting migration speeds (V_{migr}) are indicated along the speed axis for migration by flapping flight (F) and unrestricted (S1) and restricted (S2) thermal soaring flight with a climb rate (V_c) at 2 m s^{-1}.*

1.5 BMR, an osprey is expected to fly at 17 m s^{-1} to maximize migration speed. Because of the large energy costs for flapping flight it will be able to use only 7% of total migration time in flight (the rest is spent on depositing fuel reserves) giving a resulting V_{migr} at 1.1 m s^{-1} corresponding to 97 km day^{-1} (*Figure 2*). Because of its smaller size and narrow wings, flapping flight will be relatively less costly for the arctic tern. With the same P_{dep} (1.5 BMR) as the osprey, it is expected to fly during 20% of the total time, and with V_{mt} = 11 m s^{-1}, V_{migr} for the tern will reach 2.2 m s^{-1} or 187 km day^{-1} (*Figure 2*). This latter speed is in good accordance with results from ringing recoveries of arctic terns. In contrast, the migration speed estimated above for the osprey is clearly slower than that observed by satellite tracking (Kjellén *et al.*, 1997). This is because ospreys do not primarily use flapping flight but they achieve a faster V_{migr} by thermal soaring migration (see below).

Power requirements for flapping flight are expected to increase out of proportion with body mass (Pennycuick, 1989), which brings the effect of a slow migration speed in large-sized birds. *Figure 3* shows predicted migration speeds in relation to P_{dep} for a sample of different species. Whereas the tern will be able to achieve about 200 km day^{-1} with P_{dep} at 1.5 BMR, a swan will under similar conditions be able to travel at only about 50 km day^{-1}. Consequently, it will be impossible for a swan to migrate to such distant winter quarters as the tern (a swan would need 400 days to complete the 20 000 km one-way journey from northerly breeding sites to Antarctica). Only rather

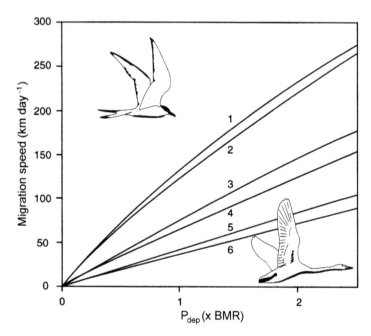

Figure 3. *Expected overall speed of migration by flapping flight in relation to rate of fuel deposition (P_{dep}) for arctic tern (1), dunlin (2), brent goose (3), osprey (4) and whooper swan (5 and 6) on the basis of flight mechanical calculations. Measurements for these species are given in Table 1. For the whooper swan migration speed has been calculated assuming a flight speed at V_{mt} (5; the flight speed giving the maximal migration speed, as also used for the other species) or at 18 m s^{-1} (6; which is the observed mean airspeed of whooper swans tracked by radar, being significantly slower than predicted V_{mt}, which is about 25 m s^{-1}; see Alerstam (1990)).*

restricted migration distances are feasible to fit within the annual time budget of the largest birds using flapping flight (Hedenström and Alerstam, 1998).

3.2 *Thermal soaring flight*

To estimate migration speed for thermal soaring migrants, the resulting cross-country speed V_x (Equation (2)) is used as V_{flight}. The mechanical energy to travel by thermal soaring migration is extracted from the rising air in thermals, so that the bird will have to spend power mainly to hold its wings outstretched in gliding position and to make the proper gliding manoeuvres. Thus, the metabolic power of a bird in gliding flight is very different and radically lower than in flapping flight. Various studies indicate that P_{flight} during gliding is close to 3 BMR (Hedenström and Alerstam, 1998). The cross-country speed depends on the strength of thermals (determining V_c; see Equation (2)), whereas P_{flight} is independent of the resulting cross-country speed.

Two cases of soaring migration are illustrated for the osprey and arctic tern in *Figure 2*—unrestricted and restricted soaring. In unrestricted soaring the assumption is that the bird can always find suitable soaring conditions when it is ready to depart on a new flight step after fuelling. However, typical thermals develop only during the day, and satellite tracking shows that, for example, ospreys restrict their migratory movements to about nine daytime hours (Kjellén *et al.*, 1997). In restricted soaring it is assumed that birds achieve an average speed that is only 9/24 of that in unrestricted soaring. Furthermore, it is assumed that the birds rest at a metabolic rate of 1 BMR when not moving (mainly during the night), so that their overall average power during travelling days with 9 h of soaring and 15 h of rest is 1.75 BMR.

Figure 2 shows that restricted soaring (with V_c between 1 and 3 m s^{-1}) gives a faster resulting migration speed than flapping flight in the osprey, but not in the arctic tern. An osprey on restricted soaring migration is expected (with P_{dep} at 1.5 BMR) to use 46% of the migration time as travelling days and the rest as fuel deposition days. The expected distance per travelling day, with thermals giving a V_c between 1 and 3 m s^{-1}, is 250–440 km, and the expected overall migration speed is 116–203 km day^{-1} (still air conditions). In two ospreys tracked by satellite-based telemetry while migrating between Sweden and tropical Africa, the average distance covered on travelling days was 274 and 263 km day^{-1}, respectively, and they used 45% and 84% of the days between departure from the breeding area and arrival at the winter destination as travelling days (Kjellén *et al.*, 1997). The latter figures do not accurately reflect the proportion of travelling time during migration, as fuel deposition before departure from the breeding area is not taken into account.

For smaller birds such as the arctic tern, thermal soaring will be a favourable alternative to flapping flight only under conditions allowing unrestricted soaring, which normally does not apply over land (*Figure 2*). Still, terns sometimes use the opportunity of gaining height by temporarily circling in thermals when climbing to high cruising altitudes (Alerstam, 1985). A more detailed analysis comparing migration by thermal soaring and flapping flight has been presented by Hedenström (1993).

3.3 *Soaring flight over the sea*

Seabirds may enjoy rather unrestricted soaring conditions (if or when they are not hindered by darkness)—winds and waves persist throughout the day. There is no theory to predict resulting 'cross-country' speeds in soaring flight over the sea waves, but radar and optical measurements show that albatrosses achieve resulting speeds of about 10 m s⁻¹ in weak or moderate winds and up to 20 m s⁻¹ in stronger winds (Alerstam *et al.*, 1993). On the basis of these speeds (V_{flight}), a flight metabolism of 3 BMR (P_{flight}) and an energy deposition rate when stopping to feed at 1.5 BMR (P_{dep}), one may expect albatrosses to reach total migration speeds of about 3.3–7 m s⁻¹, i.e. 285–605 km day⁻¹, using one-third of the time for travelling and the rest for feeding.

In fact, satellite tracking studies of wandering albatrosses on foraging trips during the breeding season over distances of 3000–15 000 km show overall mean speeds between 220 and 830 km day⁻¹ (Jouventin and Weimerskirch, 1990; Prince *et al.*, 1992). During these trips the albatrosses were travelling during both daylight and darkness (dominantly during daytime) between their feeding interludes. The longest recorded foraging trip, over a minimum distance of 15 200 km, was completed during 33 days, giving a mean travelling speed of 461 km day⁻¹. As the albatrosses achieved a positive net energy balance, accumulating energy during these trips, one must conclude that they are able to reach even higher overall travelling speeds when migrating with an equal balance between energy intake and expenditure during the non-breeding season.

One may conclude that small and medium-sized birds are expected to attain maximum migration speeds as fast as 100–300 km day⁻¹ by flapping flight. This can be matched or even surpassed by larger birds migrating over land by thermal soaring flight, even when the effect of a restricted daily thermal time is taken into account. Seabirds, especially those of larger sizes such as the albatrosses, are, by way of their soaring technique over the sea waves, probably the fastest long-distance travellers among birds. The poorest performance in this respect is expected for large-sized birds such as swans, which are confined, by their morphology and by conditions along their travelling routes, to flapping flight.

4. Range and fuel load

Aerodynamically optimal fuel loads for migrating birds may be predicted on the basis of the relationship between flight range (Y) and fuel load at departure (f). For flapping flight the slope of this positive function gradually decreases because of increased flight costs with increasing fuel burdens. The range equation for flapping flight was derived on the basis of flight mechanics (Pennycuick, 1975) by Alerstam and Lindström (1990) as

$$Y = c \left[1 - (1 + f)^{-\frac{1}{2}} \right] \qquad (4)$$

where f is the fuel load relative to lean body mass, and the coefficient c depends on, among other things, fuel composition, muscle work efficiency, bird morphology and wind. The predicted range curve for the osprey under still air conditions is shown in *Figure 4*. The utility (marginal value) of additional fuel decreases significantly with f. Hence, a given amount of additional fuel will allow a bird that already has deposited fuel reserves equal to its half ($f = 0.5$) or full ($f = 1$) lean body mass to fly an extra distance that is only 54% and 35%, respectively, of the distance attained by the same amount of additional fuel if the bird had no prior fuel reserves ($f = 0$). The range function and its derivative (which represents the instantaneous rate of migration by a bird in the process of fuel deposition) constitute a fundamental starting point for

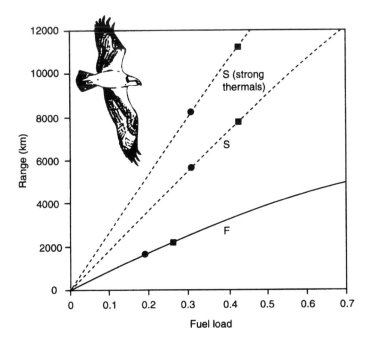

Figure 4. *Flight range in relation to fuel load (relative to lean body mass) as predicted for the osprey in flapping (F) and thermal soaring flight (S). The range curve for thermal soaring flight (Equation (6); see text) has been calculated for a climb rate in thermals of 2 m s⁻¹ (upper curve) and 1 m s⁻¹, respectively. The symbols show predicted optimal fuel loads and flight ranges with an expected energy loss of $f_o = 0.02$ (●) or a combined energy and time loss of $f_o = 0.02$ and $k\,t_o = 0.02$ (■) when settling at a new stopover site (amount of fuel is expressed relative to lean body mass; f_o is fuel cost, k is fuel deposition rate per day and t_o is days of search or settling; see Alerstam and Hedenström (1998)).*

analysing optimal departure rules and for modelling the complex process of migration with a given configuration of possible stopover sites (Gudmundsson *et al.*, 1991; Houston, 1998; Weber and Houston, 1997; Weber *et al.*, 1994, 1998).

What about the range curves for birds migrating by soaring flight? The range curve depends on the effects of increased fuel load on resulting flight speed as well as on flight metabolism. For thermal soaring flight the cross-country speed (see Equation (2)) is expected to increase with load approximately in proportion to $m^{1/6}$ (m is the total body mass). It is more difficult to tell how additional fuel affects the metabolism in gliding flight, which is assumed to be a constant multiple of BMR (about 3 BMR; see Hedenström, 1993). The extreme assumptions are that gliding flight metabolism is either unaffected by an increased load or increases at the same rate as BMR scales with body mass in between-species comparisons (proportional to $m^{3/4}$). These assumptions give the following two range equations, respectively:

$$Y = c_1\,[(1 + f)^{7/6} - 1] \tag{5}$$

$$Y = c_2\,[(1 + f)^{5/12} - 1] \tag{6}$$

In the first case, Y is a convex function of f, which means that the marginal value of fuel increases the fatter the bird becomes. This would, of course, induce birds always to accumulate before departure the full fuel reserves to cover the entire migratory journey without refuelling (provided that the rate of fuel deposition in the departure area is up to the requirements).

In the second case, Y is a decelerating function of f, but only weakly so in comparison with the range function for flapping flight. Hence, the marginal rate of gain in range at $f = 0.5$ and $f = 1$ is 79% and 67% of the corresponding rate at $f = 0$ (compare above for flapping flight). In *Figure 4*, I have calculated the predicted range curves for the osprey according to this latter case (Equation (6)) of thermal soaring flight (assuming restricted soaring conditions, i.e. 9 h of soaring flight and 15 h of rest per day; the two curves show predictions with climb rates at 1 and 2 m s^{-1}, respectively).

Apart from the difference in curvature between the range functions for flapping and soaring flight, there is another marked difference: the range on a given fuel load is much longer for soaring than for flapping flight. Optimal departure fuel loads can be predicted for energy- and time-selected migration on the basis of assumed energy and time losses at settling at new stopover sites (Alerstam and Hedenström, 1998; Alerstam and Lindström, 1990; Hedenström and Alerstam, 1997). The symbols on the range curves in *Figure 4* denote the optimal departure fuel loads and associated flight ranges for two conditions of different settling losses at stopover sites. These calculations predict that soaring migrants should be prone to put on larger fuel loads and to complete their journey by much longer flight steps (without stopping to refuel) than migrants travelling by flapping flight. Indeed, one may well suspect that some soaring migrants store enough fuel to complete their whole long-distance journey without refuelling at all (Candler and Kennedy, 1995; Smith *et al.*, 1986).

5. The triangle of velocities

Wind is an omnipresent and extremely powerful factor in bird migration. With strong enough tailwinds birds may, for example, double their performance in terms of speed and range (and reduce energy costs per distance by half), whereas strong headwinds may actually nullify migration in the desired direction. Hence, studies of bird migration must take into account the interaction between winds and flight according to the rules of trigonometry.

The resulting track vector of a flying bird (direction and speed relative to ground) is the sum of the wind vector and the birds' heading vector (flight direction and speed relative to the air). Birds migrating by flapping flight are expected to adjust both heading direction and airspeed (flight speed relative to the air) in relation to different wind conditions, and the full analysis requires the integration of trigonometry with the power equation based on flight mechanics (Alerstam and Hedenström, 1998; Liechti, 1995; Liechti *et al.*, 1994). Birds can, of course, choose any heading direction, but their choice of airspeed is limited between V_{mp} (flying slower than V_{mp} will never be favourable on migration) and V_{max}, the maximum sustained speed achievable with full muscle power capacity. The optimal solution of the triangle of velocities differs between at least four basic cases of different migratory conditions, as follows.

(i) If the bird is many flight steps from its next goal and wind is expected to vary considerably between the different flight steps, the bird will gain both time and energy by allowing itself to be drifted by wind, maintaining its heading directed towards the distant destination irrespective of wind (Alerstam, 1979a). It will be expected to adjust its airspeed in relation to the wind component along the goal direction (Alerstam and Hedenström, 1998), increasing airspeed with a headwind component and reducing it with a tailwind component. Hence, with due crosswind as illustrated in *Figure 5(a)*, this component is zero and, in energy-selected migration, the optimal airspeed is equal to V_{mr} under still air conditions.

(ii) If winds remain constant or similar throughout the journey to the next goal, complete compensation is the optimal solution (Alerstam, 1979a). The rules for optimal adjustment of the airspeed have been evaluated by Liechti *et al.* (1994). The airspeed is expected to be reduced when the resulting groundspeed exceeds the airspeed, and to be increased when the groundspeed falls short of the airspeed and, in addition, the airspeed is expected to be increased with an increasing angle of compensation, α (*Figure 5(b)*). Consequently, a fast airspeed will be instrumental, along with the proper adjustment of heading direction, in obtaining compensation under crosswinds (Liechti *et al.*, 1994).

(iii) A strategy of adapted drift implies that the bird maximizes the effective goal approach during each flight. With a migratory journey divided into several flight

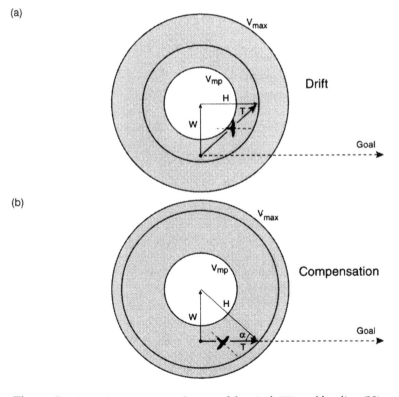

Figure 5. *The track vector (T) is the sum of the wind (W) and heading (H) vectors. In migration by flapping flight birds are expected to adopt an airspeed between V_{mp} and V_{max}. Drawing circles with radii corresponding to those speeds on top of the wind vector delimits all possible tip points of the track vector (in the shaded fields). When it is most beneficial for birds to submit to wind drift, they are under crosswinds expected to select a slower airspeed (a) than in situations when complete compensation is most favourable (b). In this latter case, birds achieve compensation most efficiently by a combined adjustment of both heading direction and airspeed.*

steps and with the wind vector varying independently between flights, the optimal solution is for the bird to allow extensive drift far away from the goal and to gradually compensate to a higher and higher degree upon approaching the goal (Alerstam, 1979a, 1990). The rules for airspeed adjustment in this strategy are intermediate between those of the two former cases, as analysed by Liechti (1995).

(iv) If winds between the departure site and the next goal vary more predictably according to some regular pattern, the 'minimum time path' will provide a favourable solution. This solution involves the calculation of successive time-fronts from the departure point according to the principles developed by Giblett (1924) and others in the early days of aviation, as reviewed by Williams (1994). To follow the path of minimum time, the traveller should adopt different degrees of partial drift and compensation depending on the expected winds during different phases of the journey.

Birds may exploit winds in still other ways, for example, by optimal vector combinations of alternating high- and low-altitude flights (Alerstam, 1979b). Thermal soaring migrants are expected to obtain the proper amount of compensation by changing their gliding heading direction only, whereas the optimal gliding speed does not vary with wind conditions (Alerstam and Hedenström, 1998).

6. Migration by vector summation

The resulting orientation and routes shown by migrating birds may be governed by the effects of simultaneous and sequential vector summation. Hamilton (1967) suggested that the orientation of a flock of birds represents a compromise between different directional preferences of the individual flock members. With a given between-individual directional scatter, the resulting flock orientation will show a decreasing variability around the population mean with increasing flock size, if the orientation of the flock is the result of simultaneous vector summation of the individuals' preferred directions. The expected reduction in angular scatter with increasing flock size has been evaluated by Hamilton (1967) and Wallraff (1978). This reduction in directional scatter is analogous to the reduction in standard error with increasing sample size for means drawn from a population with a given standard deviation, according to basic statistics. To the extent that the directional scatter reflects orientational uncertainty and inaccuracy, the orientation of all individuals will be improved by travelling in flocks and adopting the mean direction of the flock members.

The reduction in directional scatter of flocks in comparison with solitary individuals is, according to this suggestion, not the result of a 'follow the leader' effect, but rather a consequence of averaging the individual directions of all flock members (where all individual directions are given equal weight) into a common flock direction, which can be calculated on the basis of circular statistics. Whether there is any mechanism operating between individuals in a flock that could bring about such a vector summation effect is unknown. Domenici and Batty (1997) compared escape behaviour between solitary and schooling fish. They found that schooling enhanced the directionality and the co-ordination of the escape response, suggesting a process of response integration between school members.

Rabøl and Noer (1973) found that the orientational scatter decreases with flock size among skylarks on spring migration, in accordance with the expected effect of vector summation. Comparisons between the orientation of single individuals and flocks of homing pigeons have produced more equivocal results (Keeton, 1970; Tamm, 1980).

Exactly the same principles of vector summation effects as suggested for birds migrating in flocks may be applied to the successive summation of flight steps (vectors) as an individual bird or flock travels on migration. Rabøl (1978) showed how migrants drawing their orientation during each flight step from a circular probability distribution with a fairly large scatter will, by the effect of vector summation over many flight steps, produce long-distance ringing recoveries with a much more restricted angular scatter. This idea was recently elaborated by Mouritsen (1998), who claimed that simple vector orientation is sufficient to explain the distribution of ringing recoveries of juvenile night-migrating passerines (migrating individually and not in flocks).

Adding flight steps or vectors of length d, the resulting distance R from the departure point after a number of flights (n) will be

$$R = r n d \qquad (7)$$

where r is the mean vector length, which represents a measure of the angular scatter of orientation. The angular deviation (s), which is equivalent in circular statistics to the standard deviation in linear statistics, may be estimated from r according to

$$s = \sqrt{[2(1-r)]} \qquad (8)$$

where s is in radians (Batschelet, 1981).

The number of flight steps to cover a given distance is inversely proportional to r, so that birds showing large orientational scatter will make considerable detours and waste much flying time and energy to reach the same resulting distance as birds keeping their orientation perfectly constant (*Figure 6*). The graph in *Figure 6* shows how the angular deviation of the resulting migratory direction becomes reduced with an increasing number of flight steps for a population of birds with orientation varying between flight steps (around the same mean direction). This relationship is illustrated for three levels of between-step variation, corresponding to $r = 0.5$, 0.7 and 0.9. Approximately two-thirds of all observations are expected to fall within ± one angular deviation from the mean in a circular normal (von Mises) distribution, given that r is reasonably large (≥ 0.5; Batschelet, 1981). The absolute spatial scatter at the migration front (frontal spread is defined in *Figure 6*) will, of course, increase with number of flight vectors added. In fact, the resulting scatter in the plane (along and perpendicular to the axis of migration) after a large number of steps will approach a bivariate normal distribution that can be described by standard ellipses (Batschelet, 1981; K. Thorup, personal communication). For the ideal case of a population of birds migrating by vector summation, with all individuals using the same step length and completing the same total number of steps, the distribution after many steps will approach a pattern with the same scatter along as perpendicular to the migration axis, i.e. a distribution defined by standard circles (K. Thorup, personal communication).

To illustrate the consequences of migration by vector summation one may consider a population of pied flycatchers migrating 3000 km, assuming a step length of 125 km as suggested by Mouritsen (1998). All individuals migrate independently by adding vectors drawn from a circular probability distribution with the same mean. With a

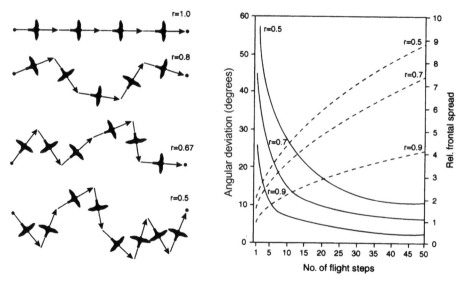

Figure 6. Vector summation of flight steps. Depending on orientational scatter (defined by the mean vector length r) birds will require different numbers of flight steps to reach the same resulting migratory distance. The angular deviation of the resulting migratory direction for a population of migrants decreases with an increasing number of flight steps (calculation based on Mouritsen (1998) and Wallraff (1978)), whereas the frontal spread (perpendicularly to the migratory direction) increases. The frontal spread is calculated as the length of the arc associated with an angle of 2 s and radius R (Equation (7)) relative to step length (see text).

between-step variation in orientation corresponding to $r = 0.5$, 48 flight steps are required to reach 3000 km (Equation (7)), giving an angular deviation of resulting migratory directions at 11° and a frontal spread of 8.8 times the step length that is, 1100 km (*Figure 6*). This means that on average two-thirds of the population have ended up within a sector of ±11° from the mean direction, corresponding to a spread over a front that is 1100 km wide perpendicularly to the migratory direction. With $r = 0.7$, 34 flight steps are required to reach the same distance (Equation (7)), giving $s = 7°$ and a frontal spread of 6.3 times the step length ($\sim$ 800 km). The corresponding values with $r = 0.9$ are 27 steps, $s = 3.5°$ and a frontal spread of about 400 km.

Mouritsen (1998) suggested, on the basis of short-distance ringing recoveries and scatter observed in orientation cage experiments, that $r = 0.67$ may be a typical between-step variation in orientation of juvenile flycatchers. If correct, the birds would follow a very circuitous route towards the destination, on average 1.5 ($= 1/r$) times longer than the straight-line distance. However, Mouritsen's (1998) estimate of between-step variation may well be inflated for a number of reasons (post-breeding dispersal, between-population and -individual differences in orientation were not taken into account). Satellite tracking is a promising possibility to investigate the process of bird migration with respect to the vector summation of successive flight steps.

Path integration, where animals integrate course angles and distances on an outward journey into a mean home vector, is an important element in, for example, insect navigation (Wehner 1998; Wehner et al., 1996), but this principle may be of less importance for the homing of migratory birds.

7. Routes on a globe

The shortest route between two points on the Earth's surface is along the great circle (orthodrome) and not along the path of constant geographic course (loxodrome or rhumbline). The shorter distance means that birds will save both time and energy for their migratory journey by following orthodromes rather than loxodromes. One may assume that it is most convenient for birds to orient along loxodromes keeping their geographic compass courses constant, whereas orientation along orthodromes requires gradually changing courses when travelling across the longitudes. Is it possible that birds have an orientation system that can meet these special requirements, allowing them to migrate along long-distance orthodrome-like routes, adapted to the principles of spherical geometry and trigonometry?

The reduction in distance along great circle routes compared with rhumblines is most pronounced for movements in east–west directions at high geographic latitudes. Considering two points on the same latitude θ, separated by a longitudinal angle λ, the distances (relative to the radius of the sphere) along the ortho- and loxodromes are

$$O = 2 \arcsin \left[\sin(\lambda/2) \cos\theta \right] \tag{9}$$

$$L = \lambda \cos\theta \tag{10}$$

where angles are in radians. The ratio O/L is plotted in relation to latitude for various longitudinal angles in *Figure 7*. Distance savings by following the orthodrome rather than the loxodrome between two points may be significant to the north of latitudes 50–60°N, or south of 50–60°S, where separation between longitudes is small, so that a journey may span a wide longitudinal angle.

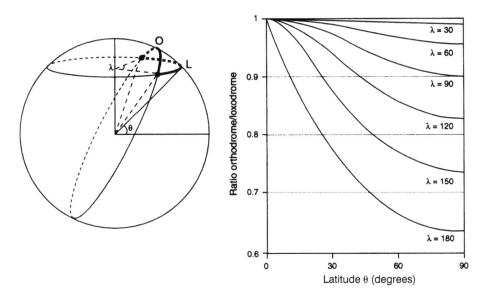

Figure 7. Ratio of distance along the orthodrome (O) to distance along the loxodrome (L) in relation to the latitude θ (same latitude for start and end points) and longitudinal angle λ between start and end points.

Several means of orientation along routes curving in a distance-saving and ortho-drome-like way have been suggested for migratory birds: (i) great circle routes may be approximated by many successive rhumblines; (ii) Kiepenheuer (1984) proposed a hypothetical magnetic compass mechanism whereby birds maintain the apparent angle of inclination constant and travel along curved 'magnetoclinic' routes; (iii) tracks with constant magnetic compass courses (magnetic loxodromes) curve in a distance-saving way in certain regions of the Earth (Alerstam and Hedenström, 1998); (iv) a time-compensated sun or celestial compass provides a possibility of great circle orientation as a consequence of the changing orientation associated with the time shift as the birds travel across longitudes (Alerstam and Pettersson, 1991).

Long-distance flight routes in polar regions are of a special interest because it is in these regions that the differences between orthodromes and loxodromes are greatest (*Figure 7*). Hence, polar flight routes may reveal if orientation of migratory birds shows any adapta-tions to the spherical geometry of the Earth. Radar studies suggest that birds may indeed accomplish approximate great circle orientation when migrating between Siberia and North America across the Arctic Ocean (Alerstam and Gudmundsson, 1999).

The ultimate performance in terms of great circle orientation would be for birds to migrate right across the geographic poles rather than making a circumflight along the parallel. The ratio O/L approaches its minimum value at $2/\pi = 0.64$ for a journey between two points on opposite sides of the pole as latitude approaches the pole (*Figure 7*). However, no transpolar migration was recorded in spite of extensive radar surveillance in the central Arctic Ocean (Gudmundsson and Alerstam, 1998a). The absence of transpolar migration is in stark contrast to the multitude of migrating birds that have been recorded by radar over more peripheral parts of the Arctic Ocean. Distances between the tundra on opposite sides of the North Pole are 2000–4000 km, well within the birds' flight range capacity. The most likely explanations for the absence of transpolar migration are that such migration is of little evolutionary advantage considering the present geographic configuration of stopover and wintering regions for tundra birds breeding in different circumpolar sectors or that orientation difficulties and the extreme daily time shift prevent successful migration across the pole (Gudmundsson and Alerstam, 1998a).

Different map projections are important tools for analysing the orientation prin-ciples in long-distance migration (Gudmundsson and Alerstam, 1998b). The geometry of migration routes may provide important clues about the orientation and navigation mechanisms guiding the birds on their global journeys.

8. Final remark

The above sections illustrate the fact that bird migration performance cannot be properly understood without incorporating basic biomechanics, trigonometry and geometry in the analysis. It is also evident that we are in the very initial phases of analysing the speed, range and precision of bird migration. We have still a long way to go before we really understand, for example, the birds' capacity of adapting their flight speeds and behaviour to different situations, the geometric principles of bird migration routes, how the birds behave with respect to wind drift and compensation, and if the principles of vector summation govern flock orientation and migration patterns. Exploring the complexity of bird migration, with its limitations, possibilities and remarkable adaptations, will remain a fascinating multi-disciplinary scientific task, and we can be certain that studies integrating biomechanics and behaviour will constitute very important elements in these future research adventures.

Acknowledgements

I wish to thank Kasper Thorup at Copenhagen University for valuable information about vector summation in circular statistics, and Paolo Domenici for valuable comments on the manuscript. I am very grateful to Gunilla Andersson for drawing the figures, and to Inga Rudebeck for typing the manuscript. This work was supported by the Swedish Natural Science Research Council.

References

Alerstam, T. (1979a) Wind as selective agent in bird migration. *Ornis Scand.* 10: 76–93.

Alerstam, T. (1979b) Optimal use of wind by migrating birds: combined drift and overcompensation. *J. Theor. Biol.* 79: 341–353.

Alerstam, T. (1985) Strategies of migratory flight, illustrated by arctic and common terns, *Sterna paradisaea* and *Sterna hirundo*. In: *Migration: Mechanisms and Adaptive Significance* (ed. M.A. Rankin). *Contrib. Mar. Sci.* 27, Suppl, 580–603.

Alerstam, T. (1990) *Bird Migration*. Cambridge University Press, Cambridge.

Alerstam, T. (1991) Bird flight and optimal migration. *Trends Ecol. Evol.* 6: 210–215.

Alerstam, T. and Gudmundsson, G.A. (1999) Bird orientation at high latitudes: flight routes between Siberia and North America across the Arctic Ocean. *Proc. R. Soc. Lond., Ser. B* 266:2499–2505.

Alerstam, T. and Hedenström, A. (1998) The development of bird migration theory. *J. Avian Biol.* 29: 343–369.

Alerstam, T. and Lindström, A. (1990) Optimal bird migration: the relative importance of time, energy, and safety. In: *Bird Migration: Physiology and Ecophysiology* (ed. E. Gwinner). Springer, Berlin, pp. 331–351.

Alerstam, T. and Pettersson, S.-G. (1991) Orientation along great circles by migrating birds using a sun compass. *J. Theor. Biol.* 152: 191–202.

Alerstam, T., Gudmundsson, G.A. and Larsson, B. (1993) Flight tracks and speeds of Antarctic and Atlantic seabirds: radar and optical measurements. *Philos. Trans. R. Soc. Lond., Ser. B* 340: 55–67.

Alexander, R.McN. (1998) When is migration worthwhile for animals that walk, swim or fly? *J. Avian Biol.* 29: 387–394.

Batschelet, E. (1981) *Circular Statistics in Biology*. Academic Press, London.

Candler, G.L. and Kennedy, P.L. (1995) Flight strategies of migrating ospreys: fasting vs. foraging. *J. Raptor. Res.* 29: 85–92.

Domenici, P. and Batty, R.S. (1997) Escape behaviour of solitary herring (*Clupea harengus*) and comparisons with schooling individuals. *Mar. Biol.* 128: 29–38.

Giblett, M.A. (1924) Notes on meteorology and the navigation of airships. *Meteorological Magazine* 59: 1–17.

Greenewalt, C.H. (1975) The flight of birds. *Trans. Am. Philos. Soc.* 65: 1–67.

Gudmundsson, G.A. and Alerstam, T. (1998a) Why is there no transpolar bird migration? *J. Avian Biol.* 29: 93–96.

Gudmundsson, G.A. and Alerstam, T. (1998b) Optimal map projections for analysing long-distance migration routes. *J. Avian Biol.* 29: 597–605.

Gudmundsson, G.A., Lindström, Å. and Alerstam, T. (1991) Optimal fat loads and long-distance flights by migrating knots *Calidris canutus*, sanderlings *C. alba* and turnstones *Arenaria interpres*. *Ibis* 133: 140–152.

Hamilton, W.J., III (1967) Social aspects of bird orientation mechanisms. In: *Avian Orientation and Navigation* (ed. R.M. Storm). Oregon State University Press, Corvallis, pp. 57–71.

Hedenström, A. (1993) Migration by soaring or flapping flight in birds: the relative importance of energy cost and speed. *Philos. Trans. R. Soc. Lond., Ser. B* **342**: 353–361.

Hedenström, A. and Alerstam, T. (1995) Optimal flight speed of birds. *Philos. Trans. R. Soc. Lond., Ser. B* **348**: 471–487.

Hedenström, A. and Alerstam, T. (1997) Optimum fuel loads in migratory birds: distinguishing between time and energy minimization. *J. Theor. Biol.* **189**: 227–234.

Hedenström, A. and Alerstam, T. (1998) How fast can birds migrate? *J. Avian Biol.* **29**: 424–432.

Houston, A.I. (1998) Models of optimal avian migration: state, time and predation. *J. Avian Biol.* **29**: 395–404.

Jouventin, P. and Weimerskirch, H. (1990) Satellite tracking of wandering albatrosses. *Nature* **343**: 746–748.

Keeton, W.T. (1970) Comparative orientational and homing performances of single pigeons and small flocks. *Auk* **87**: 797–799.

Kiepenheuer, J. (1984) The magnetic compass mechanism of birds and its possible association with the shifting course directions of migrants. *Behav. Ecol. Sociobiol.* **14**: 81–99.

Kjellén, N., Hake, M. and Alerstam, T. (1997) Strategies of two ospreys *Pandion haliaetus* migrating between Sweden and tropical Africa as revealed by satellite tracking. *J. Avian Biol.* **28**: 15–23.

Kvist, A., Klaassen, M. and Lindström, Å. (1998) Energy expenditure in relation to flight speed: what is the power of mass loss rate estimates? *J. Avian Biol.* **29**: 485–498.

Liechti, F. (1995) Modelling optimal heading and airspeed of migrating birds in relation to energy expenditure and wind influence. *J. Avian Biol.* **26**: 330–336.

Liechti, F., Hedenström, A. and Alerstam, T. (1994) Effects of sidewinds on optimal flight speed of birds. *J. Theor. Biol.* **170**: 219–225.

Lindström, Å. (1991) Maximum fat deposition rates in migrating birds. *Ornis Scand.* **22**: 12–19.

Mouritsen, H. (1998) Modelling migration: the clock-and-compass model can explain the distribution of ringing recoveries. *Anim. Behav.* **56**: 899–907.

Pennycuick, C.J. (1969) The mechanics of bird migration. *Ibis* **111**: 525–556.

Pennycuick, C.J. (1975) Mechanics of flight. In: *Avian Biology*, Vol. 5 (eds D.S. Farner and J.R. King). Academic Press, New York, pp. 1–75.

Pennycuick, C.J. (1989) *Bird Flight Performance: a Practical Calculation Manual*. Oxford University Press, Oxford.

Pennycuick, C.J. (1995) The use and misuse of mathematical flight models. *Israel J. Zool.* **41**: 307–319.

Pennycuick, C.J. (1998) Field observations of thermals and thermal streets, and the theory of cross-country soaring flight. *J. Avian Biol.* **29**: 33–43.

Pennycuick, C.J., Klaassen, M., Kvist, A. and Lindström, Å. (1996) Wingbeat frequency and the body drag anomaly: wind-tunnel observations on a thrush nightingale (*Luscinia luscinia*) and a teal (*Anas crecca*). *J. Exp. Biol.* **199**: 2757–2765.

Prince, P.A., Wood, A.G., Barton, T. and Croxall, J.P. (1992) Satellite tracking of wandering albatrosses (*Diomedea exulans*) in the South Atlantic. *Antarctic Sci.* **4**: 31–36.

Rabøl, J. (1978) One-direction orientation versus goal area navigation in migratory birds. *Oikos* **30**: 216–223.

Rabøl, J. and Noer, H. (1973) Spring migration in the skylark (*Alauda arvensis*) in Denmark. Influence of environmental factors on the flock size and correlation between flock size and migratory direction. *Vogelwarte* **27**: 50–65.

Rayner, J.M.V. (1979) A new approach to animal flight mechanics. *J. Exp. Biol.* **80**: 17–54.

Rayner, J.M.V. (1999) Estimating power curves of flying vertebrates. *J. Exp. Biol.* **202**: 3449–3461.

Smith, N.G., Goldstein, D.L. and Bartholomew, G.A. (1986) Is long-distance migration possible for soaring hawks using only stored fat? *Auk* 103: 607–611.

Tamm, S. (1980) Bird orientation: single homing pigeons compared to small flocks. *Behav. Ecol. Sociobiol.* 7: 319–322.

Thomas, A.L.R. (1996) The flight of birds that have wings and tail: variable geometry expands the envelope of flight performance. *J. Theor. Biol.* 183: 237–245.

Tucker, V.A. (1973) Bird metabolism during flight: evaluation of a theory. *J. Exp. Biol.* 58: 689–709.

Tucker, V.A. (1987) Gliding birds: the effect of variable wing span. *J. Exp. Biol.* 133: 33–58.

Tucker, V.A. (1998) Gliding flight: speed and acceleration of ideal falcons during dive and pull out. *J. Exp. Biol.* 201: 403–414.

Wallraff, H.G. (1978) Social interrelations involved in migratory orientation of birds: possible contributions of field studies. *Oikos* 30: 401–404.

Weber, T.P. and Houston, A.I. (1997) A general model for time-minimising avian migration. *J. Theor. Biol.* 185: 447–458.

Weber, T.P., Houston A.I. and Ens, B.J. (1994) Optimal departure fat loads and site use in avian migration: an analytical model. *Proc. R. Soc. Lond., Ser. B* 258: 29–34.

Weber, T.P., Ens, B. J. and Houston, A.I. (1998) Optimal avian migration. A dynamic model of nutrient stores and site use. *Evol. Ecol.* 12: 377–401.

Wehner, R. (1998) Navigation in context: grand theories and basic mechanisms. *J. Avian Biol.* 29: 370–386.

Wehner, R., Michel, B. and Antonsen, P. (1996) Visual navigation in insects: coupling of egocentric and geocentric information. *J. Exp. Biol.* 199: 129–140.

Williams, J.E.D. (1994) *From Sails to Satellites. The Origin and Development of Navigational Science.* Oxford University Press, Oxford.

Wilson, J. A. (1975) Sweeping flight and soaring by albatrosses. *Nature* 257: 307–308.

Appendix: List of symbols

α angle between track and heading

λ angle between longitudes at start and end points

θ latitude

BMR   basal metabolic rate

c proportionality coefficient in range equation for flapping flight

c_1, c_2 proportionality coefficients in range equations for thermal soaring flight

D drag

D_{ind} induced drag

D_{par} parasite drag

D_{pro} profile drag

d length of flight step

f fuel load relative to lean body mass

f_0 loss in f as search or settling cost at new stopover site

g acceleration due to gravity

H heading vector

k relative fuel deposition rate (increase in f per day)

L lift; distance along loxodrome (rhumbline)

m body mass

n number of flight steps

O distance along orthodrome (great circle)

P flight power (mechanical)

P_{dep} energy (fuel) deposition rate

P_{flight} power (metabolic) during flight, related to speed in flapping flight but not in gliding flight

R resulting distance from departure point after a number of flight steps

r mean vector length

s angular deviation

T thrust; track vector

t_0 search or settling time at new stopover site

V flight velocity

V_{bg} speed of best glide ratio in gliding flight

V_c climbing speed when circling in thermals

V_{flight} flight speed on migration

V_{max} maximum airspeed in flapping flight

V_{migr} total speed of migration, taking into account flight as well as fuel deposition

V_{min} minimum (stalling) speed in gliding flight

V_{mp} minimum power speed in flapping flight

V_{mr} maximum range speed in flapping flight

V_{ms} speed of minimum sink in gliding flight

V_{mt} speed in flapping flight for minimum migration time

V_x cross-country speed in thermal soaring flight

V_z sinking speed in gliding flight

W wind vector

Y flight range

8

Aerodynamics and behaviour of moult and take-off in birds

J.M.V. Rayner and J.P. Swaddle

1. Introduction

Moult is a costly period of the annual cycle for most birds, during which most or all of the feathers, representing up to 30% of the lean dry mass of a bird, are shed and replaced (Jenni and Winkler, 1994; Murphy, 1996). Assessment of the costs and benefits of the processes involved in moult are crucial to explaining the diversity of moult patterns (timing, duration, extent, sequence), their adaptive radiation across avian taxa, and the ways in which moult affects and interacts with other important stages of the annual cycle such as breeding and migration. In this paper we briefly review the costs of moult, and stress the importance of biomechanical considerations associated with flight when making such a cost–benefit analysis. In particular, we focus on the impact of moult on the most demanding mode of avian flight—that is, take-off from the ground—and we also indicate the factors that constrain take-off flight (*Figure 1*). The ability to take-off quickly and at a steep angle of ascent is likely to have a direct influence on fitness and survival: for example, increased take-off performance aids predatory avoidance (Cresswell, 1993; Grubb and Greenwald, 1982; Page and Whitacre, 1975; Witter *et al.*, 1994). For this reason we wish to highlight the necessity for behavioural and evolutionary biologists to consider and quantify biomechanical costs and constraints in order to gain a comprehensive view of the factors influencing important life-history strategies in birds.

2. Avian moult

2.1 *Physiology and energetics of moult*

Moult imposes both direct and indirect costs on a bird. Any moult strategy will evolve after a trade-off between these costs, a range of aspects of behaviour and life-history, and wing and body morphology. In this section we briefly review the direct

Biomechanics in Animal Behaviour, edited by P. Domenici and R.W. Blake.
© 2000 Taylor & Francis, Oxford.

(a)

(b)

Figure 1. The effect of moult on flight. (a) Lappett-faced vulture Torgos tracheliotus *making a running take-off into a strong headwind. This species is the largest African vulture, with a body mass of around 8 kg. It undergoes a very slow and irregular moult, and this individual is missing one distal primary on each wing (J.M.V.R., Kenya, 1982). (b) Lesser black-backed gull* Larus fuscus *making a sharp turn in gusty winds despite moult of secondary feathers and at least one distal primary, and considerable disruption to the trailing edge and abrasion of tail feathers (J.M.V.R., North Yorkshire, 1985).*

physiological costs of moult, which have been the subject of extensive research. Subsequently we consider the indirect costs arising from change, probably impairment, to the animal's aerodynamic performance.

Direct physiological costs of moult are well documented for numerous avian species, and have been studied and quantified from several perspectives (Lindström et al., 1993; Murphy, 1996; Murphy and King, 1992; Murphy et al., 1990; Murton and Westwood, 1977; Payne 1972). A full moult is surprisingly inexpensive, protein replacement requiring only a 2% increase in daily energetic expenditure or 6% increase in basal metabolic rate (review by Murphy (1996)). However, energy conversion into new tissues and feathers during moult is extremely inefficient: only 40–60% of metab-olized energy is actually deposited as feathers during moult (Lindström et al., 1993), and therefore estimates of energy demands of moult based on plumage and tissue renewal are likely to be gross underestimates. In addition to the increase in protein synthesis during moult (to produce feathers and skin), the rates of protein degradation and turnover also increase to up to 8.5 times greater than daily protein deposits (Murphy, 1996; Taruscio and Murphy, 1995).

Energetic expenditure during moult has predominantly been assessed by using three techniques: oxygen consumption (Dietz et al., 1992; Lindström et al., 1993), metabolized energy (Blackmore, 1969; Dietz et al., 1992), and overnight loss of body mass (Dol'nik, 1965, 1967). Oxygen consumption can rise by 9–11% of normal non-moulting levels (this varies greatly with rate of moult, body size, and the thermal environment). Metabolized energy renders similar estimates of energy expenditure to those of oxygen consumption, but the technique has been criticized as birds can alter their energy expenditure during the day by engaging in less strenuous activities (Murphy, 1996). If a constant caloric equivalent of mass loss is assumed (normally 13.8 kJ g^{-1}), then energy expenditure can be estimated from overnight body mass loss. Dol'nik (1965, 1967) used this method for 15 passerine species, and estimated energy costs to be approximately double the estimates from other methods. The assump-tions of constant energy expenditure overnight and a constant caloric equivalent to body mass have been criticized, so this method is not used very commonly (Murphy, 1996).

Although these estimates of energetic costs of moult appear relatively low, espe-cially when compared with the costs of egg production (daily energy expenditure in a typical full moult is approximately 20% of the energetic expenditure involved in producing one egg), the energetic costs are calculated without consideration of the energy expended on processing the nutrients to generate the necessary tissue (feathers and skin). Hence, there are also additional nutritional costs.

Water requirements are also greatly elevated during moult: Chilgren (1975) esti-mated that water requirements of moulting birds can be almost double those of non-moulting birds, probably because of increased metabolism and thermoregulation problems. Obtaining access to a reliable water supply may be the most challenging dietary problem posed by moult, and much of this cost will be manifest in the time and energy expended in selecting a suitable habitat. This will vary greatly among species, as do the nutritional demands on moult, and also vary geographically as some habitats will have more abundant suitable food and water sources than other habitats.

Other compositional changes during moult can include an alteration of blood constituents (DeGraw et al., 1979; Driver, 1981), which can often result from a dilution effect as a result of higher blood volumes during moult, and hypertrophy of the thymus, possibly reflecting increased production of lymphocytes as a result of expansion of blood volume (Ward and D'Cruz, 1968).

It is important to realize that moult entails not only replacement of plumage but also significant regrowth and alteration of underlying tissues, including blood vessels supplying the growing feathers, the integument surrounding the feather base, the horny covering to legs, feet and bills, and (in some cases) increased turnover of calcified bone (Murphy, 1996; Murphy *et al.*, 1992; Payne, 1972).

2.2 *Phenotypic plasticity*

These dramatic physiological changes have implications for the mass and body composition of birds during moult. There is a large and growing body of evidence that body condition and composition in birds are highly dynamic throughout the annual cycle, with not only replacement and regeneration of worn tissues, but with systematic adaptive control of body mass seasonally and in response to short-term conditions, and with organs being reduced at times when they are not required. This suite of phenomena is increasingly recognized as a major component of the avian life-history, and collectively is referred to as *phenotypic plasticity*. Moult should not be viewed simply as a restricted period of physiological and behavioural stress (although it is undoubtedly a demanding period), but also as a component of this broader pattern of phenotypic plasticity. It may be that these processes are more pronounced in birds than in other groups owing to the high energy costs of flight, and the premium of physiological and behavioural strategies that minimize total flight loads, that is, that reduce weight as far as possible. It may also be that the importance of moult is relatively emphasized owing to the unique avian adaptation of a feathered integument, which brings maintenance requirements. In birds two other periods of comparable physiological and aerodynamic demand are breeding (and raising young), and migrating. It is not surprising that in birds cycles of moult, breeding and migration are closely linked at every level from the endocrine system to behaviour and biomechanics, and that there is considerable diversity in these patterns between species in association with lifestyle, habitats and flight capabilities.

Most birds enter moult after breeding, when their body reserves of lipid and protein are often depleted (Austin and Fredrickson, 1987; Chilgren, 1975; Dhondt and Smith, 1980). In many cases, any loss of body mass will help to offset the increased costs of flying with moulting wings (Swaddle and Witter, 1997). Body mass is generally lower during moult, but often rises towards the end of moult as lipid, protein and water content of the body increase. Mass increases are especially pronounced if birds need to store fat for migration (which generally occurs immediately after post-nuptial moult) (Jehl, 1987; Morton and Morton, 1990). Protein composition of the body varies little during moult, except in cases where birds become flightless (as a result of the shedding of many flight feathers at one time) and flight muscles are atrophied (review by Murphy (1996)). In these latter cases, protein content is commonly shifted to leg muscles and the digestive organs (Ankney, 1979; Piersma, 1988) so that temporarily flightless waterfowl maintain their swimming, diving and foraging ability.

To optimize flight performance, birds hypertrophy flight muscle but down-regulate digestive tissue mass before long migration flights, and in some cases consume protein tissues during migration (Biebach, 1998; Piersma, 1998). It has been suggested that birds can down-regulate muscle mass during moult as an adaptation to reduce flight costs (Brown and Saunders, 1998). There is recent experimental evidence for this hypothesis, as European starlings *Sturnus vulgaris* exposed to endurance flight exercise training lose pectoralis mass and size, but show no decrease in take-off performance (J.P. Swaddle and A.A. Biewener, unpublished data).

Moult is extremely inefficient, and can incur substantial energetic and nutritional costs. It has long been recognized that consideration of the direct physiological costs of moult are insufficient to explain fully the common separation of moult from periods of breeding and migration, and also the vast array of moult parameters, sequences and tactics employed by different avian taxa (Earnst, 1992; Jenni and Winkler, 1994; King, 1980). To interpret the life-history implications of moult we must also consider the indirect consequences of feather replacement and its associated physiological responses. The most dramatic consequence of moult is the short-term partial or complete replacement of the flight feathers. There is therefore a pressing need to study and quantify the behavioural, biomechanical and aerodynamic consequences of moult. In the following sections we consider what is known about each of these topics in birds, and then we present recent results exploring the mechanisms by which aerodynamic performance is constrained during the moult period.

2.3 *Behaviour and moult*

Most birds alter behaviour during moult, and many reduce their level of activity (Francis *et al.*, 1991; Haukioja, 1971a; Newton, 1966; Sullivan, 1965); some species become flightless (Haukioja, 1971b; Sullivan, 1965). In birds such as waterfowl this is because their wings are already rather small and they are unable to accommodate the impact on flight speed (already high) and energy budgets (already higher power output than average for birds) from a temporary reduction in wing area (Rayner, 1988). Flightlessness may also be related to the increased physiological and energetic demands of moult: some birds may simply have insufficient energy available for aerial locomotion at the same time as replacing tissues. Naturally moulting birds reduce or alter their activity during moult, spending less time flying and preferentially seeking areas of protective cover with lower predation risk and thermal neutrality (Haukioja, 1971b; Newton, 1966; Stresemann and Stresemann, 1966; Sullivan, 1965). As it has been shown that similar responses can be induced by artificial reduction of feather lengths simulating the wing geometry observed during moult (Swaddle and Witter, 1997), we know that this behavioural response is due to flight biomechanics and not to the physiological changes occurring during moult. Birds may shift the relative timing of behavioural routines during wing moult, especially if they become flightless, to avoid increased exposure to predation (Kahlert *et al.*, 1996). Pied flycatchers *Ficedula hypoleuca* exposed to experimentally simulated moult are more susceptible to predation than are controls (Slagsvold and Dale, 1996). For all of these reasons the physiological and biomechanical costs of moult can affect behavioural routines to a large degree, resulting, for example, in birds spending less time foraging because of an increased perceived predation risk; this will have a profound influence on individual survival and fitness (Swaddle and Witter, 1997).

2.4 *Variation in moult pattern with flight ecology*

The importance of flight to moult (and vice versa) is also indicated by a brief survey of the relations between moult and flight ecology. For example, relatively slow moulting birds, such as the starling *Sturnus vulgaris*, remain capable of flight throughout moult (Jenni and Winkler, 1994; Swaddle and Witter, 1997). The moulting pattern of this bird is typical of most passerines in that it sheds one primary flight feather at a time from the wing in sequential order (proximal to distal) (Bährmann, 1964; Jenni and Winkler, 1994). This results in relatively small gaps in the wing and a maximal loss of wing area of only approximately 10% during the most intense period of moult (Swaddle and Witter, 1997). By maintaining flight capability these birds can continue to occupy similar niches as during non-moult periods, although they appear to reduce the amount of time spent flying (Feare, 1984; Swaddle and Witter, 1997).

Variation in moult among passerines is mostly the result of differences in the rate of moult (rather than in its timing, sequence or extent), and this appears to be intimately linked to their ecology (Jenni and Winkler, 1994). Species that migrate after moult often moult faster, and periods of breeding and moult generally do not often overlap (Ginn and Melville, 1983; Hemborg and Lundberg, 1998; Holmgren and Hedenström, 1995). In some cases, moult is suspended while migration occurs and then resumed once migration has finished. Alternatively, moult and migration can co-occur if few feathers are missing from the wing at any one time (Holmgren and Hedenström, 1995; Holmgren *et al.*, 1993).

In large birds with extended breeding cycles, such as albatrosses and eagle owls, it is imperative to maintain flight year round. These birds moult extremely slowly, and commonly extend their flight feather moult over a number of years (see, e.g., Prince *et al.*, 1993, 1997). Conversely, some passerines and a number of waterfowl moult very quickly and have many flight feathers missing from the wing at one time (Haukioja, 1971b; Stresemann and Stresemann, 1966) so that they become flightlessness and may hide in dense vegetation, spend a greater proportion of time than normal on the water surface (presumably to aid predator avoidance), or adopt a 'moult migration' to deep-water locations free from terrestrial predators (Salomonsen, 1968). The issue of variation in timing of moult is an important one, which may hold important clues to the diversity of moult patterns among birds. As in most such phenomena, we expect it to be a trade-off, ultimately between the need to divert increasing metabolism to tissue turnover and to any enhanced survival costs during moult, in a 'fast' moult, and a longer period during which flight patterns and behaviour are disrupted compared with normal, with consequences for other seasonal change in behaviour, in a 'slow' moult.

2.5 *Aerodynamics of moult*

It is—at least at first sight—evident that the change in wing planform during moult should have an effect on aerodynamic performance. Substantial proportions of the flight feathers, which form the lifting surface of the wings or tail, or of the covert feathers, which form the smooth surface to the body and wings, may be missing (*Figure 1*). A wing with a reduced wing area can be expected to have a reduced capacity to generate aerodynamic lift compared with the intact wing. However, little is known of the aerodynamic effect of moult in birds in flapping flight, and the change in lift may not simply be proportional to the change in area.

The aerodynamic performance of a wing is normally described by the aerofoil *polar* curve (*Figure 2*), which shows how lift and drag vary as a function of the angle of incidence of the wing section. Polar curves have a characteristic shape, which is related

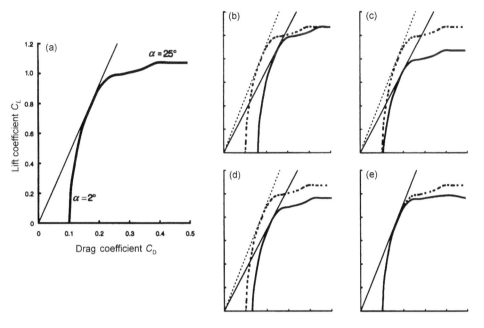

Figure 2. *Hypothetical effect of moult on wing aerodynamics. (a) Stylized polar curve for an intact wing, showing the relationship of lift and drag coefficients C_L and C_D; angle of incidence α increases moving from bottom left to top right of the curve. This curve is based on measurements on the body and wings of a swift by Nayler and Simmons (1921); the wing stalls at incidence about 30°. (b)–(e) show the possible effects of moult on the polar curve; in each case the dashed curve is the supposed polar for the intact wing. (b) Moult reduces lift with no change in drag. (c) Moult increases drag but does not affect lift. (d) Moult reduces lift and increases drag. (e) Moult reduces maximum lift coefficient but does not affect normal flight. In each of (b)–(d), the minimum lift–drag ratio, shown by the gradient of the tangent from the origin to the polar, is decreased. From measurements of gliding performance alone it may not be possible to isolate the aerodynamic mechanism for the influence of moult.*

directly to the glide polar (Tucker, 1991, 1993) and indirectly to the flapping flight power curve (Rayner and Ward, 1999). At any speed V, lift L and drag D are related to the lift and drag coefficients C_L and C_D by

$$L = \frac{1}{2} \rho S V^2 C_L \qquad (1)$$

and

$$D = \frac{1}{2} \rho S V^2 C_D \qquad (2)$$

where ρ is air density and S is wing area. If moult does not alter the force coefficients, lift and drag at a given speed should vary proportionally to the change in area. The force coefficients are likely to vary according to the morphology of the wing and feathers, the topography of the moulting feathers, and the bird's ability to compensate for their absence. We expect there will be considerable interspecific variation associated with morphology and with moult patterns, and we see potential for significant temporal variation in an individual as moult proceeds. If, for instance, just one of the wingtip primary feathers is missing, the aerodynamic load (the bound vortex circulation) normally carried by that feather may be redistributed to other wingtip feathers, with little effect on the vortex wake, and therefore little or no change in lift or induced drag, as these quantities are determined primarily by wingspan (Rayner, 1993); this might correspond to the case in *Figure 2(e)*. In a bird in which a single feather determines the wingtip and the maximum wingspan, loss of that feather may have a marked effect on lift and wake momentum: this would represent a loss of lift, with little change in drag *(Figure 2(c))*. Alternatively, when a missing secondary feather causes a gap in the trailing edge, the lift on that section of the wing is impaired; wake vortices may be shed proximally to the gap rather than more distally towards the wingtip, and either or both induced drag will rise or maximum lift will reduce *(Figure 2(b)-(d))*. A bird may compensate by adjusting the extent of overlap of the feathers, by increasing the relative speed of the wing (i.e. increasing wingbeat frequency or amplitude) or by increasing the local angle of incidence of the wing (which is very difficult to measure).

The most obvious aerodynamic effects of moult may be evident only in conditions of maximal aerodynamic performance. In cruising forward flight birds generally fly well below maximum lift coefficient, with comfortable reserves of aerodynamic performance. In slow flight and take-off a bird approaches the limits of aerodynamic force generation (Rayner, 1995; Swaddle *et al.*, 1999; this paper); it may be generating maximum lift, and therefore may be unable to maintain flight performance with even a small reduction in lifting capability. Our hypothesis is that aerodynamic lift generation is a severe constraint on take-off and slow flight, and is a prime mediator of the interaction between flight aerodynamics and behavioural strategies. We therefore expect that this constraint will be most acute at times of high energetic load such as during moult.

There is limited information about how moult affects the aerodynamic performance of a fixed, gliding, wing. Tucker (1991) recorded the change in gliding performance through moult in a Harris hawk *Parabuteo unicinctus (Figure 3)*. This species has separated primary feathers, which act to reduce induced power compared with an equivalent flat wing by spreading the vortex wake shed by the wingtip both horizontally and vertically (Tucker, 1993). In this species the loss of a distal primary forming part of the wingtip has a marked effect on gliding performance, presumably from the rise in induced drag, which the bird cannot avoid if it is to maintain lift; this corresponds to *Figure 2*(c) or (*d*). There is no information on the effect of moult of a primary feather in species with closed wingtips.

Hedenström and Sunada (1999) showed in a theoretical study of flat wings with rectangular gaps in the trailing edge simulating moult that wing gaps reduced aerodynamic performance in proportion to their size, and that gaps near the tip of the wing had a smaller effect than those created by removing proximal primary feathers. (The effect of a primary feather oriented laterally towards the wingtip was not considered.) Their analysis also indicated that an additional gap in the secondary feathers has little further influence on flight performance added to a gap in the primary feathers alone, and that the effects of wing gaps are dependent on initial aspect ratio of the wing: the effects of gaps are greater in wings of higher aspect ratio, and this may be an explanation for why birds with long, narrow wings (i.e. those with high aspect ratio) often exhibit a slow moult in which only one primary is missing from the wing at a time (Hedenström and Sunada, 1999).

Gaps or irregularity in the trailing edge of the wing during very early stages of wing moult, that is, when the most proximal primaries are being shed, could have an anal-

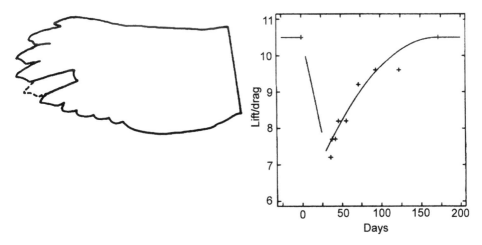

Figure 3. *Effect of moult on gliding performance in a Harris hawk* Parabuteo unicinctus *measured in a wind tunnel. Left: a fully spread wing, showing the gap left by the missing primary 6; the dotted line shows the outline of the normal feather. Right: lift–drag ratios of a gliding hawk during moulting; moult began on day 0 and was complete before day 172. Lift–drag ratio is the best (i.e. shallowest) gliding performance the bird can achieve. Immediately after moult the ratio drops sharply, and then progressively recovers. This moulting period representing almost half the year is relatively long, as may be expected for a large bird in which the absence of more than one primary at any one time may be detrimental. Adapted from Tucker (1991).*

ogous function to the notches in the trailing edges of the wings of galliforms; in a fixed wing in species with short, rounded wings this secondary notching increases the lift–drag ratio of the wing (Drovetski, 1996). It is not known how these wing gaps influence aerodynamics during flapping flight.

So far there has been less consideration of the effect of moult on the aerodynamics of bird wings during flapping flight. As we have argued above, moult may represent a particularly acute constraint in slow flight and take-off. This topic needs to be addressed before it is possible to assess the aerodynamic and biomechanical costs of moult for most birds. It is difficult to describe aerofoil action in flapping flight by polar curves, as local wing and air velocities vary in time and space, there can be considerable spanwise flow transverse to the aerofoil sections, and the flow is dominated by free vortices on, above and behind the wake; it is preferable to describe wing aerodynamics in terms of bound and trailing vorticity (Rayner 1979, 1993). We hypothesize that moult will have two effects. First, it will reduce the maximal lift that the wing can generate. This would be manifest in a lower maximum lift coefficient, and a lower maximum strength of vortex circulation bound on the wings or trailing in the wake. Second, moult may result in increased induced drag, because wake vortices are not shed as efficiently as in a normal wing; drag may not be directly influenced by reduced wing area because of the possible effects of slotting of the wingtips or of non-planarity in wake of the wingtips, neither of which are well understood in flapping flight, and may not have as marked an effect as Tucker found in a gliding bird. In normal forward flight a bird typically operates well below the maximum lift coefficient or bound circulation (Rayner, 1986, 1993), and we predict that the effect of moult may be confined to a slight increase in wing speed (i.e. wingbeat frequency and/or amplitude), and/or a small increase in local wing pitch. These changes may be so small as to be difficult to measure, or to visualize the effect on the vortex wake strength and geometry. We expect that the effects of moult will be most marked, and therefore most readily measured, during take-off and ascending flight.

2.6 *Empirical studies on moult and flight*

From the arguments above, it is to be expected that—for aerodynamic, and possibly physiological, reasons—moult will have an appreciable effect on flight performance, depending in extent on the species concerned and on their moult habits. There have, however, been relatively few studies exploring this, and most of these have artificially modified feathers to simulate moult. Noguès and Richet (1910) made dramatic mutilations to pigeon wings—far exceeding those occurring during moult—to determine how all flight capacity was lost at a critical wing area. Clement and Chapeaux (1927) showed that removal of the tips of all primary feathers of a pigeon equivalent to 3% of wingspan made all flight impossible, but that 40% of the area of secondaries could be removed with no aerodynamic effect. Wingbeat frequency was increased, and the birds were eventually forced to stop flight through fatigue. Boel (1929) trimmed the emarginated feathers of a vulture slightly, and found that although the bird could not take-off from the ground, it could still fly level; similar modifications had no effect on take-off in a pigeon. As described above, Tucker (1991) showed that the natural loss of one of the separated distal primaries in a Harris hawk depressed lift–drag ratio. Chai (1997) found that hovering flight capacity was reduced during both natural and simulated moult in the ruby-throated hummingbird *Archilochus colubris*; the effect was most marked when primary, rather than secondary, feathers were absent, and this is consistent with the mechanism outlined above by which shortening of the wingtip depresses maximum lift production. Swaddle and Witter (1997) demonstrated that a range of flight performance measures were affected by natural and experimentally simulated moult in starlings. Level flapping flight speed, ability to negotiate an aerial obstacle course and flight speed during take-off were all reduced in response to moult, but there was little indication that this reduction in performance varied through the course of moult: there was no measurable effect of moult stage and response did not depend on the size of wing gaps.

A number of workers have removed or shortened wing feathers in birds in the field, not specifically to study moult, but to explore the response of a bird to energetic stress. The argument has been that, as wings are optimized for flight, reduced wing area will impair flight performance: birds would either require more energy to fly, or would be less effective in finding food for themselves or their young. In the first such study, Harris (1971) found no change in breeding success in gulls; subsequently others have found a range of reactions, including longer foraging trips, reduced clutch sizes, lower chick growth rate, lower rates of delivery of food to young, disappearance (possibly as a result of predation), shorter duration of display, reduced mate acquisition success, and a tendency to seek protected areas (Mather and Robertson, 1993; Mauck and Grubb, 1995; Slagsvold and Dale, 1996; Slagsvold and Lijfeld, 1988, 1990; Verbeek and Morgan, 1980). These studies gave no indication of the biomechanical mechanisms involved in the compensation for modified wing profile, and reported little of changes in flight pattern of the experimental animals. In the light of the results outlined below, that starlings make an abnormal response to simulated moult, it is unlikely that these studies have much biological relevance. They do, however, confirm that wing shape is closely tuned to aspects of behaviour and ecology, and that during moult flight and foraging efficiency will be impaired, and will have consequences for avian life-history. This confirms our argument that the scheduling of moult alongside other aspects of life-history is influenced by the energetic and aerodynamic consequences of the loss of flight feathers.

2.7 *Wing shape changes and moult*

Wing shape is often used as a cue to interpret relations between flight and ecology (see, e.g. Norberg, 1990; Rayner, 1988). The patterns are complex, reflecting the diversity of trophic strategies in birds, and are overlaid by adaptation of flight muscles and of the legs, and by the phylogenetic history of individual lineages. In short, four main aspects of flight behaviour can be invoked to interpret interspecific patterns of wing design: flight speed is proportional to wing loading (and therefore is related inversely to wing area); cost of transport, or energy consumption for long distance flight, increases with aspect ratio; agility and manoeuvrability (the abilities to control the flight path) are increased by more pointed, and broader, wings (Rayner, 1988); take-off is facilitated by more rounded wings (Lockwood *et al.*, 1998).

These patterns are generally well understood, and have been extensively tested in both cross-order and interspecific studies (e.g. Rayner, 1988). However, wing shape is not stable in any bird, and varies both with growth and seasonally, and also with moult. These changes may represent a further component of phenotypic plasticity, enabling a bird to control wing shape adaptively. Changes as a result of moult may, however, be detrimental. If the most distal primary feathers are shed the wing can become shorter; depending on the change in wing area aspect ratio may also be reduced and as a result a bird may lose turning performance, have an increased cost of transport during flight, and be less able to hover or fly slowly (Lockwood *et al.*, 1998; Norberg, 1990; Rayner, 1988). Even when the feathers that form the wingtip are not moulted, gaps in the wing caused by moult of other feathers can alter the effective aspect ratio and, hence, the performance of the wing (Hedenström and Sunada, 1999).

With these considerations in mind, we feel that it would be useful to have some means of quantifying moult within a biomechanically oriented framework. Two studies have approached this issue, by converting moult scores (a subjective ranking of individual feather regrowth summed across the entire wing (Ginn and Melville, 1983)) to indices of wing raggedness and to estimates of the loss of wing area during moult (Bensch and Grahn, 1993; Hedenström, 1998). However, as indicated above, the aerodynamic influence of moult is more complex than can be assessed simply by loss of wing area. We feel that it may be more fruitful to employ an index of wingtip shape when assessing the mechanical impacts of moult (Lockwood *et al.*, 1998). This is possible with the present methods of moult scoring (Ginn and Melville, 1983) only if species exhibit little heterogeneity in moult parameters. It is more useful to compare the consequences of different moult parameters. Our proposed method takes into account the relative size and position of gaps in the wing, and also the shape of the wingtip in the non-moulting condition; all of which can affect the aerodynamic functioning of the wing. We propose that by plotting changes in the two wingtip shape indices developed by Lockwood *et al.* (1998) (wingtip roundedness and convexity), we can generate a biomechanical moult indicator that could be used to compare the flight consequences of moult among species that vary dramatically in their moult pattern, sequence and duration to explain the evolution of different moult tactics and behaviours.

To illustrate the possible application of wingtip shape as a moult indicator, we have calculated wingtip shape index scores for starlings during their complete annual moult (*Figure 4*). The two shape indices are derived from the lengths of the eight most distal primary feathers by the following equations:

$$C_2 = \log_e (3.332\ Q_1^{-3.490} Q_2^{-1.816} Q_3^{-0.893} Q_4^{-0.003} Q_5^{0.829} Q_6^{1.351} Q_7^{1.661} Q_8^{2.363}) \qquad (3)$$

and

$$C_3 = \log_e (0.0879\ Q_1^{-6.231} Q_2^{1.683} Q_3^{4.033} Q_4^{4.721} Q_5^{3.955} Q_6^{1.349} Q_7^{-3.185} Q_8^{-6.326}) \qquad (4)$$

where $Q_1 - Q_8$ are the lengths (mm) of primary feathers 1–8, respectively, numbered in ascending order from distal (i.e. nearest wingtip) to proximal on the wing. Vestigial most distal primary feathers are ignored. This unconventional numbering of primary feathers allows comparison of wingtip shapes among species with differing numbers of primary feathers (Lockwood *et al.*, 1998). In a comparative analysis of feather sizes in birds, C_2 and C_3 were shown to be effective measures of the roundedness and convexity of the wingtip, respectively, which correlated well with aspects of flight behaviour and aerodynamics (Lockwood *et al.*, 1998). When applied to moult, these two wingtip shape indices allow a quantifiable comparison of the position and size of wing gaps among species. By concentrating on the handwing (i.e. the lengths of primary feathers), this method focuses on the part of the wing that should have the largest aerodynamic effect during moult.

Although, at this stage, we do not know the aerodynamic consequences of the exaggerated values of roundedness and convexity during moult rendered by this analysis, they are useful in comparing wing geometry among species. Future experimental analyses will include simulation of these moult wing forms and study of their aerodynamic effect, both theoretically and quantitatively, on models and live birds. Exploration of the mechanical consequences of these wing forms will contribute greatly to our understanding of the flight costs of moult and so help to illuminate the selection pressures acting on both wing design and moult parameters in volant avian species.

The second way in which moult and wing shape can interact is through changes in wing shape brought about by a complete moult. Wing shape alters systematically from juvenile to adult plumage in some passerine species (Alatalo *et al.*, 1984), and as explained above these changes are clearly adaptive. Moult has the capacity to alter wing shape in a long-term manner (i.e. for the duration between successive moults). For species that show two complete moults per year, this could mean an alteration of wing shape between reproductive and migratory periods of the calendar, with the possibility of effective adaptation for differing wintering and breeding habitats. We are not aware of any studies that have explicitly explored this possibility, but it seems likely, as subtle changes in wingtip shape are associated with predictable changes in flight performance. Within-individual changes in wingtip shape in starlings are related to changes in take-off performance (Swaddle and Lockwood, unpublished): birds whose wings become more rounded because of a complete moult (i.e. comparing post-moult and pre-moult wingtip shape) take-off from the ground at a steeper angle of ascent. It is also relevant that this same measure of wingtip roundedness is related to relative predation risk, in that small passerine species with rounded wings appear to suffer less predation by sparrowhawks *Accipiter nisus* than those with relatively pointed wings (Swaddle and Lockwood, 1998).

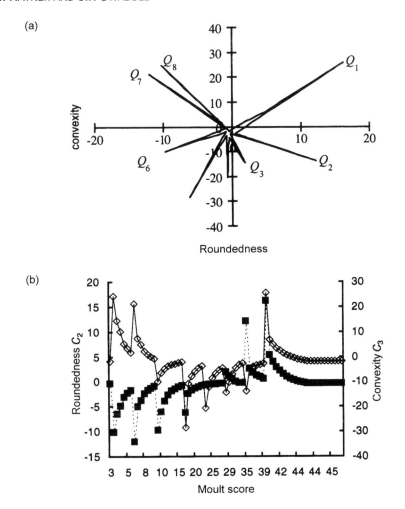

Figure 4. (a) Wingtip roundedness and convexity through a complete primary feather moult in a adult European starling Sturnus vulgaris. Q_1 *to* Q_8 *refer to the occasions when these primary feathers are shed from the wing (see text for further description). This description of moult commences with the shedding of* Q_8 *in the upper left-hand part of the graph and follows an anti-clockwise trajectory around an eight-pointed star, finishing after* Q_1 *has been replaced. Each point on the star represents the shedding of a particular primary, which results in a large gap in the wing. (b) Wingtip roundedness (■) and convexity (◊) versus moult score (Ginn and Melville, 1983) for the same individual over a single moult cycle, as above. A moult score of 45 indicates that primary feather moult has finished.*

3. Take-off in birds

Take-off is widely described as the most costly or demanding form of flight. It is generally accepted that take-off performance constrains birds, particularly in the maximum size they can reach and the loads they can carry, and that these constraints become more acute in larger birds (see, e.g., Rayner, 1995); direct experimental evidence explaining the mechanisms involved is, however, sparse (see Marden, 1987). Take-off involves high short-term energy rates associated with dynamic height gain and acceleration, and the aerodynamic lift on the wing is expected to approach maximum values. Other possible constraints on take-off performance include the contraction strain rate of the flight muscles, which may limit how much a bird can increase wing speed, the force the muscles can generate, and the power the muscles can deliver. An integrated hypothesis would explore the idea that the limits to performance set by independent aerodynamic, musculo-mechanical and physiological constraints coincide through *symmorphosis* (Weibel, 1984). In this paper we concentrate on aerodynamics, and review evidence that aerodynamic factors limit dynamic energy gain in take-off.

In a typical take-off or other extreme behaviour it may therefore not be possible to identify any one dominating constraint. Evidence of high levels of muscle activity in take-off is convincing: in pigeons, pectoralis (*pars thoracobrachialis*) EMG intensity (Dial, 1992), tail muscle activity (Gatesy and Dial, 1993) and stress developed in the pectoralis muscle (Dial and Biewener, 1993) are much higher than in normal forward flight. There are, to our knowledge, no measurements of metabolic energy consumption in take-off, but this may not be a material constraint as metabolism during brief take-off exertion is likely to be glycolytic.

The ability to take-off is vital to most birds in a trivial sense as a means of initiating flight, but it is also a major contributor to survival: many bird species escape from predators by employing short, fast flights to areas of cover (Grubb and Greenwald, 1982; Lazarus and Symonds, 1992). In particular, improved ability to take-off from the ground will reduce the risk from terrestrial-based predators (Lima, 1993), and there is considerable evidence to indicate that the speed and trajectory of escape take-off are important determinants of predation (Cresswell, 1993; Kullberg et al., 1998; Page and Whitacre, 1975; Witter et al., 1994). In this way maximized take-off performance will directly influence the probability of survival and, hence, will increase individual fitness.

Size has a marked influence on take-off performance, and it has been argued (e.g. Rayner, 1995) that take-off sets the upper limit to viable size for flying birds (but see Marden (1994)). As size increases, the margins between the power required to fly and the power available from the flight muscles, and between lift on the wing and maximum force from the flight muscles, both become narrower, and birds tend to have less scope for rapid or steep take-offs. A range of highly specialized adaptations have evolved to maintain take-off performance in relatively large species, including clap-and-fling wingbeats (in pigeons, Nachtigall and Rothe (1982)), enlarged supracoracoideus muscles associated with short, rounded and highly cambered wings (in galliforms, Rayner (1988)), or the use of extended accelerating flights often over a water surface (in gaviiforms, Norberg and Norberg (1971); in swans, Prior (1984)). Many species can take off only by running into the wind, or by dropping from a high point to convert potential into kinetic energy until speed is sufficient for aerodynamic thrust generation to be effective. Take-off has been identified as one of the major selective constraints to be overcome in the course of the evolution of flying birds from cursorial theropod dinosaurs (Rayner, 1991).

Theoretical modelling of take-off so far has been limited, possibly owing to the relative scarcity of adequate experimental information, and the practical difficulty of controlling a bird's behaviour when it is close to biomechanical or physiological limits. Norberg and Norberg (1971) modelled the generation of thrust by flapping wings during running take-offs in divers (Gaviidae), and Rayner's (1979) vortex models are applicable to take-off provided flight path dynamics are known (see below). The major limitation to formalizing such an approach has been that neither the dominant constraints on take-off performance nor the appropriate selective currency are well known.

A wing is limited in the maximum aerodynamic lift it can produce. Lift is proportional to the product of the square of the local speed of the wing and the lift coefficient (Equation (1)), which depends on the geometry of the wing cross-section and on the local angle of incidence (or pitch) of the wing section. Equivalently, lift is proportional to the strength (or circulation) of the vortices bound on the wing and trailing in the wake. Beyond a maximum pitch, the airflow over the upper surface of the wing breaks away from the wing, the wing is said to stall, and lift drops away. Stall (occurring in *Figure 2(a)* at angle of incidence of around 25°) is a major problem for fixed-wing aircraft, but is less serious for birds, for two reasons. First, stall is a dynamic pheonomenon: in flapping flight the local pitch and the magnitude of the lift force vary continuously, and instantaneous values of lift well above steady-state stall can be attained. Second, even if the wing were held still in gliding flight, a bird can readily flap to recover from a stall. In forward cruising flapping flight, the lift coefficient is well below its maximum, and a bird has considerable scope to control lift and thrust by adjusting local wing pitch and wingbeat kinematics. However, in slow flight and take-off, lift coefficient and vortex circulation are relatively high compared with forward flight, and approach the maximal levels that the wings can generate, even in flapping flight. Pigeons flying at 2–3 m s⁻¹ (Spedding *et al.*, 1984) are unable to generate sufficient wake vortex momentum to support their weight (Rayner, 1993). Mean lift coefficients as high as around five—and sometimes higher—are typical in slow flight and hovering in small birds (e.g. Norberg, 1975; Scholey, 1983), whereas lift coefficient rarely exceeds 1.5 in forward flight. In take-off flight the bird must provide additional force to gain height and to accelerate, and lift coefficient is likely to be even closer to maximal values. We hypothesize that this is the dominant constraint on take-off flight dynamics.

Physiological constraints have independently been considered to limit flight performance. Pennycuick (1969) proposed that upper limits to oxygen delivery by the respiratory and circulatory systems and to the rate of conversion of oxygen and fuel to mechanical work in the flight muscles would act as a constant upper limit to power output in any bird, independent of flight speed or behaviour, and that this power output would impose upper *and lower* limits to flight speed in steady flight. Power delivery is undoubtedly limiting, but there is little information on the magnitude of such constraints. Recent aerodynamic and mechanical evidence (Rayner, 1999; Rayner and Ward, 1999) suggests that power output at higher speeds does not rise to limiting levels, and therefore other, possibly mechanical, factors determine maximum flight speeds. It appears to us that physiological constraints setting an upper level to power output from the flight muscles are unlikely to be acute, as take-off and short slow flights are associated with glycolytic metabolism, with high bursts of power for limited duration. Indirectly, aerodynamic and physiological factors may interact to constrain take-off: to generate sufficient lift, a bird in slow flight and take-off will normally increase wingbeat kinematics and wing amplitude to maximize wing section speed during the downstroke. Force and power output from the muscles will vary with contraction rate and strain, and therefore with wingbeat kinematics, in a manner not well understood. A bird's scope for achieving high levels of performance in take-off might therefore be associated with the diversity of contractile properties of flight muscles. So far there have been no studies demonstrating substrate use or depletion in an anaerobic take-off or short-term flight in birds, and there is insufficient information to develop realistic estimates of the maximum performance.

4. Moult and take-off

In the previous sections we have described mechanisms by which moult may have an influence on flight, and have shown why take-off, climbing and accelerating flight are likely to be constrained by aerodynamic force generation. We have argued that the influence of moult will be most obvious when these two effects are combined. Indeed, it is probable that the rigorous aerodynamic constraints imposed on take-off in moulting birds are responsible for some of the changes observed in behaviour or habitat use that are commonly reported. In the following sections we report two experimental studies that have started investigation of the interaction of moult and take-off flight in starlings.

4.1 *Take-off during simulated moult*

Swaddle and Witter (1997) developed an experimental method by which the direct physiological effects of moult can be separated from the indirect effects by trimming feathers of non-moulting birds to simulate moult. A more recent experiment has expanded on these experiments (Swaddle *et al.*, 1999). In this study, manipulated starlings had their primary feathers trimmed with scissors to simulate feather lengths of a bird in mid-moult (*Figure 5*). Manipulated moult birds and control groups were filmed during take-off with a high-speed video camera (at 100 Hz) to study the changes in wingbeat kinematics during simulated moult. To quantify aerodynamic performance of the birds, Swaddle *et al.*, (1999) computed aerodynamic performance (mean wing lift coefficient by blade element theory; vortex momentum and induced power by vortex ring models (Rayner, 1979)) from measured wingbeat kinematics, and they combined measures of potential and kinematic energy gain during take-off to assess instantaneous dynamic energy per unit mass (calculated as $\frac{1}{2}(V_x^2 + V_z^2) + gz$, where V_x and V_z are the horizontal and vertical components of flight speed, respectively, g is the acceleration due to gravity, and z is height). This quantity has the advantage of combining a bird's take-off performance into a single parameter that should be independent of the bird's behavioural decisions to trade-off a gain in height (i.e. ascent angle) against a gain in speed; both are important aspects of predator avoidance, but a bird is unlikely to be able to maximize both speed and height gain.

As in the previous simulated moult study (Swaddle and Witter, 1997), birds reduced take-off speed immediately after the feather manipulations, but feather trimming had no influence on the climb angle of the take-off trajectory. Six days after the manipulations birds had regained their previous flight speed, as was also reported by Swaddle and Witter (1997). Most other measures of flight performance and kinematics (*Table 1*), including dynamic energy gain and acceleration, were significantly impaired by the manipulations, and had not recovered at least 6 days after the manipulations were performed.

Swaddle *et al.* (1999) explored possible mechanisms by which this adjustment arose. Wingbeat frequency did not change through the experiments, but wingbeat amplitude increased significantly after manipulations, although it reduced subsequently.

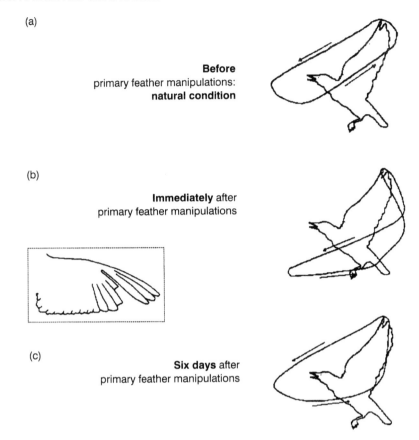

(a)

Before
primary feather manipulations:
natural condition

(b)

Immediately after
primary feather manipulations

(c)

Six days after
primary feather manipulations

Figure 5. Wingbeat kinematics in European starlings Sturnus vulgaris *during simulated moult (from Swaddle et al. (1999). The inset shows how feathers were modified to simulate an intermediate stage of a natural moult with one primary slightly shortened and the adjacent primary almost removed. All birds from this treatment group showed this consistent pattern of wingtip reversals during phase 2. Kinematics for two control groups, which experienced the same handling either with no manipulation or with trimming of the same feathers to remove less than 1% of wing area, remained similar to phase 1 throughout.*

The consequence of these changes was that the wingtip speed was increased, but because acceleration is lower the birds operated at a lower mean lift coefficient. It is expected that this was the maximum lift coefficient they were able to achieve, and this value was insufficient to maintain acceleration and rates of energy gain (= dynamic power). Immediately after manipulations there were dramatic changes in movement of the wingtip during the wingbeat (*Figure 5*). Before the manipulations the kinematics and posture are characteristic of passerines in take-off flight: the body is steeply inclined, and the wingtip follows a shallow arc anterior to the head during the upstroke, as the leading edge of the wing is held at an acute angle to the body axis. During the downstroke the wrist is moderately flexed and the wingtip follows an approximately linear path (*Figure 5(a)*). Immediately after the feather manipulation birds exhibited an abnormal wing movement not known from any bird in natural flight. The wingtip path followed a looping movement at the top of the trace, indicating the presence of a wingtip reversal during the latter phase of the upstroke, and during most of the downstroke the leading edge is at an obtuse angle to the body axis (*Figure 5(b)*). Six days after the manipulations the wingtip pattern displayed less of this disruption, but was still dissimilar to that observed during session 1 or in control birds. Wingtip reversals have been associated with lift production in the upstroke of pigeons (Brown, 1963) although flow visualization experiments do not support this interpretation (Rayner, 1995; Spedding *et al.*, 1984), and it is inconsistent with the backward-slanted orientation of the upstroke wingtip path. Upstroke lift is equally unlikely to occur in these starlings, which—like pigeons—employ a vortex ring gait during take-off (*Figure 6*). The estimated reduction in lift production during the downstroke, with parallel reductions in wake momentum, vortex ring circulation and induced power, is consistent with the decline in dynamic energy gain and acceleration (*Table 1*). Although the low value of induced power may seem advantageous, this is in practice the symptom of the birds' inability to generate more lift.

In these experiments, moult was simulated by cutting the distal portions of individual feathers (*Figure 5*). That the abnormal kinematic response arose simply from unfamiliarity with the unexpected low lift coefficient is evident from the birds' adjustment over the following 6 days. The difference from behaviour in natural moult (below) emphasizes the need for caution in employing such experimental techniques to simulate morphological variation. The 'tips' of the trimmed feathers are distinct from the true tips of a regrowing feather during moult: the shaft and vane will have different mechanical properties, and will probably be stiffer. This, in addition to the unexpected change in wing planform, are apparently responsible for the reduction in maximum lift coefficient.

There was no indication that birds generated extra lift with their tails, as tail feathers were held at similar angles among treatment groups. It does not seem that birds are exploiting novel sources of lift immediately after manipulations. Upstroke wingtip reversals are more likely to be used as a means of accelerating the wing rapidly before the downstroke, and thereby of developing lift more quickly. This mechanism is likely to be important to a starling with an impaired wing, and we propose that the increase in wingbeat amplitude and the deformation of the wingtip pattern are the immediate mechanisms of compensation for removal of feathers and loss of integrity of the distal parts of the wing. We suspect that this pattern entails some, perhaps significant, extra costs, although we have not been able to confirm this: induced power appears to decline (*Table 1*), but there may be demands on unfamiliar muscle groups associated with this flight pattern.

Subsequently, perhaps associated with adjustments to the flight muscles or simply with familiarity, the bird accommodates its wingbeat to a pattern more similar to normal, and is able to some extent to compensate for the immediate cut in lift coefficient.

4.2 *Take-off during natural moult*

Experiments with simulated moult have the advantage of separating biomechanical changes to the wing from the physiological changes to body composition, total weight and flight muscle contraction physiology, which are an essential component of the natural moult cycle. In the absence of this accommodation it is perhaps not surprising that the aerodynamic performance of birds in simulated moult is sharply depressed.

Table 1. *Mean morphology, wingbeat kinematics, flight path dynamics and computed aerodynamic performance for European starlings Sturnus vulgaris undergoing natural (left columns) and simulated (right columns) moult.*

	Natural moult			Simulated moult		
	2 weeks before moult	Mid-moult	2 weeks after moult	Before manipulation	Immediately after manipulation	6 days after manipulation
Morphology						
Body mass (kg)	0.074	0.074	0.077	0.073	0.072	0.072
Wing area (m²)	0.0188	0.0174	0.0187	0.0191	0.0176	0.0176
Wingspan (m)	0.350	0.350	0.370	0.393	0.393	0.393
Flight Path						
Speed (m s⁻¹)	2.6	2.7	3.1	2.7	2.5	2.6
Take-off angle (°)	22.8	17.4	13.8	25.4	26.0	27.9
Acceleration (g)	0.77	0.69	1.35	1.12	0.74	0.75
Dynamic energy gain in second wingbeat (J kg⁻¹)	2.0	1.7	3.3	2.5	1.7	1.8
Kinematics and aerodynamics						
Wingbeat frequency (Hz)	14.8	15.2	14.9	16.1	16.3	16.8
Wingbeat amplitude (°)	90.2	125.9	120.1	136.3	144.2	127.1
Mean downstroke wingtip speed (m s⁻¹)	8.2	11.6	11.5	15.0	16.1*	14.7
Mean lift coefficient C_L	6.6	3.7	4.9	2.9	2.3*	2.7
Wake momentum (kg m s⁻¹)	0.49	0.47	0.76	0.60	0.47	0.46
Wake vortex ring circulation (m² s⁻¹)	1.42	1.07	1.37	1.01	0.79	0.83
Power output						
Induced power (W)	4.6	3.0†	5.1	3.4	2.2†	2.4†
Dynamic power (W)	2.2	1.9	3.8	3.0	2.0	2.2
Total mechanical power (W)	6.8	4.9	8.8	6.35	4.2	4.6

Aerodynamic parameters relate to the second wingbeat after take-off ($n=7$ for natural moult, $n=7$ for simulated moult experiments). (For details of experiments see text.) Simulated moult from Swaddle et al. (1999), natural moult from Williams et al. (unpublished). Full details of experimental design, control groups, and statistical comparisons within and between groups are given in those papers.

* Assumes linear wingtip path in downstroke, and ignores wingtip reversal in upstroke; this will increase downstroke wingtip speed and reduce lift coefficient, will possibly depress induced power, but will not significantly affect wake momentum.

† Neglects change (?reduction) in aerodynamic wing efficiency (spanwise distribution of circulation) as a result of change in wing planform during moult; this might increase induced power.

(a)

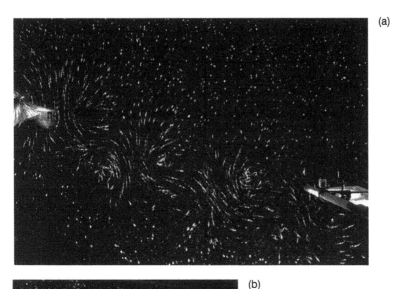

(b)

Figure 6. The vortex ring wake of starling Sturnus vulgaris *in take-off flight from a perch, seen in side (above: photo E.V. Williams) and front views (below, photo J. Martin), by helium-bubble flow visualization (methods described by Spedding* et al. *(1984)). The wake consists of a sequence of vortex rings. The rings are approximately horizontal when shed, but as they evolve they rotate to become more tilted, as seen to the bottom right of the upper figure.*

Recent observations of seven adult starlings experiencing natural moult have unveiled interesting patterns suggesting that the biomechanical consequences of moult are more complex than the simulated moult experiments indicate (Williams *et al.*, unpublished material). Take-off was filmed with a high-speed cine-camera (at 185 Hz) on three occasions: 2 weeks before moult, at mid-moult and 2 weeks after moult was completed. Wingbeat kinematics, take-off speed, angle, acceleration and dynamic energy were determined from the films, and, as described above, aerodynamic performance was estimated by vortex ring theory *(Table 1)*. Morphological measures indicate that body mass was low at the beginning of moult, remained depressed during moult, and rose when moult was completed. Swaddle and Witter (1997) have previously indicated significant mass reduction in starlings during moult. Wing area decreased during moult. Interestingly, wing length after moult was completed (16.7cm) was longer than before moult began (15.9cm), indicating the benefit of regrowing worn or abraded primary flight feathers.

During moult birds decreased their angle of ascent compared with before moult, but did not alter their speed of take-off. This is consistent with previous analyses of take-off behaviour in moulting captive starlings (Swaddle and Witter, 1997), and with aero-dynamic theory, which indicates that the faster a bird takes off, the lower the total aerodynamic work done to accelerate it. Take-off speed is less likely to be affected by a decrease in wing area, because weight support and thrust production are determined primarily by the span of the wing (Rayner, 1986, 1993). The wingspan of the birds in mid-moult in the present study remained unchanged compared with the wingspan pre-moult. Any decrease in mechanical power or aerodynamic force output may compromise ascent angle and/or acceleration, in preference to a reduction in speed, because ascent angle and acceleration are probably linearly related to energy input (Hedenström and Alerstam, 1992; Rayner, 1986), unlike take-off speed or speed in forward flight (review by Rayner and Ward (1999)). Therefore, a decrease in angle or acceleration would lead to a direct power saving, whereas a decrease in take-off speed, assuming no change in ascent angle or acceleration, would incur an increase in power requirement during take-off. We would therefore expect a bird to maintain flight speed as much as possible in an impaired take-off.

Kinematic analysis revealed that all birds exhibited uniform wingtip movements, similar to those commonly observed in this species (compare *Figure 5* pre-moult pattern), and there was no indication of the wingtip reversals observed in the simulated moult experiments (above). As in the experiments with simulated moult, wingbeat frequency remained constant, but wingbeat amplitude increased sharply in mid-moult and remaining elevated after moult had been completed, so that the wingtip was moving markedly faster. The birds were able to maintain speed, climbing angle, dynamic energy and acceleration (all fell slightly but non-significantly) with a lower lift coefficient on the wing. From this evidence we can determine that birds altered wingbeat kinematics in response to moult to maintain lift; they were not physiologi-cally or structurally constrained in their wingbeat amplitude before moult began, but during moult the maximum lift coefficient they could obtain was sharply depressed. Experiments (above) simulating moult reached the same conclusion. As birds in nature commonly enter moult in a 'physiologically challenged' state, it is possible that there are limits to the functioning of muscles involved in flight and to the ability of worn feathers to generate aerodynamic force. After moult is under way and has finished it is possible that these constraints are less severe. *In vivo* measurement of muscle and tendon performance of birds in varying physiological and photoperiodic states may shed light on these hypotheses. Birds attempting to improve take-off performance through manipulating their wingbeat kinematics may not have been able to increase downstroke wingtip speed any further owing to physiological and energetic constraints imposed by muscles and tendons, which become exacerbated by the increased demands of moult.

Because of the low lift coefficients during moult, the wake vortex circulation and the induced power were also low. It is tempting to suggest that this is an adaptive response, in reducing the costs of flight during moult, and therefore permitting energy resources to be diverted to tissue and feather synthesis. However, three points should be set against this hypothesis. First, the energy reduction is a consequence largely of the reduced acceleration rates during take-off, and any reduction in power in steady forward flight is likely to be small or absent. Second, the reduced take-off performance increases exposure of the bird to predation risk, and this is likely to outweigh any small energetic benefits. Third, if small wings with moult gaps are adaptive during moult, they should be even more beneficial at other stages of the life cycle.

The most marked evidence of accommodation of flight performance to moult comes, perhaps surprisingly, from the period after moult. Speed, acceleration, dynamic energy gain and vortex wake parameters all increase significantly, and rise to performance significantly better than that before moult. This is in contrast to the experimentally simulated moult experiment described above, in which these values did not recover. The reduction in take-off angle in response to new plumage confounds the relationship between wingtip speed and take-off speed, but this is offset by the dramatic increase in acceleration. We interpret these results to mean that before and during moult birds are constrained in lift generation by the condition of their wings (worn or abraded feathers) and by the wingbeat kinematics they can achieve. After moult, they can achieve better rates of energy gain with a lower lift coefficient than before. This may mean that the generation of aerodynamic force is a significant constraint before and during moult, but is not a constraint subsequently.

The comparison of pre-manipulation performance with that 1 week post-manipulation is probably more relevant to the effects of the mid-moult stage in the present natural moult study; that is, birds had time to adapt take-off performance to their moult plumage. Therefore, the evidence from both the natural and experimentally simulated moult experiments indicates that starlings do not experience significant decreased take-off performance as a result of prolonged exposure to a moult plumage (as happens in nature). This is a surprising finding, and has important implications for the survival and behaviour of starlings during moult. It may also help to explain why starlings (and similar passerines) have a relatively prolonged moult and shed feathers sequentially rather than simultaneously. Further experimental investigations to examine the flight costs of different moult patterns and strategies in the starling and other species may further help to explain why avian species vary so much in the timing, duration and patterns of moult. However, as our kinematic data indicated that birds increased wingbeat amplitude (and to some extent frequency) during moult, there may be a physiological and energetic cost to maintaining flight performance during moult at pre-moult levels. Therefore alteration of kinematics to minimize the apparent (to the observer) influence of moult on overall flight performance may, in itself, impose costs that could impinge on the birds' energy budget and behaviour. Hence, we advocate a life-history approach to studying moult.

There was a significant increase in dynamic energy gained per wingbeat associated with new plumage. These data further indicate the benefits of moulting into new plumage, which are important to consider when explaining the factors that give rise to the evolutionary radiation of moult parameters. Although Williams et al. (unpublished material) selected individuals with apparently undamaged plumage for their study, the data appear to indicate that the renewed plumage after moult was functionally more efficient than the year-old plumage that each bird possessed before moult. An increase in the functional efficiency of new feathers may also have important implications for other possible indirect costs and benefits of moult, such as water repellence, thermal insulation and visual appearance of display plumage. Future studies of the costs of moult could also incorporate studies of these additional indirect consequences of renewed plumage.

4.3 *Comparison of simulated and natural moult*

The mechanical realization of changes in wingform as a result of simulated and natural moult are different. The disruption of the ability of the wing to generate lift force appears comparable when feathers are physically altered to match the pattern observed in moult, or when feathers are naturally at these lengths during moult (Swaddle and Witter, 1997; Swaddle *et al.*, 1999; Williams *et al.*, unpublished). It is not possible to identify the aerodynamic mechanism, but the observed response is consistent with a sharp reduction in maximum lift coefficient C_L, and a probable reduction in lift–drag ratio (cf. *Figure 2(c)* and (*d*)). Different aerodynamic mechanisms might apply in forward cruising flight, as outlined above. The difference between natural and simu-lated flight feather moult is accentuated by the longer-term changes in flight perfor-mance in the simulated moult experiments (Swaddle and Witter, 1997; Swaddle *et al.*, 1999). In both cases, wingtip kinematics altered immediately after moult or plumage manipulations, but in the simulated moult there were dramatic changes in the geometry of the wingtip path, which represent an extreme attempt by the bird to adjust to unexpected changes. Within 1–2 weeks after the feathers had been trimmed the birds had partially regained some of their flight ability. This clearly demonstrates that birds are capable of compensating or adjusting the behavioural response of flight to a new wing geometry. As birds can do this in response to a wholly artificial manipu-lation of feather lengths, it is more than likely that they can do the same during natural moult (as the data above indicate). Possible adaptive mechanisms could include alter-ation of wingbeat amplitude, wingbeat frequency, angle of attack of the wings, altering the shape of the wing and feather overlap, increasing maximal extension of the wing during the downstroke and tighter furling of the wings to the body during the upstroke. These all deserve further investigation to help explain how birds maintain flight performance during moult.

   As natural moult entails significant physiological changes, it is also possible that the physiological alterations during moult could account for the differences between the natural and simulated moult experiments. It seems at first glance counter-intuitive that despite the additional physiological and energetic demands of moult, birds in natural moult appear to retain a greater flight performance than those that have had their feathers trimmed to simulate moult. This may simply be due to the disruption by the manipulations. We hypothesize that birds are able to reap some biomechanical benefits from the physiological changes that occur during moult. Perhaps wing weight and, hence, inertia are significantly reduced during moult: this could sharply reduce the maximum force required to move the wings (Rayner, 1986, 1993). Additionally, a loss of feathers (primaries, secondaries and coverts) from the wing may help with furling the wings tighter to the body during the upstroke, hence giving a slight reduction in profile drag (and also inertia). The expanded blood volumes during moult could increase aerobic supply to the flight muscles. The most important physiological effect on biomechanical performance may be the reduction in body mass commonly observed during moult. This will help the bird to achieve more with the reduced lift force from the wing. This loss of mass can even comprise loss of flight muscle mass without a detrimental effect on flight (J.P. Swaddle and A.A. Biewener, unpublished data). In simulated moult, birds are immediately exposed to a sub-optimal wing geometry and lower wing area without any alteration of body mass or other aspects of body composition. In natural moult, birds have often lost mass before wing moult commences, and therefore their wing loading is preserved at a lower, less costly, level. The experimental data we have presented in this chapter suggest that this mass loss is the primary mechanism by which starlings can reduce flight costs or compensate for reduced aerodynamic performance during natural moult. However, this does not preclude the action of alternative and additional mechanisms, nor can it explain flight performance in species that do not lose mass. In further investigations it will be necessary to take account of the physiological state of birds before moult begins, as the data presented in this chapter indicate that these initial conditions are vital in determining the influence of the wing geometry of moult on biomechanical performance.

5. Discussion

Moult frequently occurs between periods of breeding and migration (or overwintering for non-migratory species) and there tends to be little overlap among these periods for most avian species (Ginn and Melville, 1983; Hemborg and Lundberg, 1998; Holmgren and Hedenström, 1995; Langston and Rohwer, 1996; Slagsvold and Lifjeld, 1989). Hence, moult (especially in those species that moult once per year) often delimits the end of breeding and also the beginning of migration, and we must adopt a broad, life-history view of the factors that influence and drive the evolutionary radiation of moult parameters. These factors (i.e. the relative costs and benefits of moult as well as evolutionary and developmental constraints) are not only important in shaping 'moult' as a life-history trait, but they will also have indirect influences on breeding, migration and overwintering strategies. Therefore, quantification of cost–benefit trade-offs and identification of developmental and evolutionary constraints acting on moult parameters are essential in assessing many life-history elements of bird species. The review and presentation of novel data in this chapter further indicate the importance of biomechanical considerations to such analyses. We cannot obtain a full understanding of the factors that shape the evolution of moult without empirical and theoretical studies of biomechanical costs, benefits and constraints.

A life-history approach to the study of moult has often indicated the interplay between moult parameters and features of the ecology and behaviour of particular species (e.g. Ashmole, 1963, 1965; Furness, 1988; Langston and Rohwer, 1996; Slagsvold and Lifjeld, 1989). An intriguing example is that of Laysan and black-footed albatrosses (*Diomedea immutabilis* and *D. nigripes*, respectively), which appear to display trade-offs between the time available for moult versus breeding (Langston and Rohwer, 1996). Because of their extended breeding period and the lack of overlap between moult and breeding, there is often insufficient time available to complete a full moult every year; hence Laysan and black-footed albatrosses exhibit various patterns of incomplete moult. In years when breeding starts earlier or when the breeding period is shorter or abandoned, individuals are able to moult more feathers after their breeding attempt. In addition, when feathers become too abraded (as they may not be replaced every year), breeding may be sacrificed so that birds have time available to complete a full moult and renew their worn feathers (see discussion by Langston and Rohwer (1996)). Hence, moult not only influences breeding during the same year, but can also influence future breeding attempts, as albatrosses may be able to trade-off long-term investment in moult versus breeding. Ultimately, these birds are constrained in their rate of moult by the cost of regrowing many primary feathers at one time (as feather growth rates are very conservative in these species (Langston and

Rohwer, 1996)). This may entail energetic and physiological costs of feather replacement (as indicated earlier in this chapter), but will also incur significant biomechanical cost, as having more feathers missing from the wing will increase the size and number of wing gaps (these birds show split sequences of primary moult in which the outer primaries are shed proximal to distal, but the inner primaries are shed distal to proximal). Therefore, the biomechanical consequences of moult can have profound influences on the breeding tactics of individuals.

These forms of life-history trade-offs are not particular to slow moulting seabirds with prolonged breeding periods, such as albatrosses. Other large birds have extended moult cycles, sometimes taking more than 1 year to complete a full wing moult. Similar observations have also been reported in many passerine species. For example, breeding pairs that invest more during the breeding season (i.e. raise a larger brood, or raise more broods) tend to start their post-nuptial moult later (e.g. Dhondt, 1973; Newton, 1966; Slagsvold and Lifjeld, 1989). Additionally, in cases where the male tends to expend less parental effort than the female, the male will often moult earlier than the female (e.g. Hemborg and Merilä, 1998; Orell and Ojanen, 1980; Slagsvold and Lifjeld, 1989). Hence, breeding, migration and moult strategies can be intimately linked in many avian species.

Biomechanical considerations are intrinsic in determining the relative costs of these different strategies, and so are important in mediating life-history trade-offs. For example, moult could overlap with breeding (or migration) in species that have short rounded wings, as the aerodynamic cost of wing gaps may be relatively small for species with low aspect ratio (Hedenström and Sunada, 1999). Also, wing gaps may be smaller in species with short, rounded wings, as there is more overlap among neighbouring feathers than in species with longer, more pointed wings. Slow moulting species with short rounded wings may be able to minimize the costs of moult and, therefore, have more flexibility to overlap moult with periods of breeding or migration. These observations could explain why species such as the dunlin *Calidris alpina* moult during their migration (Holmgren et al., 1993).

Many large soaring or gliding birds have wingtips with separated primary feathers. The geometry of these wingtips is critical in determining the structure of the wing's vortex wake, and therefore the drag. It is to be expected that loss of primaries in species with separated wingtips could be aerodynamically more costly than for birds with closed wingtips. Not only will the gaps in the wing be larger for birds with separated primaries, but these species will also lose (to some degree) the benefit of possessing wingtip slots, which will result in a relative increase in induced drag (Tucker, 1993). In addition, mechanical aerodynamic loads must be redistributed over the remaining feathers, increasing the incidence of abrasion and wear. Hence, biomechanical factors may constrain such species to slow moult in which only one feather is missing from the wingtip at any one time: Harris hawks display this form of slow moult sequence (Tucker, 1991), which is typical of raptors. Vultures are commonly observed flying with primaries missing (*Figure 1(a)*), indicating a very slow moult (Fischer, 1962).

Biomechanically derived hypotheses may also help explain natural variation in moult patterns. For example, Laysan albatrosses show an incomplete primary feather moult, but the most commonly replaced feathers are the outer primaries (Langston and Rohwer, 1995). These are the longest primaries and, hence, are the most energetically costly feathers to regrow (there is more feather to replace). So why moult the longest feathers in preference to shorter inner primaries? There are three mechanical explanations. First, distal feathers may be more likely to become abraded or damaged than proximal feathers, as the distal feathers form the wingtip. Hence, these feathers may need to be replaced more often. Second, loss of aerodynamic function of outer primaries during non-moult periods may be more costly than loss of function in inner primaries. So even if abrasion rates were the same across the wing, it may be more important to replace distal feathers than proximal feathers for flight mechanics reasons. Third, a gap created during moult at the wingtip (i.e. by shedding of distal feathers) may be aerodynamically less costly than a gap at a more proximal section of the wing (Hedenström and Sunada, 1999). For this reason replacing outer primaries may be mechanically less costly than replacing inner primaries.

The data that we have presented and reviewed in this chapter further support the life-history perspective that we advocate. In natural moult, birds appear to undergo several adaptations that reduce the flight costs of missing feathers from the wing. The most notable of these is a loss of body mass. Regulating a lower body mass during moult will impose its own suite of costs and benefits (see Witter and Cuthill, 1993). Two of the most obvious costs of reduced mass may be an increased probability of starvation (which has obvious fitness consequences) and also increased nutritional stress during moult, which is known to increase the occurrence of deformities and abnormalities in regrowing feathers (e.g. Swaddle and Witter, 1994; see below), which are also likely to have fitness consequences. Hence, although birds appear to reap flight benefits from a reduced mass during moult, there will also be costs and, ultimately, constraints to this mass loss, all of which can only be fully interpreted by adopting a life-history approach. Experiments in which moult is simulated may appear to isolate specific biomechanical or behavioural aspects of the problem; our comparison of aerodynamic responses to simulated and natural moult in starlings indicates that the response to simulation of moult for which the bird is unprepared induces abnormal behaviour, approaching the limits of aerodynamic performance. We advocate caution in experiments employing artificial manipulation of wing or tail feathers to simulate natural inter- or intra-individual morphological variation.

A possible integrated approach to study many of the costs and benefits associated with moult is in terms of energetics (as described earlier). The increase in wingbeat amplitude and wingbeat frequency observed during natural moult are bound to impose an increased physiological and energetic burden on the birds. Not only is the wingtip moving faster, but it is also moving further. In terms of flight, these responses help to maintain take-off performance at pre-moult levels, but the costs of these kinematic changes may be difficult to ascertain without measuring energetics. This form of study also needs to be integrated with knowledge of physiological and anatomical constraints of muscle, tendon and joint operation during moult. It is possible that these are altered by moult, or that birds do not perform maximally when outside moult periods. Once we have gained a more complete understanding of the energetic costs and constraints imposed by both the direct physiological implications and also the indirect biomechanical and behavioural consequences of moult, these elements can be combined with existing models of moult energetics. Without incorporating such elements we will not be able to fully explain the adaptive radiation of moult parameters or the interaction of moult with other important periods of the avian annual cycle, such as breeding and migration.

Studying the energetic cost of various elements of moult could create a number of methodological difficulties. Implicit throughout our discussion we have assumed that birds trade-off costs with benefits, and there is no reason to think that birds would not do this in experimental conditions, and would even alter the dynamics of these trade-offs among experimental protocols. Studies using free-flying birds could be limited as birds could compensate, energetically, by shifting behavioural patterns and reducing energy expended in particular behaviours (as was indicated explicitly by Swaddle and Witter (1997)). Hence, we would need to know the energetic implications of all behaviours exhibited by birds during the experiment and also the amount of time spent performing each of these behaviours. In most cases this would be an onerous (and probably unfeasible) task. Conversely, bringing birds into captive conditions in which their behaviour can be monitored more closely or their environment controlled to reduce confounding environmental factors, also poses problems. The studies reported here are limited in that we do not know, ultimately, if birds can trade-off any loss of take-off performance with any other part of their behavioural repertoire. We also do not know whether birds' motivation to escape from an experimenter in an enclosed aviary is the same as the performance birds would show in a real predatory event. What we suggest is the explicit integration of field observations and controlled laboratory experimentation. This is something that many behavioural ecologists are used to but appears less prevalent in the biomechanical literature, as birds often have to be in controlled conditions to give accurate data. Hence, not only can biomechanics make substantial contributions to the study of behaviour, but also behavioural ecological approaches can be incorporated into biomechanical studies to increase the ecological and evolutionary relevance of any data that are collected.

In this chapter we hope to have provided some initial ideas and empirical evidence for the importance of biomechanics to the moult tactics and behaviour of birds. Crucial to our arguments are identification of biomechanical constraints acting during moult, and also quantification of the mechanical costs and benefits of moult. We have discussed the aerodynamic consequences of moult at length, but there may be additional costs that should also be incorporated. These include assessment of the thermoregulatory costs of feather loss. This should be studied not only in terms of feather loss, but also in terms of the physiological alterations and demands of natural moult. As water requirements are greatly elevated during moult, there may be significant thermoregulatory consequences independent of the insulatory effects of plumage loss. Other factors to be considered are water repellence (which will affect not only thermoregulation but also buoyancy) and also alteration of the appearance of the plumage in terms of visual communication. The appearance of plumage colours that are derived from dietary sources (e.g. coloration produced by carotenoids (Gray, 1996)) is mediated by moult, and the coloration of such plumage areas can have profound influences on natural and sexual selection processes (e.g. Hill, 1990). This latter issue does not solely apply to gross colour or morphological changes observed in birds that switch between nuptial and overwintering plumage, but also to more subtle changes in structural colours, which could be influenced by feather wear and ultrastructural properties of feathers. There are also numerous studies that could be performed to assess biomechanical constraints acting during moult. It is possible that muscle, tendon and joint functional limits are different during moult because of the gross physiological changes that birds experience during this period.

A life-history approach also demands that we consider the long-term effects of moult. These could include annual alteration of wing form, such as changes in wing dimensions (e.g. aspect ratio and wingtip shape). These are factors that are known to influence flight costs (Lockwood *et al.*, 1998; Norberg, 1990; Pennycuick, 1975; Rayner, 1988), and so will impinge on numerous behavioural, energetic and physiological traits. Conditions during moult have also been demonstrated to affect the left–right symmetry of growing feathers (i.e. fluctuating asymmetry (Swaddle and Witter, 1994, 1998)). Changes in asymmetry of feathers have been demonstrated to influence a range of important behaviours, such as angle of escape take-off (Swaddle, 1997), speed of take-off (Swaddle *et al.*, 1996), aerial manoeuvrability and agility (Evans *et al.*, 1994; Swaddle and Witter, 1998; Thomas, 1993), level-flight speed (Swaddle, 1997; Thomas, 1993) and mate preferences (Swaddle and Cuthill, 1994). All of these behaviours are likely to have a direct influence on fitness and survival. Environmental conditions during moult may also influence the overall condition and performance of feathers, hence influencing flight, abrasion and breakage resistance, thermoregulation and water repellence.

It is important also to take account of the condition of birds before moult. For example, Laysan albatrosses with a higher parasitic load of oesophageal nematodes replaced fewer feathers in their moult than those with a relatively lower number of parasites (Langston and Hillgarth, 1995). Hence, pre-moult condition may place constraints on moult, which may in turn influence behaviour and survival. Condition of feathers before moult is also an important consideration. Williams *et al.*, (unpublished) explicitly demonstrated that the flight costs of bearing year-old feathers can be as significant as the flight costs of moult itself. Future studies could include assessment of pre-moult feather condition in terms of a range of physical and mechanical properties (e.g. abrasion, hardness, stiffness, thermoregulation, water repellence, visual appearance). These properties may vary among species and within individuals, and may help explain the timing of feather replacement. For example, species occupying harsh environments and that experience high levels of feather abrasion may have to moult more than once a year.

The preceding discussion has indicated the importance of moult to most birds and, in particular, that moult can have profound life-history influences, including indirect influences on other life-history traits (such as breeding, migration and overwintering). Biomechanical costs and constraints are important elements of the factors that shape moult and so will have life-history implications. Therefore, we urge evolutionary biologists to consider and quantify the mechanical effects of moult so we can achieve a better understanding of the vast array of moult parameters observed in nature.

Acknowledgements

To reduce the number of references we have cited several review papers in the text. This is not intended to detract or deflect credit from the original authors. J.P.S. was funded by a Natural Environment Research Council Postdoctoral Fellowship during the collection of experimental data and a Royal Society of London University Research Fellowship while preparing the manuscript. J.M.V.R.'s research on flight aerodynamics has been funded by grants from BBSRC, EPSRC and the Royal Society of London. We are grateful to Jim Martin and Emma Williams for the images in *Figure 6*, to Emma Williams for permission to quote unpublished results, and to Thomas Alerstam for comments on the manuscript.

References

Alatalo, R.V., Gustafsson, L. and Lundberg, A. (1984) Why do young passerine birds have shorter wings than older birds? *Ibis* **126**: 410–415.

Ankney, C.D. (1979) Does the wing molt cause nutritional stress in Lesser Snow Geese? *Auk* **96**: 68–72.

Ashmole, N.P. (1963) Molt and breeding in populations of the Sooty Tern *Sterna fuscata. Postilla Yale Peabody Mus.* **76**: 1–18.

Ashmole, N.P. (1965) Adaptive variation in the breeding regime of a tropical sea bird. *Proc. Natl Acad. Sci. USA* **53**: 311–318.

Austin, J.E. and Fredrickson, L.H. (1987) Body and organ mass and body composition of post-breeding female Lesser Scaup. *Auk* **104**: 694–699.

Bährmann, U. (1964) Über die Mauser des europäischen Stars *Sturnus vulgaris. Zool. Abh. Dresden* **27**: 1–9.

Bensch, S. and Grahn, M. (1993) A new method for estimating individual speed of molt. *Condor* **95**: 305–315.

Biebach, H. (1998) Phenotypic organ flexibility in Garden Warblers *Sylvia borin* during long-distance migration. *J. Avian Biol.* **29**: 529–535.

Blackmore, F.H. (1969) The effect of temperature, photoperiod and molt on the energy requirements of the House Sparrow, *Passer domesticus. Comp. Biochem. Physiol.* **30**: 433–444.

Boel, M. (1929) Scientific studies of natural flight. *Trans. Am. Soc. Mech. Engrs* **51**: 217–242.

Brown, R.E. and Saunders, D.K. (1998) Regulated changes in body mass and muscle mass in molting Blue-winged Teal for an early return to flight. *Can. J. Zool.* **76**: 26–32.

Brown, R.H.J. (1963) The flight of birds. *Biol. Rev. Cambridge Philos. Soc.* **38**: 460–489.

Chai, P. (1997) Hummingbird hovering energetics during moult of primary flight feathers. *J. Exp. Biol.* **200**: 1527–1536.

Chilgren, J.D. (1975) Dynamics and bioenergetics of postnuptial molt in captive White-crowned Sparrows (*Zonotrichia leucophrys gambelii*). PhD thesis, Washington State University, Pullman.

Clement, H. and Chapeaux, E. (1927) Quelques modifications expérimentales des surfaces portantes du pigeon et leurs effets sur le vol. *C. R. Assoc. Fr. Avanc. Sci.* **50**: 415–416.

Cresswell, W. (1993) Escape responses by redshanks, *Tringa totanus*, on attack by avian predators. *Anim. Behav.* **46**: 609–611.

DeGraw, W.A., Kern, M.D. and King, J.R. (1979) Seasonal changes in the blood composition of captive and free-living White-crowned Sparrows. *J. Comp. Physiol.* **129**: 151–162.

Dhondt, A.A. (1973) Postjuvenile and postnuptial moult in a Belgian population of Great Tits, *Parus major*, with some data on captive birds. *Gerfaut* **63**: 187–209.

Dhondt, A.A. & Smith, J.N.M. (1980) Postnuptial molt of the Song Sparrow on Mandarte Island in relation to breeding. *Can. J. Zool.* **58**: 513–520.

Dial, K.P. (1992) Activity patterns of the wing muscles of the pigeon (*Columba livia*) during different modes of flight. *J. Exp. Zool.* **262**: 357–373.

Dial, K.P. and Biewener, A.A. (1993) Pectoralis muscle force and power output during different modes of flight in pigeons (*Columba livia*). *J. Exp. Biol.* **176**: 31–54.

Dietz, M.W., Dann, S. and Masman, D. (1992) Energy requirements for molt in the Kestrel, *Falco tinnunculus. Physiol. Zool.* **65**: 1217–1235.

Dol'nik, V.R. (1965) Bioenergetics of the molts of fringillids as adaptations to migration. *Nov. Ornithol.* **4**: 124–126 [In Russian].

Dol'nik, V.R. (1967) Annual cycles of the bioenergetical adaptations for life condition in some passerines. *Dokl. Akad. Nauk SSSR* **40**: 115–163 [In Russian].

Driver, E.A. (1981) Hematological and blood chemical values of Mallard, *Anas p. platyrhynchos*, drakes before, during and after remige [sic] moult. *J. Wildl. Dis.* **17**: 413–421.

Drovetski, S.V. (1996) Influence of the trailing-edge notch on flight performance of galliforms. *Auk* **113**: 802–810.

Earnst, S.L. (1992) The timing of wing molt in Tundra Swans: energetic and non-energetic constraints. *Condor* **94**: 847–856.

Evans, M.R., Martins, T.L.F. and Haley, M. (1994) The asymmetrical cost of tail elongation in red-billed streamertails. *Proc. R. Soc. Lond. Ser. B* **256**: 97–103.

Feare, C. (1984) *The Starling*. Oxford University Press, Oxford.

Fischer, W. (1962) *Die Geier*. Franckh'sche Verlag, Stuttgart.

Francis, I.S., Fox, A.D., McCarthy, J.P. and McKay, C.R. (1991) Measurement and moult of the Lapland Bunting *Calcarius lapponicus* in West Greenland. *Ringing Migr.* **12**: 28–37.

Furness, R.W. (1988) Influences of status and recent breeding experience on the moult strategy of the yellow-nosed albatross *Diomedea chlororhynchos*. *J. Zool.* **215**: 719–727.

Gatesy, S.M. and Dial, K.P. (1993) Tail muscle activity patterns in walking and flying pigeons (*Columba livia*). *J. Exp. Biol.* **176**: 55–76.

Ginn, H.B. and Melville, D.S. (1983) *Moult in Birds*. British Trust for Ornithology, Tring.

Gray, D.A. (1996) Carotenoids and sexual dichromatism in North American passerine birds. *Am. Nat.* **148**: 453–480.

Grubb, T.C. and Greenwald, L. (1982) Sparrows and a bushpile: foraging responses to different combinations of predation risk and energy costs. *Anim. Behav.* **30**: 637–640.

Harris, M.P. (1971) Ecological adaptations to moult in some British gulls. *Bird Study* **18**: 113–118.

Haukioja, E. (1971a) Summer schedule of some subarctic passerine birds with reference to post-nuptial moult. *Rep. Kevo. Subarctic Res. Stat.* **7**: 60–69.

Haukioja, E. (1971b) Flightlessness in some moulting passerines in Northern Europe. *Ornis fennica* **48**: 13–21.

Hedenström, A. (1998) The relationship between wing area and raggedness during moult in the willow warbler and other passerines. *J. Field Ornithol.* **69**: 103–108.

Hedenström, A. and Alerstam, T. (1992) Climbing performance of migratory birds as a basis for estimating limits for fuel carrying capacity and muscle work. *J. Exp. Biol.* **164**: 19–38.

Hedenström, A. and Sunada, S. (1999) On the aerodynamics of moult gaps in birds. *J. Exp. Biol.* **202**: 67–76.

Hemborg, C. and Lundberg, A. (1998) Costs of overlapping reproduction and moult in passerine birds: an experiment with the pied flycatcher. *Behav. Ecol. Sociobiol.* **43**: 19–23.

Hemborg, C. and Merilä, J. (1998) A sexual conflict in collared flycatchers, *Ficedula albicollis*: early male moult reduces female fitness. *Proc. R. Soc. Lond. Ser. B* **265**: 2003–2007.

Hill, G.E. (1990) Female house finches prefer colourful males: sexual selection for a condition-dependent trait. *Anim. Behav.* **40**: 563–572.

Holmgren, N. and Hedenström, A. (1995) The scheduling of moult in migratory birds. *Evol. Ecol.* **9**: 354–368.

Holmgren, N., Ellgren, H. and Pettersson, J. (1993) The adaptation of moult pattern in migratory Dunlins *Calidris alpina*. *Ornis Scand.* **24**: 21–27.

Jehl, J.R. (1987) Moult and moult migration in a transequatorial migrating shorebird: Wilson's phalarope. *Ornis Scand.* **18**: 173–178.

Jenni, L. and Winkler, R. (1994) *Moult and Ageing of European Passerines*. Academic Press, London.

Kahlert, H., Fox, A.D. and Ettrup, H. (1996) Nocturnal feeding in moulting Greylag Geese *Anser anser*—an anti-predator response? *Ardea* **84**: 15–22.

King, J.R. (1980) Energetics of avian moult. *Acta Int. Ornithol. Congr.*, **27**: 312–317.

Kullberg, C., Jakobsson, S. and Fransson, T. (1998) Predator-induced take-off strategy in Great Tits (*Parus major*). *Proc. R. Soc. Lond. Ser. B* **265**: 1659–1664.

Langston, N. and Hillgarth, N. (1995) Molt varies with parasites in Laysan albatrosses. *Proc. R. Soc. Lond. Ser. B* **261**: 239–243.

Langston, N.E. and Rohwer, S. (1995) Unusual patterns of incomplete primary molt in Laysan and black-footed albatrosses. *Condor* **97**: 1–19.

Langston, N.E. and Rohwer, S. (1996) Molt–breeding tradeoffs in albatrosses: life history implications for big birds. *Oikos* **76**: 498–510.

Lazarus, J. and Symonds, M. (1992) Contrasting effects of protective and obstructive cover on avian vigilance. *Anim. Behav.* **43**: 519–521.

Lima, S. (1993) Ecological and evolutionary perspectives on escape from predatory attack: a survey of North American birds. *Wilson Bull.* **105**: 1–47.

Lindström, Å., Visser, G.H. and Daan, S. (1993) The energetic cost of feather synthesis is proportional to basal metabolic rate. *Physiol. Zool.* **66**: 490–510.

Lockwood, R., Swaddle, J.P. and Rayner, J.M.V. (1998) Avian wingtip shape reconsidered: wingtip shape indices and morphological adaptations to migration. *J. Avian Biol.* **29**: 273–292.

Marden, J.H. (1987) Maximum lift production during take-off in flying animals. *J. Exp. Biol.* **130**: 235–258.

Marden, J.H. (1994) From damselflies to pterosaurs: how burst and sustainable flight performance scale with size. *Am. J. Physiol.* **266**: R1077–R1084.

Mather, M.H. and Robertson, R.J. (1993) Honest advertisement in flight displays of bobolinks (*Dolichonyx oryzivorus*). *Auk* **109**: 869–873.

Mauck, R.A. and Grubb, T.C.J. (1995) Petrel parents shunt all experimentally increased reproductive costs to their offspring. *Anim. Behav.* **49**: 999–1008.

Morton, G.A. and Morton, M.L. (1990) Dynamics of postnuptial molt in free-living mountain White-crowned Sparrows. *Condor* **92**: 813–828.

Murphy, M.E. (1996) Energetics and nutrition of molt. In: *Avian Energetics and Nutritional Ecology* (ed. C. Carey). Chapman and Hall, New York, pp. 158–198.

Murphy, M.E. and King, J.R. (1992) Energy and nutrient use during moult by White-crowned sparrows *Zonotrichia leucophrys gambelii*. *Ornis scand.* **23**: 304–313.

Murphy, M.E., King, J.R. and Taruscio, T.G. (1990) Amino acid composition of feather barbs and rachises in three species of pygoscelid penguins: nutritional implications. *Condor* **92**: 913–921.

Murphy, M.E., Taruscio, T.G., King, J.R. and Truitt, S.G. (1992) Do molting birds renovate their skeletons as well as their plumages—osteoporosis during the annual molt in sparrows. *Can. J. Zool.* **70**: 1109–1113.

Murton, R.K. and Westwood, N.J. (1977) *Avian Breeding Cycles.* Clarendon Press, Oxford.

Nachtigall, W. and Rothe, H.J. (1982) Nachweis eines 'clap-and-fling-Mechanismus' bei der im Windkanal fliegenden Haustaube. *J. Ornithol* **123**: 439–443.

Nayler, J.L. and Simmons, L.F.G. (1921) A note relating to experiments in a wind channel with an alsatian swift. *ARC Rep. Memo.* **708**.

Newton, I. (1966) The moult of the bullfinch *Pyrrhula pyrrhula*. *Ibis* **108**: 41–67.

Noguès, P. and Richet, C.R. (1910) Expériences sur le vol des pigeons à ailes rognées. *Trav. Inst. E.J. Marey* **2**: 217–224.

Norberg, R.Å. and Norberg, U.M. (1971) Take-off, landing, and flight speed during fishing flights of *Gavia stellata* (Pont.). *Ornis Scand.* **2**: 55–67.

Norberg, U.M. (1975) Hovering flight in the pied flycatcher. In: *Swimming and Flying in Nature* (eds T.Y.-T. Wu, C.J. Brokaw and C. Brennen). Plenum, New York, pp. 869–880.

Norberg, U.M. (1990) *Vertebrate Flight.* Springer, Heidelberg.

Orell, M. and Ojanen, M. (1980) Overlap between breeding and moulting in the great tit *Parus major* and willow tit *P. montanus* in northern Finland. *Ornis Scand.* **11**: 43–49.

Page, G. and Whitacre, D.F. (1975) Raptor predation on wintering shorebirds. *Condor* **77**: 73–83.

Payne, R.B. (1972) Mechanics and control of moult. In: *Avian Biology*, Vol. 2 (eds D.S. Farner & J.R. King). Academic Press, New York, pp. 103–155.

Pennycuick, C.J. (1969) The mechanics of bird migration. *Ibis* **111**: 525–556.

Pennycuick, C.J. (1975) Mechanics of flight. In: *Avian Biology*, Vol. 5 (eds D.S. Farner & J.R. King). Academic Press, New York, pp. 1–75.

Piersma, T. (1988) Breast muscle atrophy and constraints on foraging during flightless periods of wing moulting in the great crested grebes. *Ardea* **76**: 96–106.

Piersma, T. (1998) Phenotypic flexibility during migration: optimization of organ size contingent on the risks and rewards of fueling and flight? *J. Avian Biol.* **29**: 511–520.

Prince, P.A., Rodwell, S., Jones, M. and Rothery, P. (1993) Moult in Black-browed and Grey-headed Albatrosses *Diomedea melanophris* and *D. chrysostoma*. *Ibis* **135**: 121–131.

Prince, P.A., Weimerskirch, H., Huin, N. and Rodwell, S. (1997) Molt, maturation of plumage and ageing in the wandering albatross. *Condor* **99**: 58–72.

Prior, N.C. (1984) Flight energetics and migration performance in swans. Ph.D. Thesis, University of Bristol.

Rayner, J.M.V. (1979) A new approach to animal flight mechanics. *J. Exp. Biol.* **80**: 17–54.

Rayner, J.M.V. (1986) Vertebrate flapping flight mechanics and aerodynamics, and the evolution of flight in bats. In: *Biona Report 5, Bat Flight—Fledermausflug* (ed. W. Nachtigall), Gustav Fischer, Stuttgart, pp. 27–74.

Rayner, J.M.V. (1988) Form and function in avian flight. *Curr. Ornithol.* **5**: 1–66.

Rayner, J.M.V. (1991) Avian flight evolution and the problem of *Archaeopteryx*. In: *Biomechanics in Evolution. Seminar Series of the Society for Experimental Biology*, Vol. 36. (eds J.M.V. Rayner and R.J. Wootton). Cambridge University Press, Cambridge, pp. 183–212.

Rayner, J.M.V. (1993) On aerodynamics and the energetics of vertebrate flapping flight. In: *Fluid Dynamics in Biology. Contemporary Maths*, Vol. 141 (eds A.Y. Cheer and C.P. van Dam). American Mathematical Society, Providence, RI, pp. 351–400.

Rayner, J.M.V. (1995) Flight mechanics and constraints on flight performance. *Israel J. Zool.* **41**: 321–342.

Rayner, J.M.V. (1999) Estimating power curves for flying vertebrates. *J. Exp. Biol.* **202**: 3449–3461.

Rayner, J.M.V. & Ward, S. (1999) On the power curves of flying birds. In: *Proc. 22nd Int. Ornithol. Congr., Durban* (eds N.J. Adams and R.H. Slotow). Johannesburg: Bird Life South Africa, pp. 1786–1809.

Salomonsen, F. (1968) The moult migration. *Wildfowl* **19**: 5–24.

Scholey, K.D. (1983) Developments in vertebrate flight. Ph.D. Thesis, University of Bristol.

Slagsvold, T. and Dale, S. (1996) Disappearance of female pied flycatchers in relation to breeding stage and experimentally induced molt. *Ecology* **77**: 462–471.

Slagsvold, T. and Lijfeld, J.T. (1988) Ultimate adjustment of clutch size to parent feeding capacity in a passerine bird. *Ecology* **69**: 1918–1922.

Slagsvold, T. and Lifjeld, J.T. (1989) Hatching asynchrony in birds: the hypothesis of sexual conflict over parental investment. *Am. Nat.* **134**: 239–253.

Slagsvold, T. & Lijfeld, J.T. (1990) Influence of male and female quality on clutch size in tits (*Parus* spp.). *Ecology* **71**: 1258–1266.

Spedding, G.R., Rayner, J.M.V. and Pennycuick, C.J. (1984) Momentum and energy in the wake of a pigeon (*Columba livia*) in slow flight. *J. Exp. Biol.* **111**: 81–102.

Stresemann, E. and Stresemann, V. (1966) Die Mauser der Vögel. *J. Ornithol.* **107**: 3–448.

Sullivan, J.O. (1965) 'Flightlessness' in the Dipper. *Condor* **67**: 535–536.

Swaddle, J.P. (1997) Within-individual changes in developmental stability affect flight performance. *Behav. Ecol.* **8**: 601–604.

Swaddle, J.P. and Cuthill, I.C. (1994) Female zebra finches prefer males with symmetric chest plumage. *Proc. R. Soc. Lond. Ser. B* **258**: 267–271.

Swaddle, J.P. and Lockwood, R. (1998) Morphological adaptations to predation risk in passerines. *J. Avian Biol.* **29**: 172–176.

Swaddle, J.P. and Witter, M.S. (1994) Food, feathers and fluctuating asymmetries. *Proc. R. Soc. Lond. Ser. B* **255**: 147–152.

Swaddle, J.P. and Witter, M.S. (1997) The effects of molt on the flight performance, body mass, and behavior of European starlings (*Sturnus vulgaris*): an experimental approach. *Can. J. Zool.* **75**: 1135–1146.

Swaddle, J.P. and Witter, M.S. (1998) Cluttered habitats reduce wing asymmetry and increase flight performance in European starlings. *Behav. Ecol. Sociobiol.* **42**: 281–287.

Swaddle, J.P., Witter, M.S., Cuthill, I.C., Budden, A. and McCowen, P. (1996) Plumage condition affects flight performance in starlings: implications for developmental homeostasis, abrasion and moult. *J. Avian Biol.* **27**: 103–111.

Swaddle, J.P., Williams, E.V. and Rayner, J.M.V. (1999) The effect of simulated flight feather moult on escape take-off performance in starlings. *J. Avian Biol.* **30**: 351–358.

Taruscio, T.G. and Murphy, M.E. (1995) 3-Methylhistidine excretion by molting and non-molting sparrows. *Comp. Biochem. Physiol.* **IIIA**: 397–403.

Thomas, A.L.R. (1993) The aerodynamic costs of asymmetry in the wings and tail of birds: asymmetric birds can't fly round tight corners. *Proc. R. Soc. Lond. Serv. B* **254**: 181–189.

Tucker, V. A. (1991) The effect of moulting on the gliding performance of a Harris' hawk (*Parabuteo unicinctus*). *Auk* **108**: 108–113.

Tucker, V.A. (1993) Gliding birds—reduction of induced drag by wing tip slots between the primary feathers. *J. Exp. Biol.* **180**: 285–310.

Verbeek, N.A.M. and Morgan, J.L. (1980) Removal of primary remiges and its effect on the flying ability of the glaucous-winged gull. *Condor* **82**: 224–226.

Ward, P. and D'Cruz, D. (1968) Seasonal changes in the thymus gland of a tropical bird. *Ibis* **110**: 203–205.

Weibel, E.R. (1984) *The Pathway for Oxygen*. Harvard University Press, Harvard, MA.

Witter, M.S. & Cuthill, I.C. (1993) The ecological costs of avian fat storage. *Philos. Trans. R. Soc. Lond., Ser. B* **340**: 73–90.

Witter, M.S., Cuthill, I.C. and Bonser, R.H.C. (1994) Experimental investigations of mass-dependent predation risk in the European starling, *Sturnus vulgaris*. *Anim. Behav.* **48**: 201–222.

9

The physics of chemoreception revisited: how the environment influences chemically mediated behaviour

Robb W.S. Schneider and Paul A. Moore

1. Introduction

Many organisms move through environments in a directed fashion to locate patchily distributed resources. To effectively orient or navigate, organisms need to acquire information about their environment using their sensory systems. This information can include the location of predators, food resources, potential mates, and habitats (e.g. Busdosh et al., 1982; Hargrave, 1985; Hessler et al., 1972; Ingram and Hessler, 1983; Peckarsky, 1980; Petranka et al., 1987; Sainte-Marie, 1986; Wilson and Smith, 1984). For a number of organisms, this information is obtained from chemical signals. For example, chemical signals are important for tracking scents (Hamner and Hamner, 1977), aggregating microorganisms (Bell and Mitchell, 1972), and migration in *Daphnia* (Dodson et al., 1997). In addition, algal exudates (e.g. Maier and Müller, 1986), pheromones (e.g. Atema and Engstrom, 1971; Gleeson et al., 1987), sperm attractants (e.g. Miller, 1979), and predatory chemicals (Peckarsky, 1980; Petranka et al., 1987) play crucial roles in the feeding, aggregation, and spawning of widely dispersed organisms. These signals serve as guideposts in orientation behaviour. In all of these instances, it has been shown that the chemical signals play an important role in the behaviour of animals.

To clarify the role that stimulus patterns play in an organism's behaviour, it is critical to quantify the stimulus patterns and types of information available to animals during orientation to chemical sources. When studying chemoreception, we need to understand how the physical processes that transport chemicals through environments structure the information in signals. At macroscopic scales this requires understanding

Biomechanics in Animal Behaviour, edited by P. Domenici and R.W. Blake.
© 2000 Taylor & Francis, Oxford.

the hydrodynamics involved in fluid flow, and quantifying chemical signal distributions at the same spatial and temporal scales as used by organisms.

Information in chemical signals is contained in four components: quality, intensity, spatial distribution, and temporal fluctuation. Quality is the chemical composition of the stimulus signal. For example, in aquatic systems amino acids serve as feeding stimulants for shrimp (Carr and Derby, 1986), whereas in terrestrial systems blends of aldehydes, esters, and alcohols serve as mating signals for moths (Agosta, 1996). Intensity is defined as the number of molecules that excite a receptor per unit of time. The third component is the spatial distribution of chemicals in the environment and the fourth is the temporal fluctuation of chemicals across receptor surfaces. Chemical composition of the signal can be manufactured directly from an organism itself, as in mating situations (Agosta, 1996), or indirectly as body fluids released as chemicals signaling a predation event (Hazlett, 1985, 1990). This aspect of the chemical signal is dependent on the signaling organism. Conversely, fluid dynamics and the physics of the environment affect the information contained in the last three components (intensity, spatial distribution, and temporal fluctuations). To characterize the importance of signal intensity, spatial distribution, and temporal fluctuations it is imperative to understand how fluid dynamics structures information available to organisms.

In 1977, Berg and Purcell wrote a seminal paper on the physics of chemoreception. That paper outlined the role that molecular diffusion plays in structuring information in chemical signals for microscopic organisms. Although critical for organisms and processes that occur at smaller spatial scales (less than millimeter scale), the physics outlined in that paper has little bearing on chemoreceptive processes and organisms that live or function at larger spatial scales (greater than millimeter scale). This is primarily due to the temporal dynamics associated with molecular diffusion over larger spatial scales and to the increasing importance of fluid flow. Given the physics of dispersion at these scales, chemical gradients are radially symmetrical around the female gamete. As a result of the diffusion-dominated dispersion, concentration gradients at this scale are gradual and predictable, and are easily followed by organisms (Bell and Mitchell, 1972; Berg, 1983; Berg and Purcell, 1977; Maier and Müller, 1986). The male gamete performs a 'random walk' strategy (kinesis) up the concentration gradient to locate the female gamete.

The purpose of this chapter is two-fold: first, to demonstrate how the physics of chemical dispersion (physical environment) can significantly influence an organism's movement through different environments at scales larger than those covered by Berg and Purcell (1977); second, to illustrate some general principles and lessons learnt from current studies on fluid dynamics of chemoreception.

2. Qualitative measures of the physical environment

At larger spatial scales (greater than millimeter scale), the chemical senses are unique among the senses in that the physical process of transmission through a medium (i.e. fluid flow) is largely independent of any inherent excitatory properties of the receptor-activating signal (i.e. chemical composition of the signal). This is not true for light, as the quality of light (spectral frequency or wavelength) directly influences the various transmission impediments such as scattering, absorption, and attenuation (Jerlov, 1976; Shifrin, 1988). Similarly, the frequency (or wavelength) of sound is a critical factor in determining propagation through various media (Clay and Medwin, 1977) and also in the reflection and diffraction involved in echolocation (Griffin, 1986; Nachtigall and Moore, 1986). Conversely, two physical processes distribute chemical signals: fluid flow and molecular diffusion. These two processes are not independent of each other, but rather intertwined in chemical signal dispersion. To characterize which of these processes is more important for structuring information in chemical signals, we can use a dimensionless number (Peclet number) that is the ratio of these two forces.

The Peclet number (Pe) is a dimensionless number describing the relative importance of fluid flow and molecular diffusion in chemical dispersion. As it is dimensionless, it allows one to scale the physical processes involved in chemical dispersion independently of the medium involved, organisms under investigation, or behavioural situation. Thus, it can provide a global view of the physical processes that structure information in chemical signals. Pe is the ratio of dispersion by fluid flow (*ul*) and molecular diffusion (D_m), and can be calculated using the formula

$$\text{Pe} = \frac{u\,l}{D_m} \tag{1}$$

where *u* is the free stream fluid velocity (cm s⁻¹), *l* is a characteristic length scale, typically taken as the length from the source to the receptor (cm), and D_m is the molecular diffusion coefficient (cm² s⁻¹).

If the Pe values are below one, molecular diffusion dictates chemical dispersion (*Figure 1*). In these environments, the physics of chemoreception described by Berg and Purcell (1977) are important for structuring the intensity, spatial distribution, and temporal fluctuations associated with chemical signals and organismal behaviour (such as the algal attraction system discussed above). For environments with Pe numbers

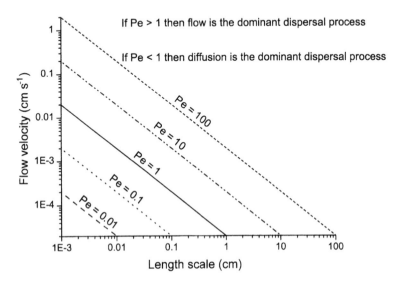

Figure 1. *The Peclet number is the ratio between the two forces involved in chemical dispersion, molecular diffusion and fluid flow. This figure shows ranges of Peclet value for different length and velocity scales (Peclet value shown within a range is the minimum for that band). These values are determined using the diffusion coefficient of an amino acid in sea water.*

larger than one, fluid flow is the dominant dispersal process and is more important for structuring information in chemical signals. To understand how fluid flow influences chemical signal structure, it becomes important to understand the dynamics of flow.

The Reynolds number (Re) provides a quantitative measure of the hydrodynamics of a flow system (Vogel, 1981). The Reynolds number is the ratio of inertial forces to viscous forces in a moving fluid (or a solid body moving in a stationary fluid). Re is calculated using the following formula

$$\text{Re} = \frac{u\,l}{v} \tag{2}$$

where v is the kinematic viscosity ($m^2\,s^{-1}$), u is the fluid velocity ($m\,s^{-1}$) and l is a characteristic length scale (m). At a broad scale the Reynolds number is an indication of the level of turbulence in a flow-based system (*Figure 2*). (Other factors, such as bed roughness and shape of objects in flow, can alter when the different transitions from laminar to turbulent occur. Therefore, it is important to use the Reynolds number as a broad indicator of typical flow. As a general rule for a given hydrodynamic situation, when the Reynolds number increases the level of turbulence will also increase.) At low Re (<1), viscous forces dominate the fluid mechanics of a system. At these Re, the typical flow structure that dominates fluid motion at spatial scales with which we are familiar is absent. Eddies are extremely short lived and motions tend to be dampened. Streamlines, which are the paths along which water particles move, do not cross each

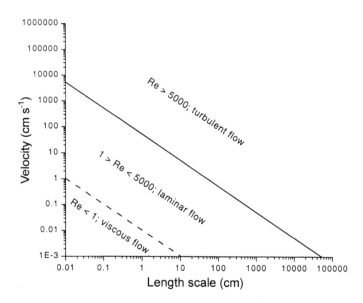

Figure 2. *For flowing systems, the Reynolds number describes the type of flow patterns present. The Reynolds number is the ratio between inertial and viscous forces. For large Reynolds numbers (Re > 5000), turbulence causes rapid chemical transport and mixing. For intermediate Reynolds numbers (Re = 1–5000), flow is laminar, and mixing is slower than turbulence. For low Reynolds numbers (Re < 1), viscous forces dominate fluid flow, and mixing is inhibited. The values presented in the figure are only estimates of the flow type (laminar vs. turbulence). Exact transitions between the types of flow will be dependent on other factors, such as bed roughness or objects within the flow.*

other. Because streamlines are parallel, mixing rates are slow and dominated by diffusive processes. At higher Re, inertial forces dominate the fluid mechanics. At these scales, large eddies can be generated and sustained within the flow field. Flow can become turbulent and streamlines will cross or become chaotic. As turbulence increases, mixing as a result of fluid flow (rather than dispersion as above) increases and this mixing has profound effects for information in chemical signals. Any change within the environment of flow (such as different substrates in streams, or moving from a field to a forest in the case of airborne signals) will result in a change in the structure of turbulence. Again, a change in turbulence will result in a change in chemical signal information.

The Reynolds number is a rudimentary measure of fluid dynamics and more elaborate parameters are needed for a more detailed view of the fine-scale turbulence structure that determines chemical signal fluctuations. These measures include friction velocity (U_*), roughness Reynolds number (Re_*), turbulent energy dissipation rate (ϵ), the Kolmogorov microscale (η), and the Batchelor microscale (η_s). For an excellent review of these measures in relation to ecology and behaviour, readers are referred to Sanford (1997).

Of particular importance for chemoreception are the Kolmogorov microscale and the Batchelor microscale. These scales are measures of the smallest eddy sizes under specific flow conditions (Kolmogorov) and the smallest distance for differences in chemical concentrations (Batchelor). In a broad sense, these measures give the researcher some insight into the temporal spectrum of chemical fluctuations within a flowing system. In a more detailed analysis, the Batchelor scale gives an indication of the small-scale changes of intensity within a signal and provides some insight into the temporal frequency of these fluctuations. As these scales become smaller, there are two associated changes in information in chemical signals. First, the variance associated with the mean concentration increases. This may present a problem to organisms that need accurate estimates of concentration. Second, a spectral analysis of chemical signals shows an increase in the high-frequency fluctuations within an odour plume. Again, this may present a problem to sensory systems that are temporally tuned to specific fluctuations within habitats. A more detailed discussion of this phenomenon in relation to sensory coding has been presented elsewhere (Schneider et al., 1998).

It is possible to calculate both the Kolmogorov microscale (η) and Batchelor microscale (η_s; Mann and Lazier 1996) from the turbulent energy dissipation rate,

$$\eta = 2\pi \left(\frac{v^3}{\epsilon} \right)^{1/4}, \text{ and } \eta_s = 2\pi \left(\frac{vD_m^2}{\epsilon} \right)^{1/4}, \tag{3}$$

where D_m is the molecular diffusion coefficient of the signal molecule, ϵ is the turbulent energy dissipation rate and v is the kinematic viscosity. In reality, it is probably these last two numbers (Kolmogorov microscale and Batchelor microscale) that play the most important role in explaining chemosensory behaviour. These values indicate the smallest spatial scales (and thus temporal scales) over which differences in eddy structure (Kolmogorov) and chemical fluctuations (Batchelor) can be maintained by fluid flow. Indeed, the summary table (*Table 1*) would be greatly enhanced by replacing the Reynolds number with the Kolmogorov scale and adding a composite of

Table 1. *Properties of chemical signals at different Peclet and Reynolds numbers*

Peclet number	Reynolds number	Intensity	Spatial information	Temporal information	Example organisms
Low	Low	Smooth predictable concentration gradients	Predictable spatial patterns of equal spatial dispersion	Slow onset of signal intensity, longer-lasting signals	Bacteria, olfactory mucosa, snails and flounders near solid substrates
Intermediate	Low	High concentrations that vary little over time	Somewhat predictable patterns with unequal spatial dispersion	Slow onset and decay of signal intensity, low-frequency fluctuations	Copepods, boundary layers around appendages
High	Intermediate	Variable concentrations with high fluctuations in intensity	High degree of variability in spatial distribution	Rapid signals with slow onset of intensity, increasingly higher-frequency fluctuations	Small fish, benthic invertebrates in slow flow
High	High	Highly variable concentrations	Heterogeneous and unpredictable spatial patterns	Rapid signals with high-frequency fluctuations	Moths, crayfish, lobsters, sharks

both the Peclet and Batchelor scale. Unfortunately, we could not find enough detailed flow information in the behavioural literature to add these values. In the future, we would hope that researchers concerned with the biomechanics of chemoreception be aware of these numbers.

3. Fluid flow and chemical signals

The physical forces governing odour dispersal at Pe > 1 and high Re will depend on the flow regime into which the chemical stimulus is introduced (for review, see Westerberg (1991)). For example, odour plumes introduced in the free stream area of flow (far away from any solid surface) are dispersed by differential fluid velocities of the surrounding medium, producing irregular patterns of patches with a wide range of size and time scales. Large-scale turbulent eddies, at the size of or larger than the length of the odour plume, produce shifts of the plume as a whole, causing it to meander through a large time and space frame. Medium-scale eddies (size of the plume diameter) break up the plume into large patches. Small-scale eddies (smaller than the plume diameter) cause the larger patches to break up into smaller ones and to decrease the concentration gradient (slope) at patch boundaries (Aylor, 1976; Aylor et al., 1976; Miksad and Kittredge, 1979). Patch gradients in the smallest patches decline gradually as a result of molecular diffusion. Increasing time and distance from the source causes patches to spread over a larger area, thus decreasing internal concentration (Moore and Atema, 1988, 1991; Murlis, 1986; Murlis and Jones, 1981).

These odour parameters allow a moving organism to compare several patches encountered, which could indicate the direction to the source. For example, crayfish live in a number of hydrodynamically distinct habitats. These habitats vary in both flow velocity and substrate type. Changes in substrate or flow velocity will affect the types of turbulence that are generated within those habitats, which will, in turn, affect the dynamics of odour patches in odour plumes. Moore and Grills (1999) showed that crayfish located an odour source faster, walked faster, and spent less time stationary during orientation in habitats with higher turbulence. In addition, Moore et al. (2000) showed that specific odour parameters associated with patch dynamics are different in habitats with higher turbulence. Odour concentrations in plumes are heterogeneous when measured at fast temporal and small spatial scales (Atema, 1985; Moore and Atema, 1988, 1991; Moore et al., 1992; Murlis and Jones, 1981; Zimmer-Faust et al., 1988). The patchy structure of odour plumes is due to the irregular flow produced by the mechanical forces acting on a moving fluid. The magnitude of concentration fluctuations is dependent on the interaction between the size of the turbulent eddies and the size of the odour plume. This size dependence implies that estimates of odour concentration are determined in part by the temporal and spatial sampling scale of the chemical measurements (Aylor, 1976; Aylor et al., 1976; Miksad and Kittredge, 1979). As a result of this turbulence, animals located down-current of an odour source will experience periods during which odour concentrations are well above or below the mean odour concentration, and which exhibit unpredictable temporal variation. Thus, mean concentrations and time-averaged distributions (i.e. 5–10 min) may not be indicative of the information available for many macroscopic animals attempting to orient towards an odour source (Elkinton et al., 1984; Moore and Atema, 1988, 1991; Zimmer-Faust et al., 1988).

Odour transport is also influenced significantly by the interaction between a moving fluid (air or water) and a solid surface, such as when air flows over the Earth's surface or water flows over the sea floor. The interface between a stationary solid and a moving fluid is called a 'boundary layer', and is a region through which fluid velocities steadily increase from zero with increasing distance from the solid surface. Odour signals introduced into a boundary layer have significantly different properties from odour plumes introduced away from solid surfaces (Westerberg, 1991).

Independent of where an odour plume is introduced, the degree of turbulence is important for structuring chemosensory information. As a rough indicator, animals larger than about 1–10 cm (or Re ≫ 1000; flow velocities ≥0.1 cm s^{-1}) live in turbulent environments dominated by inertial forces. At these size scales (or Re scales), distances associated with molecular diffusion and viscous forces are negligibly small. Any initially homogeneous odour distribution is broken up by turbulent flow into a dynamic pattern of odour patches of varying sizes and concentrations. The exact dynamics of temporal fluctuations, spatial distribution, and intensity (three main components of information for organisms) will be determined by the hydrodynamics specific to the environment. As organisms use these signal components to make behavioural decisions about their environments, we can now make a broad-based prediction regarding animal behaviour and chemosensory information: we predict that organisms across a number of different taxa and spatial scales should have habitat-specific chemosensory mediated behaviour, because the information in chemical signals is habitat specific.

To demonstrate this phenomenon, we will review the physical environment of organisms living at different spatial and temporal scales and provide some insight into the dynamics of chemical signals in those habitats (*Table 1*). We focus primarily on the spatial scales because it is this, and not the temporal scale, that occurs in all of the dimensionless numbers presented above. Finally, we will present what is known about their chemosensory behaviour in relation to the chemical signal structure for that habitat.

4. Chemosensory behaviour in different environments

4.1 *Low Peclet and low Reynolds number environments*

In environments with low Peclet numbers (*Table 1*), thermal motion of molecular diffusion is the dominant process of odour dispersion. Consequently, fluid flow plays no role in structuring intensity, spatial distribution, or the temporal fluctuations of chemicals. In addition, viscous forces dominate over inertial forces and the environment is undisturbed by turbulence. As a result of molecular diffusion, chemical signals are slow with expanding concentric concentration gradients emanating from the source. Thus, odour signals have predictable concentration gradients and predictable spatial gradients. Intensity gradually increases as organisms move towards the source and signals have very slow onsets and decays (on the order of hours). This is the world that Berg and Purcell (1977) illuminated in their paper, and readers are referred to that paper for a further discussion on the physics and signal properties at these scales.

4.2 *Intermediate Peclet and low Reynolds number environments*

In environments with intermediate Peclet numbers (*Table 1*), molecular diffusion and fluid flow have approximately equal impact on dispersing chemical signals. Molecular diffusion acts to disperse signals in a symmetrical fashion generating circular (in two dimensions) and spherical (in three dimensions) areas of equal concentrations. In these environments, smooth concentration gradients exist and always increase as organisms move towards the source. The low Reynolds number flow works to stretch or shear these concentration gradients. In low Reynolds number environments, viscous forces overwhelm inertial forces. As stated above, eddies are nonexistent and streamlines are parallel to each other. Mixing is extremely difficult and chemicals will remain within streamlines (*Figure 3*). As inertial forces are negligible, fluid motion stops as soon as the source of the generating motion stops.

Diffusion and shear act on chemicals released into these creeping flows. Shear is created by different relative motions of adjacent fluid particles, such as the shear in a copepod feeding current. Shear can occur along a single streamline, differing fluid velocities along a streamline, or across streamlines, differing fluid velocities between streamlines. Shear fields cause relative motion between the source and its surrounding chemical gradients. Shear along the streamline will stretch the concentration gradients established by molecular diffusion and will increase the distance between the leading edge of the chemical gradient and the source, whereas shear across streamlines will cause the concentration gradients to spread and stretch laterally. Small aquatic zooplankton or larger benthic organisms located within boundary layers inhabit these types of habitats. In these habitats, chemical signals have predictable spatial patterns (generated by molecular diffusion) that are stretched either laterally or longitudinally by shear fields. Temporal fluctuations, so common in the habitats described below, are absent.

Copepods provide the best-studied example of an intermediate Peclet number and low Reynolds number environment. Copepods are small crustaceans (smaller than a few millimeters) found primarily in the open ocean. A number of recent studies have shown that chemical signals are important for copepods. Female copepods use hops

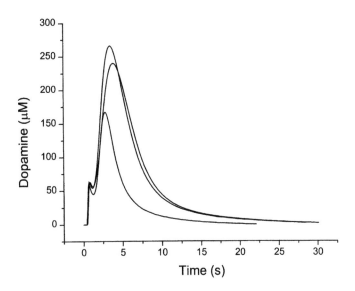

Figure 3. *Typical chemical signal structure in an intermediate Peclet number and low Reynolds number environment. These recordings are taken in the feeding current of a* Calanus *copepod* (Pe ~ 10; Re ~ 100). *Lines are replicates taken at the same location within the boundary layer.*

(that are triggered by male odours) to increase encounter probabilities and to induce flow (Van Duren *et al.*, 1998), and recent work has shown that copepod mating is probably chemically mediated (Yen *et al.*, 1998). Although these are excellent examples of chemosensory behaviour, the majority of research on copepod behaviour has focused on foraging decisions. A copepod feeds by generating a current that draws potential food particles from the surrounding water to maxillipeds. As food approaches, the copepod reaches out and captures the prey item. It is presumed that copepods rely on chemoreception for foraging decisions (Koehl and Strickler, 1981; Poulet and Ouellet, 1982). On the basis of behavioural observations, Andrews (1983) provided a physical model for the remote detection of particles that suggested differences in feeding currents of copepods (their environment) would change the organisms' chemoreceptive abilities. Empirical studies (Moore *et al.*, 1999) support this idea and show that the spatial and temporal distributions of the chemical signals are strongly influenced by the structure of the feeding current. Different species of copepods have different feeding currents, and thus, live in different sensory environments.

The spatial distribution of chemicals, temporal fluctuation, and intensity of chemical signals is highly dependent on the hydrodynamics of a copepod's feeding current (Moore *et al.*, 1999). At these Peclet numbers, diffusion defines the initial gradient, and low Reynolds number flow within the laminar feeding current distorts the shape of the odour field. Chemical signals located in high shear feeding currents (characteristic of the copepod *Pleuromamma xiphias*) have different spatial and temporal distributions from signals located in lower shear flow fields (characteristic of the copepod *Euchaeta rimana*; Moore *et al.*, 1999). Chemical signals in high shear are more longitudinally stretched, creating a longer distance from the leading edge of the chemical field to the center of the odour source. This longer distance provides the copepod with a greater lag time between detection and the time that the prey item enters its capture zone. Copepods with low shear feeding currents do not have these same capabilities. Thus, differences in the physical environment of feeding currents will result in different chemosensory behaviours, such as foraging or mate location. These predictions of chemoreceptive behaviours are based on the physics of chemical dispersion and direct measurement of chemical signals, and have yet to be empirically tested.

4.3 *High Peclet and intermediate Reynolds number environments*

In environments with high Peclet numbers (*Table 1*), molecular diffusion has little impact on structuring information in chemical signals. Fluid flow is the dominant dispersal force. Quantifying the exact nature of the hydrodynamics within the environment becomes critical for understanding how flow influences the intensity, spatial distribution and temporal fluctuations in chemical signals. For intermediate Reynolds number environments, fluid flow can be laminar or quasi-laminar, or exhibit boundary layer conditions, depending on various factors such as the roughness and the morphology of the substrate. In these environments, turbulence tends to be lower, eddies dissipate at larger sizes, and streamlines remain roughly parallel to each other. It must be kept in mind that the Reynolds number is a rough estimate of flow conditions and it should not be used as an exact estimate of the microscale flow structure. Given the right conditions, an intermediate Reynolds number environment can become highly turbulent. Still, in a broad sense, these environments have lower turbulent energy than higher Reynolds number environments.

Chemical signals in these types of environments are intermediate in structure between the previous example and the next environment (*Figure 4*). Signal intensity begins to fluctuate and smooth concentration gradients are nonexistent. The spatial distribution is somewhat patchy and has lost most of the predictability seen in the previous environment. Chemical signals will have faster onsets (on the order of seconds) and will decay along the same temporal scales. When compared with the previous habitat, signals are more heterogeneous in space and time.

Many different organisms live in these types of habitats. In particular, benthic invertebrates (crayfish, blue crabs, aquatic insect larvae) or fish (flounders) are often located in benthic boundary layers with laminar or quasi-laminar flows. These environments are often characterized by long-lasting unidirectional flow that allows boundary layers to be established. Even within a single habitat, topographical differences in habitat structure influence the level of turbulent flow. These include riffles (faster flow), pools (slower flow), sandy substrates (smaller roughness elements), and cobble areas (larger roughness elements). Hart *et al.* (1996) have shown that these small changes in habitat structure can lead to large changes in the hydrodynamics of that habitat. These hydrodynamic changes will then lead to differences in the intensity, spatial distribution, and temporal fluctuation of chemical signals (Westerberg, 1991). Recent measurements of these components of chemical signals on different substrate types have shown profound differences in chemical signal structure, which will subsequently influence animal behaviour (Moore *et al.*, 2000).

Crayfish (*Orconectes rusticus*) and blue crabs (*Callinectes sapidus*) can use chemical signals in these habitats to locate food sources (Moore *et al.*, 2000; Weissburg and Zimmer-Faust, 1993). Both organisms respond well and are capable of locating odour

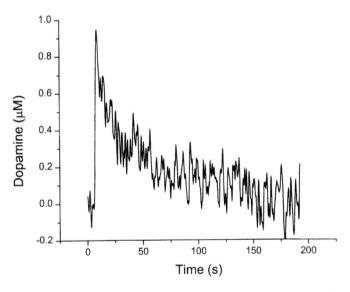

Figure 4. *Typical chemical signal structure in a high Peclet number and intermediate Reynolds number environment. Chemical recordings are taken in the boundary layer of a unidirectional artificial stream (Pe ~ 1000; Re ~ 1000).*

sources in these types of flow. In fact, both organisms showed similar orientation behaviours. Although the exact spatial structures of orientation paths were different, both crayfish and blue crabs oriented while in the middle of the plume and tended to orient in fairly direct fashion towards the odour source. Neither of these organisms showed any of the characteristic zigzag pattern found in higher Reynolds number environments (see below). We hypothesize that the similarity in behaviours is a result of the similar chemical signal structure in the two experiments. When the hydrodynamics of the experimental trials were changed, both organisms showed specific changes in their behavioural responses. For example, crayfish can locate the source of the odour with greater than 90% success rate on two distinct substrates: sand (small roughness elements) and cobble (larger roughness elements). A detailed analysis of the spatial structure of the orientation path showed that they were similar on the two substrates (Moore and Grills, 1999). However, crayfish orienting over the sand substrate took twice as long to find the source, walked twice as slowly, and spent more time not moving. Thus, crayfish in higher turbulent environments (cobble substrates) can locate odour sources more efficiently. Conversely, blue crabs are more efficient in locating the odour source in lower turbulent flows. In higher turbulent conditions, blue crabs were less successful in finding the source. In trials in which they were successful in locating the source, they took significantly longer than in the lower turbulence conditions (Weissburg and Zimmer-Faust, 1993). These results show that the chemosensory behaviour in an individual organism can be habitat dependent.

4.4 *High Peclet and Reynolds number environments*

High Peclet number environments are flow dominated and molecular diffusion plays a very limited role (*Table 1*). In addition, at high Reynolds numbers (>5000), the environment is very turbulent and chaotic. Within this environment streamlines cross and are mixed, and the thickness of boundary layers is greatly reduced. An odour plume in this environment is no longer stretched by the flow but rather ripped into filaments and patches by the chaotic nature of turbulence. A further consequence of turbulence is eddy circulation patterns. Large eddies of varying size can move an entire plume, medium eddies distribute patches within a plume, and small eddies swirl the odour within a patch.

As a result of turbulence effects, odour plumes at these scales lack any predictable concentration gradients. The spatial distribution is dominated by a patchy nature and the odour signal is heterogeneous in space and time. Signals have very rapid onsets of concentration with fast decays (on the order of milliseconds; *Figure 5*). Spectral analysis of these odour signals shows relatively high-frequency fluctuations in concentration (>10–20 Hz).

Because of the chaotic nature of turbulent odour signals, organisms are faced with a complex behavioural problem to solve; how do sensory systems extract reliable information from chaotic signals? Moths, tsetse flies, lobsters, salmon, and sharks are examples of organisms that rely heavily on chemical signals for various behaviours. All of these organisms exist in very turbulent environments where relevant odour signals are highly complex. Studies have shown that the behavioural mechanisms involved in orientation are just as complex as the odour signals. Although behaviours are complex, several similarities can be seen across these diverse taxa and habitats. Orientation behaviours exhibited by these organisms have three similarities: they

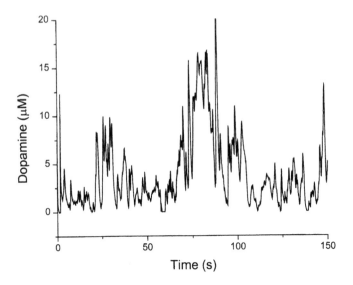

Figure 5. *Typical chemical signal structure in a high Peclet number and Reynolds number environment. Recordings show the typical fluctuations of a turbulent odour plume in a unidirectional flow field (Pe ~ 1000; Re ~ 10 000).*

involve both rheotactic (anemotactic) and chemotaxic components, and generally involve some aspect of an internally generated search behaviour. Rheotactic or anemotactic behaviour involves turning into current and heading upstream. Bilateral sampling (chemotaxis) allows organisms to make discriminatory decisions based on some aspect of the odour signal, and internally generated search behaviours, such as casting in moths or circling in bonnet-head sharks, provide the organism with an evolutionarily successful solution for the problem at hand.

The hydrodynamics within the environment that the tsetse fly occupies span a wide range. The differences in hydrodynamics are a function of host size, substrate flora, and wind conditions. The primary odour source might be small (bird), intermediate (wildebeest, buffalo, elephant), or extremely large (herd of mammals) (Brady et al., 1989), and the environment ranges from grassland to shrub based. In addition, wind velocities can range from 0.01 to over 3 m s^{-1}. The host location behaviour consists of a series of short parabolic flights about 30–60 cm above the substrate towards a host (Bursell, 1984). While airborne, the tsetse fly evaluates the presence of the chemical signal and modifies its course. If plume contact is lost, the tsetse fly executes an in-flight turn that will bring it back into contact with the plume. Also, when resting, tsetse flies tend to be located on the upwind side of the horizontal branches, thus avoiding the turbulent mixing at the downwind side and receiving the benefits of additional exposure to odour-laden wind (Brady et al., 1989). In woodland habitats where wind speeds are low (<0.5 m s^{-1}) odour plumes tend to meander. Thus, tsetse fly success is reduced; conversely, host location increases when the wind speed is about 0.75 m s^{-1}.

For nocturnal moths, evolutionary success is based in part on locating a 'calling' female. The female releases a species-specific sex pheromone into the environment, to which the male moth orients. The Peclet number for this system is in the scale of 10, whereas the Re number of this system is approximately 10^4 (Schneider et al., 1998).

These benchmarks testify that the chemical signal is distributed by fluid flow, and is turbulent.

Moth orientation behaviour has been extensively studied over the past three decades (Bell and Cardé, 1984). The behaviour of the male moth has been defined as an odour-guided optomotor anemotaxis. The finer aspects of the orientation strategy used while flying differ from species to species; nevertheless, the underlying role of the chemical signal is to mediate the behaviour. Upon contact with a pheromone plume while in flight, an optomotor anemotaxis is triggered, causing the male to turn upwind and steer his flight with respect to wind direction. As the concentration of the pheromone increases as he approaches the pheromone source, the male alters his ground speed and track angles using the visual feedback. Generally, upon loss of the odour signal the male moth begins the classic casting behaviour: forward progress is reduced and the moth moves back and forth across the windline to sample more area so as to relocate the plume.

Changes in the turbulent structure of an odour plume alter the moth's orientation behaviour. Mafra-Neto and Cardé (1995a) noted that when a moth is orienting up a ribbon plume versus a turbulent plume the flight track was wider and more crosswind than upwind. These changes in structure of the plume also have more influence than changes in concentration. Mafra-Neto and Cardé (1995b) found that increased frequency of pulses in the plume structure and volume of pheromone introduced increased air and ground speeds as well as resulting in a narrower flight track. Willis *et al.* (1994) showed that greater track angles were associated with point sources whereas lower track angles were exhibited with wide diffuse plumes. Baker *et al.* (1985) showed that *Grapholeta molesta* will not fly in a homogeneous cloud of pheromone, but rather prefers a patchy heterogeneous pheromone plume.

Salmon are known to migrate great distances (≈ 4000 km) to home streams to spawn. There have been two behaviours proposed for these long distance migrations and home stream localization: nearshore migration and stream migration. Both of these behaviours use odour as part of the mechanism of orientation. Additional sensory information is probably gathered by sun compass, polarized light, and/or magnetic fields (Dittman and Quinn, 1996).

When the salmon is in the open ocean, it is unable to use spatial references to orient. The primary mechanism for migration is olfaction (Døving *et al.*, 1985). In the nearshore environment there are many layers in vertical stratification as a result of density differences. By moving through these layers, the salmon is able to determine its overall direction of movement. The resulting behaviour of the salmon is to perform a vertical zigzag maneuver between the density layers (Johnsen, 1983).

Salmon migrate up the streams and rivers to spawn in the stream of their birth. At an early age, the salmon are imprinted with the odours from the home stream, which provide the olfactory cue necessary for the behaviour. To return to their home stream to spawn the salmon use a straight positive rheotaxis (up current) as long as odours are present. When the odours are not present, the salmon exhibit negative rheotaxis and actively move downstream to locate the home stream compound (Johnsen, 1983).

Although rheotaxis is the primary behavioural mechanism for guiding the salmon to a chemical source, tropotaxis in this case is not used (presumably because the bulk of the stream has the signature compound). Tropotaxis is seen in several chondrichthyes, however. The typical zigzagging behaviour seen in moths has also been found in nurse sharks (Johnsen, 1983). In addition to head zigzag behaviour, bonnet-head sharks have been found to use an internally generated local search program upon detection of a relevant chemical compound. This circling behaviour halts the sharks' forward movement and allows them to relocate the odour signal (Johnsen and Teeter, 1985).

In all of these orientation behaviours, we see organisms exhibiting searching patterns that are spatially more complex. By increasing the spatial complexity of orientation, organisms may be sampling a larger area of their environment and using these larger samples to reconstruct a more reliable estimate of the spatial or temporal distribution of chemical signals. Given the high level of heterogeneity in chemical signal structure at high Peclet and Reynolds number scales, this may be a necessity for efficient orientation.

5. Conclusions

Environments have unique physical features that influence the type of hydrodynamics. Given that information in chemical signals is heavily influenced by the hydrodynamics, different habitats will have habitat-specific signal composition with regard to intensity, spatial distribution, and temporal fluctuations, as demonstrated for the habitats above. It follows that the habitat-specific hydrodynamics sets constraints on the types of information available to organisms and sensory systems. For example, copepods with higher transverse shear have more temporal and spatial information available to them when compared with copepods with lower shear rates. These differences in sensory information, in turn, shape their chemosensory mediated behaviour.

Evolutionary theory would predict that if sensory systems are to extract environmentally and behaviourally relevant information from odour signals, both the sensory system and subsequent behaviour should be adapted to the particular constraints of different habitats. Further, if we assume that organisms are adapted to their sensory environments, we can accept as a working hypothesis that sensory physiology and sensory behaviour are 'matched' to the most biologically important signals (Wehner, 1987). This hypothesis states that the properties of a sensory system are 'matched' or tuned to that part of the sensory environment that carries the most relevant information. Perceiving the environment through a matched filter helps limit the amount of extraneous and irrelevant information arriving at the nervous system and 'frees up the CNS from intricate computations to extract information needed for fulfilling a particular task' (Wehner, 1987). We are now extending this hypothesis to include chemosensory mediated behaviour and are proposing that behaviours will be 'matched' to the hydrodynamics that structure chemosensory information. Thus, to begin to understand evolutionary adaptations of chemosensory mediated behaviours, it is imperative to study the physics of signal dispersion.

References

Agosta, W. (1996) *Bombardier Beetles and Fever Trees: A Close-up Look at Chemical Warfare and Signals in Animals and Plants.* Addison Wesley, Reading, MA.

Andrews, J.C. (1983) Deformation of the active space in the low Reynolds number feeding current of calanoid copepods. *Can. J. Fish. Aquat. Sci.* **40**: 1293–1302.

Atema, J. (1985) Chemoreception in the sea: adaptation of chemoreceptors and behavior to aquatic stimulus conditions. *Symp. Soc. Exp. Biol.* **39**: 387–423.

Atema, J. and Engstrom, D.G. (1971) Sex pheromone in the lobster, *Homarus americanus.* *Nature* **232**: 261–263.

Aylor, D.E. (1976) Estimating peak concentrations of pheromones in the forest. In: *Perspectives in Forest Entomology* (eds J.E. Anderson, and M.K. Kaya), Academic Press, New York, pp. 177–188.

Aylor, D.E., Parlange, J.-Y. and Granett, J. (1976) Turbulent dispersion of disparlure in the forest and male gypsy moth response. *Environ. Entomol.* **10**: 211–218.

Baker, T.C., Willis, M.A, Haynes, K.F. and Phelan, P.L. (1985) A pulsed cloud of sex pheromone elicits upwind flight in male moths. *Physiol. Entomol.* **10**: 257–365.

Bell, W.J. and Cardé, R.T. (1984) *Chemical Ecology of Insects.* Sinauer, Sunderland, MA.

Bell, W. and Mitchell, R. (1972) Chemotactic and growth responses of marine bacteria to algal extracellular products. *Biol. Bull.* **143**: 265–277.

Berg, H.C. (1983) *Random Walks in Biology.* Princeton University Press. Princeton, NJ.

Berg, H.C. and Purcell, E.M. (1977) The physics of chemoreception. *J. Biophys.* **20**: 193–219.

Brady, J., Gibson, G. and Packer, M.J. (1989) Odour movement, wind direction, and the problem of host-finding by tsetse flies. *Physiol. Entomol.* **14**: 369–380.

Bursell, E. (1984) Observations on the orientation of tsetse flies (*Glossinia pallipdipes*) to wind-borne odours. *Physiol. Entomol.* **9**: 133–137.

Busdosh, M., Robillard, G.A., Tarbox, K. and Beehler, C.L. (1982) Chemoreception in an arctic amphipod crustacean: a field study. *J. Exp. Mar. Biol. Ecol.* **62**: 261–269.

Carr, W.E.S. and Derby, C.D. (1986) Behavioral chemoattractants for the shrimp, *Palaemonetes pugio*: identification of active components in food extracts and evidence of synergistic mixture interactions. *Chem. Senses* **11**: 49–64.

Clay, C.S. and Medwin, H. (1977) *Acoustical Oceanography.* John Wiley, New York.

Dittman, A.H. and Quinn, T.P. (1996) Homing in Pacific salmon: mechanisms and ecological basis. *J. Exp. Biol.* **199**: 83–91.

Dodson, S.I., Ryan, S., Tollarian, R. and Lampert, W. (1997) Individual swimming behavior of *Daphnia*: effects of food, light, and container size in four clones. *J. Plankt. Res.* **19**: 1537–1552.

Døving, K.B., Westerberg, H. and Johnsen, P.B. (1985). Role of olfaction in the behavioral and neuronal responses of atlantic salmon, *Salmo salar*, to hydrographic stratification. *Can. J. Fish. Aquat. Sci.* **42**: 1658–1667.

Elkinton, J.S., Cardé, R.T. and Mason, C.J. (1984) Evaluation of time-average dispersion models for estimating pheromone concentration in a deciduous forest. *J. Chem. Ecol.* **10**: 1081–1108.

Gleeson, R.A., Adams, M.A. and Smith, A.B. (1987) Hormonal modulation of pheromone-mediated behavior in a crustacean. *Biol. Bull.* **172**: 1–9.

Griffin, D.R. (1986) *Listening in the Dark.* Comstock, Ithaca, NY.

Hamner, P. and Hamner, W.M. (1977) Chemosensory tracking of scent trails by the planktonic shrimp *Acetes sibogae australis*. *Science* **195**: 886–888.

Hargrave, B.T. (1985) Feeding rates of abyssal scavenging amphipods (*Eurythenes gryllus*) determined *in situ* by time-lapse photography. *Deep-Sea Res.* **32**: 443–450.

Hart, D.D., Clark, B.D. and Jasentuliyana, A. (1996) Fine-scale field measurement of benthic flow environments inhabited by stream invertebrates. *Limnol. Oceanogr.* **41**: 297–308.

Hazlett, B.A. (1985) Disturbance pheromones in the crayfish *Orconectes virilis*. *J. Chem. Ecol.* **11**: 1695–1711.

Hazlett, B.A. (1990) Source and nature of disturbance-chemical system in crayfish. *J. Chem. Ecol.* **16**: 263–275.

Hessler, R.R., Isaacs, J.D. and Mills, E.L. (1972) Giant amphipods from the abyssal Pacific Ocean. *Science* **175**: 636–637.

Ingram, C.L. and Hessler, R.R. (1983) Distribution and behavior of scavenging amphipods from the central North Pacific. *Deep-Sea Res.* **30**: 683–706.

Jerlov, N.G. (1976) *Marine Optics.* Elsevier, Amsterdam.

Johnsen, P.B. (1983) New directions in fish orientation studies. In: Signposts in the Sea (eds W.F. Hernnkind and A.B. Thistle), Florida State University Press, Tallahassee, FL.

Johnsen, P.B. and Teeter, J. (1985) Behavioral responses of bonnethead sharks (*Sphyrna tiburo*) to controlled olfactory stimulation. *Mar. Behav. Physiol.* **11**: 283–291.

Koehl, M.A.R. and Strickler, J.R. (1981) Copepod feeding currents: food capture at low Reynolds number. *Limnol. Oceanogr.* **26**: 1062–1073.

Mafra-Neto, A. and Cardé, R.T. (1995a) Influence of plume structure and pheromone concentration on the upwind flight of *Cadra cautella* males. *Physiol. Entomol.* **20**: 117–133.

Mafra-Neto, A. and Cardé, R.T. (1995b) Effect of the fine-scale structure of pheromone plumes: pulse frequency modulates activation and upwind flight of almond moth males. *Physiol. Entomol.* **20**: 229–242.

Maier, I. and Müller, D.G. (1986) Sexual pheromones in algae. *Biol. Bull.* **170**: 145–175.

Mann, K.H. and Lazier, J.R.N. (1996) *Dynamics of Marine Ecosystems: Biological–Physical Interactions in the Oceans*. Blackwell Scientific, Cambridge, MA.

Miksad, R.W. and Kittredge, J. (1979) Pheromone aerial dispersion: a filament model. *14th Conference on Agriculture and Forest Meteorology, Vol. 1*. American Meteorological Society, Boston, MA, pp. 238–243.

Miller, R.L. (1979) Sperm chemotaxis in the Hydromedusae. II. Some chemical properties of the sperm attractants. *Mar. Biol.* **53**: 115–124.

Moore, P.A. and Atema, J. (1988) A model of a temporal filter in chemoreception to extract directional information from a turbulent odor plume. *Biol. Bull.* **174**: 355–363.

Moore, P.A. and Atema, J. (1991) Spatial information in the three-dimensional fine structure of an aquatic odor plume. *Biol. Bull.* **181**: 408–418.

Moore, P.A. and Grills, J. (1999) Chemical orientation to food by the crayfish, *Orconectes rusticus*: influence by hydrodynamics. *Anim. Behav.* **58**: 953–963.

Moore, P.A., Zimmer-Faust, R.K., BeMent, S.L., Wiessburg, M.J., Parrish, J.M. and Gerhardt, G.A. (1992) Measurement of microscale patchiness in a turbulent aquatic odor plume using a semiconductor-based microprobe. *Biol. Bull.* **183**: 138–142.

Moore, P.A., Fields, D.M. and Yen, J. (1999) Physical constraints of chemoreception in foraging copepods. *Limnol. Oceanogr.* **44**: 166–177.

Moore, P.A., Grills, J. and Schneider, R.W.S. (2000) Habitat specific signal structure for olfaction: an example from artificial streams. *J. Chem. Ecol.* (in press).

Murlis, J. (1986) The structure of odour plume. In: *Mechanisms in Insect Olfaction* (eds T.L. Payne, M.C. Birch and C.E.J. Kennedy), Clarendon Press, New York, pp. 27–38.

Murlis, J. and Jones, C.D. (1981) Fine-scale structure of odour plumes in relation to insect orientation to distant pheromone and other attractant sources. *Physiol. Entomol.* **6**: 71–86.

Nachtigall, P.E. and Moore, P.W.B. (1986) Animal sonar: processes and performance. In: *Proceedings of a NATO Advanced Study Institute Conference on Animal Sonar Systems*, vol. 156, pp. 1–852 (eds P.E. Nachtigall and P.W.B. Moore). Plenum, New York.

Peckarsky, B.L. (1980) Predator–prey interactions between stoneflies and mayflies: behavioural observations. *Ecology* **61**: 932–943.

Petranka, J.W., Kats, L.B. and Sih, A. (1987) Predator–prey interaction among fish and larval amphibians: use of chemical cues to detect prey. *Anim. Behav.* **35**: 420–425.

Poulet, S.A. and Ouellet, G. (1982) The role of amino acids in the chemosensory swarming and feeding of marine copepods. *J. Plankt. Res.* **4**: 341–361.

Sainte-Marie, B. (1986) Effect of bait size and sampling time on the attraction of the lysianassid amphipods *Anonyx sarsi* (Steele and Brunel) and *Orchomenella pinguis* (Boeck). *J. Exp. Mar. Biol. Ecol.* **99**: 63–77.

Sanford, L.P. (1997) Turbulent mixing in experimental ecosystem studies. *Mar. Ecol. Prog. Ser.* **161**: 265–293.

Schneider, R.W.S., Lanzen, J. and Moore, P.A. (1998) Boundary layer effect on chemical signal movement near the antennae of the Sphinx moth, *Manduca sexta*: temporal filters for olfaction. *J. Comp. Physiol. A.* **182**(3): 287–305.

Shifrin, K.S. (1988) *Physical Optics of the Ocean Water*. American Institute of Physics, New York.

van Duren, L.A., Stamhuis, E.J. and Videler, J.J. (1998) Reading the copepod personal ads: increasing encounter probability with hydromechanical signals. *Philos. Trans. R. Soc. Lond. B* **353**: 691–700.

Vogel, S. (1981) *Life in Moving Fluids: The Physical Biology of Flow*. Princeton University Press, Princeton, NJ.

Wehner, R. (1987) 'Matched filters'—neural models of the external world. *J. Comp. Physiol. A* **161**: 511–531.

Weissburg, M.J. and Zimmer-Faust, R.K. (1993) Life and death in moving fluids: hydrodynamic effects on chemosensory-mediated predation. *Ecology* **74**: 1428–1443.

Westerberg, H. (1991) Properties of aquatic odour trails. In: *Proceedings of the 10th International Symposium on Olfaction and Taste* (ed. K. Døving). Graphic Communication System, Oslo, pp. 45–65.

Willis, M.A., David, C.T., Murlis, J. and Cardé, R.T. (1994) Effects of pheromone plume structure and visual stimuli on the pheromone-modulated upwind flight of male gypsy moths (*Lymantria dispar*) in a forest (Lepidoptera: Lymantriidae). *J. Insect Behav.* **7**: 385–409.

Wilson, R.R. and Smith, K.L. Jr. (1984) Effect of near-bottom currents on detection of bait by the abyssal grenadier fishes *Coryphaenoides* spp. recorded *in situ* with a video camera in a free vehicle. *Mar. Biol.* **84**: 83–91.

Yen, J., Weissburg, M.J. and Doall, M.H. (1998) The fluid physics of signal perception by mate tracking copepods. *Philos. Trans. R. Soc. Lond. B* **353**: 787–804.

Zimmer-Faust, R.K., Stanfill, J.M. and Collard, S.B., III (1988) A fast multichannel fluorometer for investigating aquatic chemoreception and odor trails. *Limnol. Oceanogr.* **33**: 1586–1595.

10

Passive orientation, hydrodynamics, acoustics and behaviour in swash-riding clams and other sandy beach invertebrates

Olaf Ellers

1. Introduction

Biomechanics can provide context and tools for studying behaviour. Knowledge of the physical world is essential in studying mechanical, chemical, and thermal sensory capabilities. In addition, biomechanics can define optima and constraints relative to which behavioural choices and boundaries can be understood. In this paper I present an integrative case study in which biomechanics is used to understand a locomotory behaviour common to a diverse assemblage of animal species that live on sandy beaches.

Many diverse animals that live on sandy beaches locomote by riding in flows from waves. Such wave riding is often part of a tidal migration; but is also used for seasonal shifts into the intertidal or subtidal habitats, for pursuing prey or escaping from predators, or for movement of larvae into or out of the habitat.

In this paper, I will concentrate first on describing some physical and sensory aspects of the motion of one clam, *Donax variabilis*, that lives on beaches in the southeastern USA. This work is described in more detail in three papers (Ellers, 1995a,b,c). The behaviour, sensory capabilities, mode of locomotion, and physical characteristics of this clam provide a suite of characteristics that are common to many sandy beach inhabitants. In addition, towards the end of this paper, I will describe physical characteristics of the wave-swept shore (e.g., dimensionless fall velocity) that are relevant to habitat choice of all sandy beach animals, to biodiversity of sandy beaches, and to size-specific predictions for migratory behaviour on beaches.

Biomechanics in Animal Behaviour, edited by P. Domenici and R.W. Blake.
© 2000 Taylor & Francis, Oxford.

2. Habitat

Waves and the flows they generate are the most important part of the physical environment in the context of the behaviours described here. When waves approach the shore, as the water becomes progressively shallower, waves form into breakers. The breaker moves shoreward, and either spills directly onto the shore (typical on more steeply sloped beaches), or collapses into a bore (typical on less steeply sloped beaches). A bore is a cliff of water, several centimeters high, that travels beachward, and itself collapses and forms a sheet of water that moves shoreward. The sheet of water formed either by breakers spilling onto the shore or by a collapsing bore is called the swash or run up. Swash flows shoreward, slows down, stops and then starts to flow seaward. That seaward flow is called the backwash.

3. Behaviour

On beaches in the southeastern USA, the clam *Donax variabilis* maintains position at the edge of the sea by migrating tidally. To migrate, a clam of this species occasionally jumps out of the sand, rides a wave and digs in again at a new position on the beach. During incoming tides, net movement is shoreward and during outgoing tides net movement is seaward. During incoming tides jumping occurs before a wave reaches the position of the clams on the beach whereas during outgoing tides jumping occurs directly into seaward moving flow.

I have named this method of locomotion 'swash-riding'. As with other methods of locomotion such as running or swimming, it is possible to discover underlying physical principles that describe and categorize the locomotion. Such a description of underlying principles and characteristics makes possible meaningful comparisons of behaviour between and within species.

4. Orientation

After jumping in front of incoming flow from waves, a clam is carried shoreward. The flow slows and starts moving seaward; it is usually at this time that a clam will orient with the anterior end upstream. Siphons protrude posteriorly whereas the foot protrudes anteriorly. The tip of the foot is used to anchor in the sand and by repeated thrusts and pulls of its foot a clam digs into the sand again. As a consequence of being oriented with the foot upstream, a clam does not twist about the anchor point. Rather, it is able to dig into a groove formed in the sand just upstream of the shell where sand is scoured out around the shell by flow. The formation of the groove in the sand thus facilitates the digging behaviour.

Orientation in wave flows is passive and does not involve direct behavioural control. Indeed, dead clams (but with both valves still in place) orient the same way as do live ones in a flow tank. Therefore, orientation is a physical characteristic of the clam shell. Such passive orientation is not a merely a general characteristic of bivalve shells; rather it is a specific feature of swash-riding species of *Donax*. For instance, other haphazardly chosen bivalve species do not orient in flow and do not ride waves (Ellers, 1995c).

Passive orientation occurs for the same reasons that a weathervane orients to show the wind's direction. The mechanically relevant features of a weathervane are: (i) the vane must have a vertical rod, which forms a fixed pivotal axis about which the weathervane rotates; (ii) the surface area of the vane must be larger on one side of the pivotal axis. Wind causes the vane to experience drag, which is a force parallel to the direction of flow and has a magnitude given by

$$D = 0.5\rho u^2 C_d A \tag{1}$$

where ρ is the density of the medium, u is freestream speed, C_d is the coefficient of drag, and A is the surface area projected perpendicular to flow. The drag on the side with the larger area is higher, which causes the vane to spin into a stable position with the larger area downstream (*Figure 1*). Similarly, the features of a clam that are important in causing orientation are: (i) the shell must have a pivotal point, where the clam touches the sand, with the clam rotating about a vertical axis through the pivotal point; (ii) the shell must have a wedge-like shape, with more projected surface area on one side of the pivotal axis than on the other; (iii) as a corollary to (i), the shell must be sufficiently heavy that it keeps the pivotal point in contact with the sand despite flow forces tending to lift the clam. The third requirement increases downward forces that tend to keep the shell on the sand countering upward lift exerted by flow on the clam shells in most orientations. A 'heavy' shell is reflected in a high density (mass per volume) relative to clam species that do not swash-ride (Ellers, 1995c).

Once a clam is sliding with the anterior end upstream, there are several consequences to the motion of the clam. These can be quantified by measuring the distribution of pressures on the surface of a clam in various orientations (*Figure 2*). The pressure distribution drives passive orientation.

In addition to providing an explanation for passive orientation, these measurements provide information about total force experienced in various orientations. Most importantly, a clam with the anterior end upstream experiences downward lift that tends to keep the clam in contact with the sand. Staying in contact with the sand increases the likelihood that a clam will be able to dig in at a chosen location rather than being swept further out to sea with the backwash. In contrast, when forces on the clam are measured in other orientations, lift is upward, which tends to destabilize the clam's motion in flow.

Finally, a clam that is oriented with the anterior end upstream experiences a distribution of pressures that tends to push the anterior end into the sand while lifting the

Pivotal axis

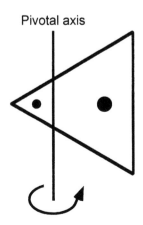

Figure 1. A weathervane rotates because of a turning moment exerted about the pivotal axis by the wind. Moments, which tend to cause rotation, are products of force and distance from the pivotal axis. In a weathervane the wind exerts pressure (force per area) on the area of the weathervane. The inequality of total areas on either side of the pivotal axis and the inequality of the distribution of area contribute to the total moment generated, which causes the weathervane to spin until it reaches an oriented position where the wind's direction is parallel to the surface of the weathervane.

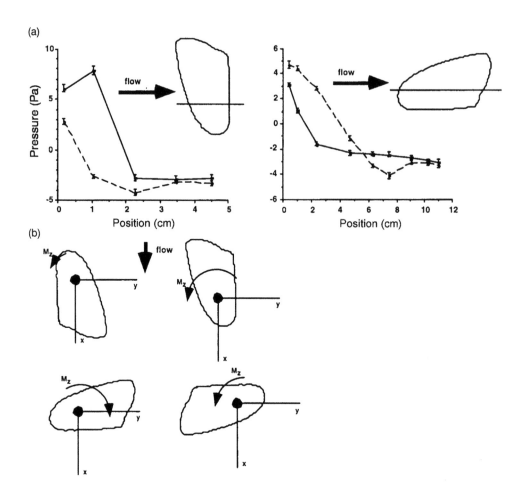

Figure 2. *Pressure distribution (a) along the indicated transects on the surface of a (five-times life size) scale model of a clam and the resulting moments (b). Positive pressures indicate forces pushing perpendicularly into the surface of the clam; negative ones outward. Pressures at points on the upper surface (dashed lines) and pressures on the lower surface (continuous lines) along many transects similar to those shown can be used to calculate (Ellers, 1995c) the total moments on the clam in various orientations (b). The calculation involves: (1) calculation of a distribution of forces exerted at each point on the clam by multiplication of each measured pressure by an associated projected area; (2) calculation of a distribution of moments by multiplication of each force by the distance from the location of the force to the pivotal point about which the clam rotates; and (3) summation of all the moments. The rotations implied by those calculations are shown in (b). The tendency of clams to orient in flows can thus be understood in terms of the shape of, and consequent pressure distribution on, the clams. The pressure distribution in (a) also illustrates that there is downward lift when oriented with the anterior end upstream (left) and upward lift when oriented in other ways (right). This is reflected in the low pressure on the underside of the clam when anteriorly oriented (left). Reprinted from Ellers (1995c) Biol. Bull. **189**, pp. 138–147.*

posterior end. Such a distribution of force tends to contribute moments that are appropriate for digging.

5. Choice of waves

Locomotion often has significant energetic costs and we can expect to find behaviour and adaptations that tend to reduce such costs. Migration by swash-riding is no different. Pushing out of the sand, and digging in again is energetically costly in *Donax* (Ansell and Trueman, 1973). A clam may reduce the cost of migration by reducing the number of swash-rides, which can be achieved by maximizing the distance traveled in each swash-ride. Maximizing the distance traveled could be achieved by preferentially riding waves that create swash that has a large excursion. The excursion of a swash is the distance between the point where that swash passes the previous backwash and the maximum shoreward reach of that swash (*Figure 3*). Do clams jump for and ride only the largest waves?

I surveyed jumping responses and recorded the excursion of waves on four summer days on an outer bank island in North Carolina. For each incoming swash, either many clams jumped, or no clams jumped, so it was possible to determine which swash clams choose. Clams jumped preferentially for the largest swash (*Figure 4*). They showed no

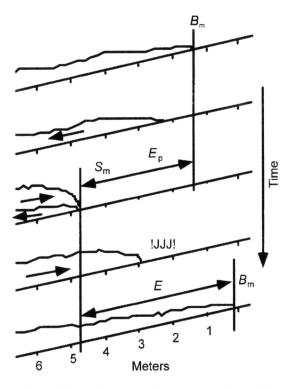

Figure 3. *Definition of variables used to assay whether clams preferentially jumped in response to waves. E is the excursion of a swash; E_p is the receding excursion of the previous swash; B_m is the beachward maximum and S_m is the seaward minimum extent of the swash. These variables can be assayed relative to the timing of observed jumping events. The results are summarized in Figure 4. Reprinted from Ellers (1995a) Biol. Bull. 189, pp. 120–127.*

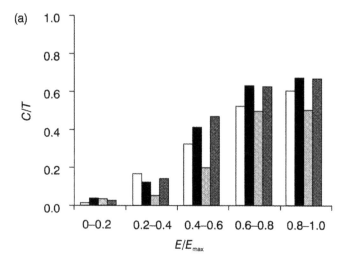

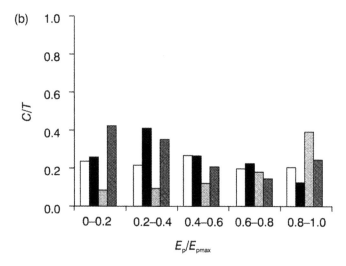

Figure 4. *(a) Clams jumped for about half of the largest 20% of excursions whereas they jumped for less than one-tenth of the smallest 20% of excursions. They are choosing to jump for the largest waves. (b) In contrast, the receding excursion of the previous swash showed no preferential pattern of jumping. Thus clams are not cueing on any size-related feature of the wave preceding their jump. C/T is the fraction of waves for which clams jumped in each category, E/E_{max} is the fractional size of a wave's excursion compared with the maximum excursion seen on a given day. E_p/E_{pmax} is a similar ratio applied to the immediately preceding receding excursion. Columns with different shading are from observations on different days. (For details and statistics see Ellers (1995a)). Reprinted from Ellers (1995a) Biol. Bull. **189**, pp. 120–127.*

preferences with regard to the excursion of the previous backwash. Thus clams appear to be able to determine which swash will carry them the furthest. Furthermore, on rising tides, clams jump out ahead of incoming swash; thus they are able to determine the timing of an incoming wave. What cues are clams using to determine the timing of their jump and to predict the excursion of the swash?

6. Acoustic cues from waves

Ocean waves generate sound that could cue clams to jump. Sound from ocean waves is a possible candidate because the sound signal, traveling at the speed of sound, would arrive at the clam's location before the ocean wave, thus enabling clams to anticipate the arrival of the ocean wave. Furthermore, incoming waves that generate swash of large excursion also generate loud sounds. The loudness of generated sound depends on the energy input at the source, which for breakers or bores depends on their height. Large bores create large excursions of the swash. Excursions can be predicted (Bradshaw, 1982) using

$$E = \frac{u_0^2}{g \tan \beta} \tag{2}$$

where E is the excursion of the swash, g is the acceleration due to gravity, β is the beach slope, and u_0 is the speed of the swash immediately after bore collapse and is given by

$$u_0 = u' + 2\sqrt{gh'} \tag{3}$$

where u' is the speed of the bore and h' is the height of the bore.

I recorded sounds that were made by incoming waves and that were potentially available to clams as cues. These sounds were recorded by placing a hydrophone into the sand amidst a population of *D. variabilis*. The sounds are reminiscent of thunder, with loud rumbling noises containing a mixture of low frequencies (*Figure 5*). The noises become louder as a wave approaches the hydrophone (*Figure 6*) and the loudness is dependent on the size of the incoming bore. It is possible for a person to judge the time of arrival and size of incoming swash from these noises. Do the clams detect and respond to such sounds?

Clams subjected to wave sounds or to similar low-frequency sounds in an aquarium do respond by jumping out of the sand. Low-frequency sounds were either generated by knocking with knuckles on the table on which rested the aquarium containing the clams, or by presenting synthesized sounds to the clams using an underwater loudspeaker. Clams are more responsive to low-frequency sounds than to high-frequency sounds.

Responsiveness can be judged by considering the loudness of sounds that elicit a given level of response. Loudness of sound as perceived by an organism is subjective. However, physical aspects of the sound wave that may relate to perceived loudness can

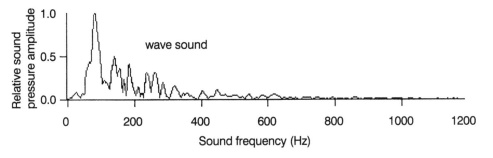

Figure 5. *Sound frequency content of incoming waves. This sounds like rolling thunder. Reprinted from Ellers (1995b) Biol. Bull. **189**, pp.128–137.*

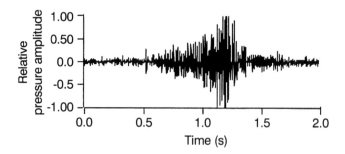

Figure 6. *Sound pressure versus time showing increase in loudness of an approaching wave.*
Reprinted from Ellers (1995b) Biol. Bull. 189, pp.128–137.

be measured. Two conventional measures of loudness are the intensity level IL and the
sound pressure level SPL (Kinsler *et al.*, 1982) given by the equations

$$\text{IL} = 10\log_{10}\left(\frac{I}{I_{\text{ref}}}\right) \qquad (4)$$

where I is the sound intensity $I = \frac{P_e^2}{\rho c}$, P_e^2 is the root mean square sound pressure, ρ is the
density of the medium, and c is the speed of sound in that medium, and

$$\text{SPL} = 20\log_{10}\left(\frac{P_e}{P_{\text{ref}}}\right) \qquad (5)$$

For analyzing clam responses I used the root mean square sound pressure (which is the
basic variable used in the two conventional measures above) as a measure of loudness.
The clams I observed were most responsive to lower-frequency sounds. For instance,
an 832 Hz sound that was 10 times louder than a 72 Hz sound nevertheless elicited
jumping responses from only half as many clams (Ellers, 1995b).

7. Tidal rhythm of responsiveness

Clams respond to such low-frequency sounds in the laboratory on a tidal rhythm
(*Figure 7*). Two sets of experiments were conducted several weeks apart, such that daily
and tidal rhythms were shifted relative to each other, thus allowing distinction
between the two rhythms. In both sets, clams were maximally responsive at high tide
and did not respond at all within an hour on either side of low tide, with gradients of
responsiveness in the intervening periods. This pattern of activity was observed while
the clams were in an aquarium in a laboratory, in the absence of direct tidal cues, indicating that there is an internal rhythm of responsiveness to sound. This pattern of
activity in the laboratory corresponds to observations of activity patterns on the
beach. Specifically, clams stop migrating seaward about 1 h before low tide, and resume
migrating shoreward about 1 h after low tide.

Although the existence of tidal rhythms in marine invertebrates is known, little
specific work has been done to document them, or to study the genetic or physiological basis of the clocks controlling the rhythms. Naylor (1985) gave a review of
rhythmic behaviour in crustaceans, molluscs, and polychaetes. The most closely

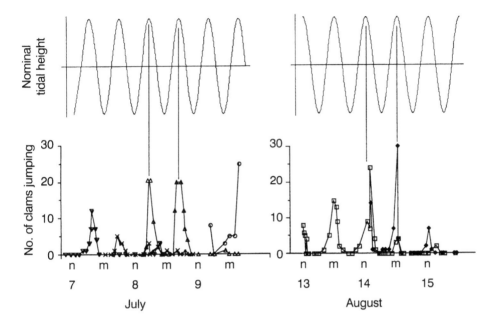

Figure 7. *Tidal rhythm of responsiveness to sound. Upper traces show tidal state at the beach from which clams were gathered. Lower traces show the number of clams responding to loud artificial noise in an aquarium. Clams were maximally responsive at high tide. (For details and statistics, see Ellers (1995b)). Reprinted from Ellers (1995b) Biol. Bull. **189**, pp.128–137.*

related organism to *Donax* in which an endogenous tidal rhythm has been demon-strated is a venerid clam that has a tidal rhythm of shell gaping (Williams *et al.*, 1993). The clock controlling such tidal rhythms is suggested, at least in crabs, to be not a 12.4 h clock but rather two 24.8 h clocks that match the lunar day, and that run coupled out of phase with each other to match the two tidal peaks per day (Palmer, 1996).

8. Generality of results: similarity with behaviour of other sandy beach swash-riders

There are numerous other animals that migrate or move on sandy beaches by riding flows from waves. These wave-riders include several genera of gastropods, mole crabs, and amphipods. Swash-riding gastropods include *Terebra saleana* (Kornicker, 1961), *Hastula inconstans* (Miller, 1979), *Olivella biplicata* (Johnson, 1966), and *Bullia digi-talis* (Odendaal *et al.*, 1992). Swash-riding mole crabs include *Emerita analoga* (Cubit, 1969) and *Remipes truncatifrons* (Mori, 1938). Indeed, although literature reports are rare, swash-riding is common on sandy beaches.

Wave-riding is used by both predatory and filter-feeding animals. For instance, *Donax* are filter feeders, and the edge of the sea might be a location that enhances the availability of suspended material because waves continually suspend material off the sea floor. For mole crabs and some gastropods, the backwash is the important flow for filtering food. Mole crabs place their antennae on the surface of the sand and filter the backwash, occasionally retracting their antennae and wiping them through their mouth appendages to retrieve filtered food. Similarly, some olivellid gastropods have flanges on the anterior end of their feet that are deployed at the surface of the sand to filter food from the backwash. Using a movement very reminiscent of mole crabs, these olivellid gastropods flick the foot flange past their mouth periodically to pass filtered material from the foot flange to their mouth. Other swash-riders use swash-riding to locate prey. For instance, *Bullia digitalis* uses chemotaxis while swash-riding to locate prey (Odendaal *et al.*, 1992).

There are several patterns of behaviour, of which *D. variabilis* is an example, that all such wave-flow riders have in common. Swash-riders: (i) orient in flow; (ii) respond to pressure changes such as are present in sound, infrasound and hydro-static pressures as waves move shoreward over a location; (iii) have rapid burrowing abilities; (iv) tend to have a wedge-like shape. I have surveyed orientation behaviours in several swash-riders. I have observed hastulid and olivellid gastropods orienting with the pointed end (apex) of the shell upstream and the expanded, umbrella-shaped foot downstream. Furthermore, all swash-riding *Donax* species that I have observed (*D. assimilis*, *D. denticulatus*, *D. variabilis*, *D. fossor*, *D. texasianus*, *D. gouldii*, *D. ecuadoriensis*) orient in flow, whereas those that swash-ride less frequently, such as *D. californicus*, do not orient in flow. Finally, mole crabs swim actively in the swash and, just before re-entry into the sand, they orient with the pointed, posterior end upstream. Comparative information on responses to rapid pressure changes is rare; but some swash-riding amphipods respond to pressure changes (Enright, 1961, 1963; Forward, 1986). Surveys of burrowing suggest that beach dwellers on shores with large waves are relatively rapid burrowers (Ansell and Trevallion, 1969). Finally, swash-riders tend to be wedge-shaped with respect to flow. The profile of a mole crab is wedge-shaped with the posterior end being the thin edge of the wedge. The terebrid and hastulid gastropods have wedge-shaped spires. An everted foot, as in swash-riding olivellid, terebrid and hastulid gastropods, also contributes to a functional wedge-shape.

That density and a wedge-like shape are important variables has been demonstrated in a comparative study of six populations of *Donax serra* that had variable morphology and differing zonation patterns on beaches (Donn, 1990). Wedge-shaped individuals and those with thicker shells (and hence heavier per unit volume), tended to be located higher on the beach than did more blade-shaped and thinner shelled individuals, and differences in these variables among populations were related to their position on the beach. As will be detailed below, individuals with less dense shells and less wedge-like shape will tend to be swept off the beach, or at least further seaward while riding waves.

Similar effects have been observed (McLachlan *et al.*, 1995) among 12 bivalve species found on a variety of beach types ranging from dissipative to reflective. Species found on reflective beaches, with harsher wave climates, were denser (higher weight per volume) and uniformly wedge-shaped. In contrast, species from dissi-pative beaches, with milder wave climates, were less dense and varied in shape from blade-like to spherical. Thus, both density and a wedge-like shape are important in determining the nature of the beach habitat in which beach-dwelling bivalves are found.

9. Generality of results: physical effects on diversity and habitat choice

9.1 *Context*

Species that live on beaches are profoundly affected by flows from waves. That flow is important to the lives of marine animals has been documented in numerous books (Denny, 1988, 1993; Vogel, 1994). Flow effects have been characterized in terms of drag in swimming animals, drag, lift and dislodgment in sessile rocky shore creatures, and various fluid dynamic effects important to filter feeders. Sandy beaches provide a new category of flow effects that are important – specifically, the effects of transport of animals and sand by waves.

9.2 *Diversity and wave climate*

The wave climate of beaches is very important in determining the richness of species on a beach. Species richness increases from reflective to dissipative beaches (Borzone *et al.*, 1996; Hacking, 1998; McLachlan *et al.*, 1993). Reflective beaches have higher slopes, larger waves and coarser sand than do dissipative beaches.

Wave climate and beach type are characterized by an index called the dimensionless fall velocity (Masselink and Short, 1993). This index compares the wave period (time between successive waves) with how long it takes an object to sink a distance equal to the wave height. This index thus indicates whether a given object will tend to remain suspended in wave flows and be washed off beaches or whether it will tend to sink to the sand and stay on beaches. The turbulent flow of incoming breaking waves tends to lift animals and sand off the beach and into the wave flow; whether the sand grains and animals are washed off the beach depends on whether they sink to the sand before the next incoming wave arrives. The correlation of species richness with dissipative beaches is a result of the wave climate as expressed by the dimensionless fall velocity. On reflective beaches with very large waves, many species are swept off the beach and thus unable to maintain a position on the beach, thus adversely affecting species richness.

The dimensionless fall velocity, also called the dimensionless fall time, is given by

$$F = \frac{H_b}{VT} \tag{6}$$

where H_b is the breaker height from the top of the breaker to the sand, V is the sand fall speed and T is the wave period. Fall speed is the speed at which sand grains on a particular beach fall through water, and that speed is dependent on grain size. For instance, very fine sand, 35 μm in diameter, sinks at 1 mm s^{-1}; whereas coarse sand, 1.5 mm in diameter, sinks at 20 cm s^{-1} at 20°C. There will typically be a size range of particles on a given beach, so there will be a range of fall velocities corresponding to that particle size range.

The dimensionless fall velocity can be used to understand continuing patterns of onshore–offshore sediment transport on a beach, and to understand the development of typical beach profiles and features such as bars and berms. Approximately, if F is less than one, then sand tends to accrete on the beach above Mean Low Water, whereas if F is greater than one then sand tends be deposited offshore. There are other factors involved, such as the deep-water wave steepness and the wave height to grain size ratio, but to a first approximation the dimensionless fall velocity is useful in understanding the development of beach profiles and the movement of sand (US Army Corps of Engineers, 1977).

Both species diversity and richness are correlated with dimensionless fall velocity. Another index, beach state index, BSI, a variant of the dimensionless fall velocity that also accounts for relative tidal range, has a slightly better correlation with species richness than does F alone (McLachlan et al., 1993).

9.3 *Clam size and wave climate*

Sinking speeds of objects in water depend on size and density. Smaller objects sink less quickly and thus are washed off beaches more readily than are larger objects. This phenomenon explains the typical distribution of sand on beaches, with finer sands being found on beaches with smaller waves and longer periods, and coarser sand being found on beaches with larger waves and shorter periods.

For spherical objects of given density, the sinking speed can be calculated as follows. We equate the gravitational and drag forces on the sphere and solve for speed

$$(\rho_s - \rho_w)g\,\frac{4\pi a^3}{3} = \frac{1}{2}\,\rho_w u_t^2 C_a \pi a^2 \tag{7}$$

where ρ_s and ρ_w are the density of the sphere and of sea water, respectively, g is the gravitational acceleration, a is the radius of the sphere, C_d is the drag coefficient of the sphere, and u_t is the terminal speed. In a Reynolds number (Re) range over which the coefficient of drag is constant, the formula above can be rearranged to give

$$u_t = \sqrt{\frac{8ag(\rho_s - \rho_w)}{3C_d \rho_w}} \tag{8}$$

For a sphere, the coefficient of drag is 0.4 over a Reynolds number range from 10^5 to 2×10^5. Below Reynolds number of one, the drag of a sphere is given by Stokes' law (Vogel, 1994) and the terminal speed is given by

$$u_{t,lowRe} = \frac{2a^2 g(\rho_s - \rho_w)}{9\mu} \tag{9}$$

where μ is the viscosity of sea water. Between a Reynolds number of one and 10^5, there is no explicit function that gives the falling speed as a function of sphere radius because the empirically derived formula that gives the coefficient of drag in that range does not lend itself to rearrangement as a function. It is, however, possible to write and plot an implicit function by using an empirical formula for the coefficient of drag that is useful between Reynolds numbers of one and 2×10^5. That empirical formula (described by Vogel (1994)) for the coefficient of drag of a sphere is

$$C_d = \frac{24}{Re} + \frac{6}{1 + Re^{0.5}} + 0.4 \tag{10}$$

where Re is given by

$$Re = \frac{\rho_w l u_t}{\mu} \tag{11}$$

where l is the characteristic length, which I have taken here as the diameter. Then, combining Equations (7) and (10), the implicit formula linking the sphere radius, a, and the terminal sinking speed in that middle Re range, v_t, is

$$0 = \frac{-4\pi a^3 g(\rho_s - \rho_w)}{3} + 0.5\pi a^2 u_t^2 \rho_w \left[0.4 + \frac{12\mu}{au_t \rho_w} + \frac{6}{1 + \left(\frac{2au_t \rho_w}{\mu}\right)^{0.5}} \right] \quad (12)$$

and using an implicit function plotting algorithm, implemented via a computer, the falling speed is plotted for various sphere densities and radii (*Figure 8*).

I have measured the falling speed of several species of *Donax* and they are typically less than, but within a factor of two of, the calculated falling speeds of spheres of similar size (using height or width of clam compared with diameter of sphere) and density. Estimating falling speeds from size and density, *Figure 8* suggests that clam genera with lower densities such as *Pecten*, *Mya*, and *Mytilus* (around 1100 kg m⁻³) have relatively low fall speeds. Lower falling speeds enhance the likelihood of being washed off beaches. Not coincidentally, these less dense genera do not ride waves on beaches. *Figure 8* also suggests that smaller individuals of a given species and density would have lower falling speeds. Size-specific effects can be more specifically quantified for *Donax*

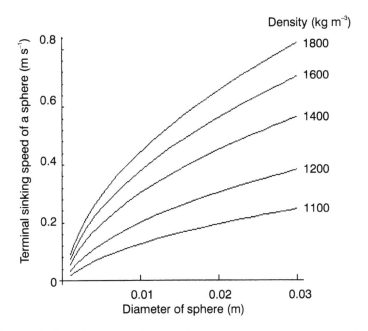

Figure 8. *The calculated falling or sinking speeds as a function of sphere density and radius. The ranges of speeds and sphere sizes shown correspond to Reynolds numbers in the range 70–70 000. Bivalve species have densities from 1100 kg m⁻³ (e.g. Tagellus plebeius) to 1700 kg m⁻³ (e.g. Donax variabilis). Range of radii shown reflects approximate sizes common in bivalves. Falling speed, together with wave height and period, can be combined to produce the dimensionless fall index, which determines which objects will tend to be washed off beaches. The higher the falling speed, the higher the tendency to remain on a beach. This parameter can be used to explain the size distribution of* Donax *on a beach (Figure 9).*

(and other species) by considering the fall speed of a specific species and the dimensionless fall velocity at a given beach at a given time.

For a sinking clam, the terminal speed is determined by a balance between the net downward force of gravity minus buoyancy and the upward force as a result of drag

$$gk_1 lwh(\rho_c - \rho_w) = 0.5\rho_w v_t^2 C_d k_2 wh \tag{13}$$

where g is the acceleration due to gravity; k_1 is a dimensionless constant that expresses volume of the clam relative to a rectanguloid of the same length l, width w, and height h as the clam; ρ_c is the density of the clam; ρ_w is the density of sea water; v_t is a clam's terminal sinking speed; C_d is the coefficient of drag; and k_2 is a constant that expresses frontal projected area of the clam relative to a rectangular area of the same length, width and height as the clam. The drag formula used above is that usually used at higher Reynolds numbers, at which drag is a function of speed squared. At low Reynolds number, however, drag is a function of speed, and in a middle Reynolds number range, drag gradually changes from the square function to the linear function of speed. In that middle Reynolds number range, that change can sometimes be modeled using the equation

$$C_d = k_4 v_t^{k_3} \tag{14}$$

(A more complex function similar to that used for spheres, Equation (10) above, or for cylinders, as in Vogel (1994), would be necessary to fit all points from Reynolds numbers close to one to high Reynolds numbers where the coefficient of drag is a constant, but for the Reynolds number range (70–20 000) of interest for *Donax*, the function above provides a good fit.) Using Equations (13) and (14) and solving for terminal speed gives

$$v_t = \left[\frac{2gk_1 l(\rho_c - \rho_w)}{k_2 k_4 \rho_w} \right]^{(\frac{1}{2 + k_3})} \tag{15}$$

Assuming that ρ_c is not a function of clam length (Ellers, 1995c), Equation (15) reduces to

$$v_t = (k_5 l)^{(\frac{1}{2 + k_6})} \tag{16}$$

where k_5 and k_6 are constants that can be obtained by a nonlinear curve-fitting algorithm using data on clam length and sinking speed.

I collected data for *D. assimilis*, a wave-riding species from exposed beaches on the Pacific coast of Panama. Using 15 individuals ranging in size from 12 to 40 mm in length, the fitted function above explained 94% of the variance. The fitted fall speed function, Equation (16), can be used in place of a sand fall speed, V, in the formula for dimensionless fall velocity, Equation (6), to provide dimensionless fall velocities for various sizes of *Donax*

$$F_D = \frac{H_b}{Tv_t} = \frac{H_b}{T(k_5 l)^{[1/(2 + k_6)]}} \tag{17}$$

Using this function, I predicted the specific wave conditions under which variously sized *D. assimilis* would be expected to be deposited onshore, and thus remain on a beach or be deposited offshore and thus be located subtidally (*Figure 9*).

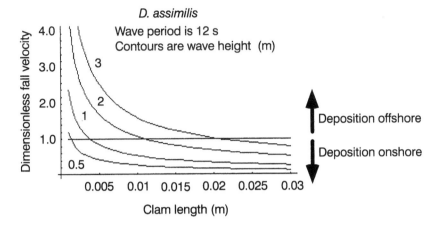

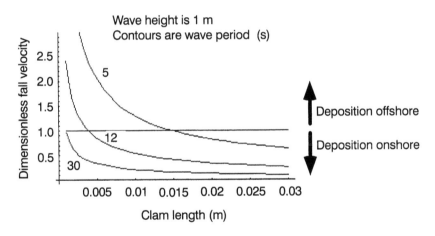

Figure 9. *Predictions of offshore or onshore deposition of* Donax *depend on clam size and wave conditions. Predictions indicated by the dimensionless fall velocity are shown given wave height and period conditions indicated. The horizontal line shows the break point between onshore and offshore deposition at an F of one. Smaller clams are more likely to be deposited offshore, and all clams are more likely to be deposited offshore when waves are higher and periods are shorter.*

Size-specific location of *Donax*, such as that predicted in *Figure 9*, has been reported in the literature. For two sympatric species in the southeastern USA, there is a smaller species, *D. parvula* that is maximally about 1 cm long and a larger species, *D. variabilis*, that is up to 3 cm long. *D. parvula*, now known as *D. fossor* (Ademkewicz and Harasewych, 1996), the smaller species, tends to be distributed more subtidally than does *D. variabilis*, although that distribution trend varies with season (Nelson *et al.*, 1993), as does wave climate. In addition, small individuals of both *D. variabilis* and *D. parvula* were shown to be present primarily subtidally, moving onshore as they grew larger (Bonsdorff and Nelson, 1992)—specimens less than 3 mm long were always subtidal. It is typical of all intertidal *Donax* species that they recruit into the intertidal zone only after reaching some minimum size. Similarly, on the coast of Texas, the larger *D. variabilis* tends to be distributed in the intertidal zone during the summer months, whereas the smaller *D. texasianus* (then called *D. tumida*) tends to be distributed subtidally or in the lower intertidal zone (Loesch, 1957). These size-specific effects are explained by the dimensionless fall velocity.

10. Conclusion

An integrative, biomechanical, behavioural case study of one species of clam, *D. variabilis*, revealed the influence of the physical environment on the behaviour of these clams. A mechanical stimulus (sound) is used by the clams to cue their swash-riding behaviour, and is also the likely basis for locomotion in all or most other swash-riding species.

The wave climate of a beach determines the size range of clams on that beach and provides an indispensable background for explaining an observed behaviour; namely, that small clams tend to be subtidal or low in the intertidal zone. Wave climate also provides an indispensable background for explaining the distribution of other clam species and other invertebrate beach dwellers. Finally, the ultimate importance of wave climate is that it determines the biodiversity of sandy beaches.

References

Ademkewicz, S.L. and Harasewych, M.G. (1996) Systematics and biogeography of the genus *Donax* (Bivalia: Donacidae) in eastern North America. *Am. Malacol. Bull.* **13**: 97–103.

Ansell, A.D. and Trevallion, A. (1969) Behavioural adaptations of intertidal mollusks from a tropical sandy beach. *J. Exp. Mar. Biol. Ecol.* **4**: 9–35.

Ansell, A.D. and Trueman, E.R. (1973) The energy cost of migration of the bivalve *Donax* on tropical sandy beaches. *Mar. Behav. Physiol.* **2**: 21–32.

Bonsdorff, E. and Nelson, W.G. (1992) The ecology of coquina clams *Donax variabilis* Say, 1822, and *Donax parvula* Philippi, 1849, on the east coast of Florida. *Veliger* **35**: 358–365.

Borzone, C.A., Souza, J.R.B. and Soares, A.G. (1996) Morphodynamic influence on the structure of inter- and subtidal macrofaunal communities of subtropical sandy beaches. *Rev. Chilena Hist. Nat.* **69**: 565–577.

Bradshaw, M. (1982) Bores and swash on natural beaches. *Coastal Studies Unit Technical Report No. 82/4.* Coastal Studies Unit, Department of Geography, the University of Sydney, Sydney, N.S.W.

Cubit, J. (1969) Behaviour and physical forces causing migration and aggregation of the sand crab *Emerita analoga* (Stimpson). *Ecology* **50**: 118–123.

Denny, M.W. (1988) *Biology and the Mechanics of the Wave-Swept Environment,* (1st Edn). Princeton University Press, Princeton, NJ.

Denny, M.W. (1993) *Air and Water.* Princeton University Press, Princeton, NJ.

Donn, T.E. Jr. (1990) Morphometrics of *Donax serra* Röding (Bivalvia: Donacidae) populations with contrasting zonation patterns. *J. Coastal Res.* **6**: 893–901.

Ellers, O. (1995a) Behavioral control of swash-riding in the clam *Donax variabilis. Biol. Bull.* **189**: 120–127.

Ellers, O. (1995b) Discrimination among wave-generated sounds by a swash-riding clam. *Biol. Bull.* **189**: 128–137.

Ellers, O. (1995c) Form and motion of *Donax variabilis* in flow. *Biol. Bull.* **189**: 138–147.

Enright, J.T. (1961) Pressure sensitivity of an amphipod. *Science* **133**: 758–760.

Enright, J.T. (1963) Responses of an amphipod to pressure changes. *Comp. Biochem. Physiol.* **7**: 131–145.

Forward, R.B. (1986) Behavioral responses of a sand–beach amphipod to light and pressure. *J. Exp. Mar. Biol. Ecol.* **102**: 55–74.

Hacking, N. (1998) Macrofaunal community structure of beaches in northern New South Wales, Australia. *Mar. Freshwater Res.* **49**: 47–53.

Johnson, P.T. (1966) On *Donax* and other sandy beach inhabitants. *Veliger* **9**: 29–30.

Kinsler, L.E., Frey, A.R., Coppens, A.B. and Sanders, J.V. (1982) *Fundamentals of Acoustics*, 3rd Edn. John Wiley, New York.

Kornicker, L.S. (1961) Observations on the behaviour of the littoral gastropod *Terebra salleana*. *Ecology* **42**: 207.

Loesch, H.C. (1957) Studies of the ecology of two species of *Donax* on Mustang Island, Texas. *Publ. Inst. Mar. Sci. Univ. Texas* **4**: 201–227.

Masselink, G. and Short, A.D. (1993) The effect of tide range on beach morphodynamics and morphology—a conceptual beach model. *J. Coastal Res.* **9**: 785–800.

McLachlan, A., Jaramillo, E., Donn T.E. and Wessels, F. (1993) Sandy beach macrofauna communities and their control by the physical environment: a geographical comparison. *J. Coastal Res.* Special Issue **15**: 27–38.

McLachlan, A., Jaramillo, E., Defeo, O., Dugan, J., Ruyck A.D. and Coetzee, P. (1995) Adaptations of bivalves to different beach types. *J. Exp. Mar. Biol. Ecol.* **187**: 147–160.

Miller, W. (1979) The biology of *Hastula inconstans* (Hinds, 1984) and a discussion of life history similarities among other *Hastulas* of similar proboscis type. *Pacific Sci.* **33**: 289–306.

Mori, S. (1938) Characteristic tidal rhythmic migration of a mussel, *Donax semigranosus* Dunker, and the experimental analysis of its behaviour at the flood tide. *Dobutsugaku Zasski*, (*Zool. Mag. Jpn*) **50**: 1–12.

Naylor, E. (1985) Tidally rhythmic behaviour of marine animals. *Symp. Soc. Exp. Biol.* **39**: 63–93.

Nelson, W.G., Bonsdorff, E. and Adamkewicz, L. (1993) Ecological, morphological, and genetic differences between the sympatric bivalves *Donax variabilis* Say, 1822, and *Donax parvula* Philippi, 1849. *Veliger* **36**: 317–322.

Odendaal, F.J., Turchin, P., Hoy, G., Wickens, P., Wells, J. and Schroeder, G. (1992) *Bullia digitalis* (Gastropoda) actively pursues moving prey by swash-riding. *J. Zool.* **228**: 103–113.

Palmer, J.D. (1996) Time, tide and the living clocks of marine organisms. *Am. Sci.* **84**: 570–578.

US Army Corps of Engineers (1977) *Shore Protection Manual*. US Government Printing Office, Washington, DC.

Vogel, S. (1994) *Life in Moving Fluids*, 2nd Edn. Princeton University Press, Princeton, NJ.

Williams, B.G., Palmer, J.D. and Hutchinson, D.N. (1993) Comparative studies of tidal rhythms XIII. Is a clam clock similar to those of other intertidal animals? *Mar. Behav. Physiol.* **24**: 1–14.

Appendix: List of symbols

a	radius of a sphere
A	projected area, perpendicular to the direction of drag, of an object in flow
β	beach slope
c	the speed of sound in that medium
C_d	coefficient of drag
D	drag
E	excursion of the swash
F	the dimensionless fall velocity
F_D	the dimensionless fall velocities of a set of *Donax* clams as a function of clam length (equation 17)
g	acceleration due to gravity

h	height of the bore
H_b	breaker height from the top of the breaker to the sand
I	sound intensity
IL	intensity level
k_1	dimensionless constant that expresses volume of a clam relative to a rectanguloid of the same length, l, and width, w, and height, h, as the clam
k_2	dimensionless constant that expresses frontal projected area of a clam relative to a rectangular area of the same length, width and height as the clam
k_3	the exponent relating the drag coefficients of a set of clams to their terminal speeds (equation 14)
k_4	the coefficient relating the drag coefficients of a set of clams to their terminal speeds (equation 14)
k_5	the coefficient relating the terminal speeds of a set of clams to their lengths (equation 16)
k_6	a constant in an expression that is an exponent in equation 16 which relates terminal speeds of a set of clams to their lengths
l	characteristic length of an object used in the calculation of Re
μ	the dynamic viscosity of sea water
P_e^2	the root mean square sound pressure
ρ	density of medium
ρ_c	density of the clam
ρ_s	density of a sphere
ρ_w	density of sea water
Re	Reynolds number
SPL	sound pressure level
T	wave period
u	flow speed
u'	speed of the bore
u_0	the speed of the swash immediately after bore collapse
u_t	terminal velocity of a sphere
$u_{t,low\ Re}$	terminal velocity of a sphere at low Re
V	sand fall speed
v_t	clam's terminal sinking velocity

11

Crab claws as tools and weapons

R.N. Hughes

1. Introduction

Crab claws assume major importance as tools and weapons used in the essential processes of feeding and reproductive behaviour. As crab claws are amenable to bio-mechanical investigation and crabs themselves make good behavioural subjects, the system has become an important experimental model for studies of optimality in behaviour and morphology. The purpose of this review is to illustrate these principles using appropriate examples, rather than to be comprehensive.

A relatively simple evolutionary development is required to convert a crustacean walking limb into a claw. The distal segment (dactylus) of the walking leg articulates about a pivot on the adjacent, penultimate segment (propodus). Flexion and extension are achieved by two antagonistic muscles, and nervous control involves stretch receptors lying in an elastic membrane spanning the articulation. Modification to a claw is achieved by growth of a finger-like protuberance of the propodus to lie alongside the dactylus. Flexion and extension now move the dactylus against and away from the rigid protuberance (pollex) in the manner of a pincer.

2. Claws as tools

As tools, claws serve principally for gripping, squeezing, cutting and tearing, but in specialized cases also for scraping, digging, or resisting displacement by hydrodynamic forces (Blake, 1985). Biomechanical characteristics of gripping, squeezing and cutting can be understood in terms of muscular and skeletal morphology of the claw. Matching of these characteristics to tasks that in most natural situations entail risks of failure (Hughes, 1980) or even damage (Juanes, 1992) involves neurological, physiological and behavioural control, which also therefore need to be understood.

2.1 *Biomechanical characteristics*

Muscular morphology. Because crustacean muscles are enclosed within a rigid exoskeleton, they must not swell on contraction and so they are of the pinnate form

Biomechanics in Animal Behaviour, edited by P. Domenici and R.W. Blake.
© 2000 Taylor & Francis, Oxford.

where, instead of running in parallel between two planes of insertion, the fibres are inserted at an angle and staggered along an axial plane (apodeme) in a feather-like arrangement. As the fibres contract, the angle of pinnation increases, without concomitant swelling of the muscle. Moreover, pinnate muscles can exert twice as much force as parallel-fibred muscles of similar size (Alexander, 1968). The flexor muscle of the crustacean walking limb therefore is preadapted for the role of a powerful, claw closer muscle, requiring only an exaggeration in size and an associated increase in volume of the surrounding exoskeleton (manus) (*Figure 1(a)*). The extensor muscle retains a relatively small size consistent with the modest force needed to open the claw. The closer muscle contains fast and slow fibres in a ratio that varies according to the precise function of the claw.

Exoskeletal morphology. For a given size of closer muscle, squeezing potential depends on the mechanical advantage of the lever system within the claw. This mechanical advantage can be approximated as the ratio of the distance ($L1$) between the pivot and point of attachment of the closer-muscle apodeme to the distance ($L2$) between the pivot (fulcrum) and the point where the force is applied along the dactylus (*Figure 1(a)*). The point of force application may be located at the tip or more proximally along the dactylus. Distal location results in lower mechanical advantage and lower applied force, but achieves greater and faster movement that could be useful in catching mobile prey, cutting soft material or fending off enemies. Among families of molluscivorous crabs, 'distal mechanical advantage' ($L2$, i.e. distance between pivot and tip of claw, *Figure 1(a)*) is in the order of 0.2–0.5 (Seed and Hughes, 1995). Proximal location ($L2$, i.e. distance between pivot and apex of proximal molar, e.g. *Figure 1(c)*) has the opposite results, with mechanical advantage in the order 0.7–1.3 (Blundon and Kennedy, 1982; Hughes and Elner, 1989) and so is useful in crushing hard-shelled prey. These divergent properties are reflected by exoskeletal morphology, which sometimes differs between left and right claws (claw dimorphism) and frequently differs among species of crab.

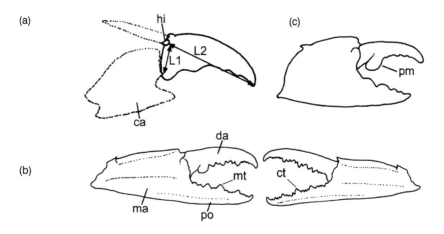

Figure 1. (a) *Skeletal morphology of the crab claw. ca, closer apodeme; pi, pivot (fulcrum). Mechanical advantage at the claw tip is L1/L2. (b) Claw dimorphism in* Callinectes sapidus. *Right, minor claw; left, major claw; ct, cutting teeth; da, dactylus; ma, manus; mt, molariform teeth; po, pollex. (c) Major claw of* Ozius verreauxii. *pm, proximal molar.*

The blue crab, *Callinectes sapidus* Rathbun, exemplifies claw dimorphism. As is usually the case among species showing dimorphism, the right claw is more massive than the left (*Figure 1(b)*). The greater volume of the manus of the right (major) claw accommodates a larger closer muscle and the concomitantly greater distance between pivot and attachment of the closer muscle apodeme increases mechanical advantage. The major claw therefore is capable of squeezing with greater force than the left (minor) claw. Appropriately, the dactylus and pollex of the major claw are relatively stout and armed with directly opposing molariform teeth suitable for crushing hard-shelled prey (*Figure 1(b)*). The slenderer shape of the minor claw is associated with a smaller closer muscle and a distal mechanical advantage some 23% lower than that of the major claw (Seed and Hughes, 1997). Greater movement and speed of closure facilitate snapping and cutting. Accordingly, teeth of the minor claw form slightly offset, opposing blades, suitable for shearing (*Figure 1(b)*). Degree of claw dimorphism, however, varies among species, from virtual absence as in the edible crab, *Cancer pagurus* (L.), through moderate expression as in swimming crabs of the family Portunidae, to the exaggerated state seen for example in fiddler crabs, *Uca* spp. Within species, claw dimorphism tends to be more pronounced in males than females, indicating that its expression may reflect behaviourally driven sexual selection additionally, or preferentially, to specialized mechanical functions (Lee, 1995; see also Section 3.2).

In addition to relative size, the major claw also varies among species in shape and dentition, reflecting differing degrees of specialization for crushing. A feature variously expressed among species within several families of crab and therefore showing repeated, independent evolution, is an exaggerated blunt tooth (proximal molar) near the base of the pollex (*Figure 1(c)*). Mechanical advantage at the proximal molar is considerably greater than at the tip of the claw, allowing correspondingly greater force to be applied. In *C. sapidus*, for example, proximal mechanical advantage is about four times the distal mechanical advantage. Accordingly, blue crabs make great use of the proximal molar for crushing molluscan prey (Blundon and Kennedy, 1982). In more specialized molluscivores, the proximal molar may become exaggerated and used with precision to break well-armoured shells (see Section 2.2).

Muscular activity pattern. Continuous recordings of forces applied by the major claw to objects furnished with strain gauges reveal characteristic patterns of muscular activity. Elner's classic (1978) study showed that *C. maenas* attempts to break open mussels, *Mytilus edulis* L., by repeatedly applying sub-maximal forces in bouts separated by periods during which the mussel is rotated to new positions (*Figure 2(a)*). Shell material consists of microscopic crystals of calcium carbonate embedded in an organic matrix that interrupts the spreading of fracture lines, making the shell resistant to compressive forces (Currey and Kohn, 1976). Consequently, for any but smaller molluscan prey (Blundon and Kennedy, 1982), maximum forces that could be generated by the claw usually are less than the critical force necessary to break the shell in a single application (static strength) as measured by a materials-testing machine (Boulding and LaBarbera, 1986; Currey and Hughes, 1982; Elner, 1978). On the other hand, the critical force sufficient to break the prey's shell may be of the same order as that sufficient to break the claw itself (Juanes, 1992) and, indeed, breakage of the dactylus occasionally has been witnessed in laboratory trials. In this regard, mechanical strength of the claw requires further study, using materials-testing technology. Coupled with periodic rotation of the shell, repeated application of weaker forces

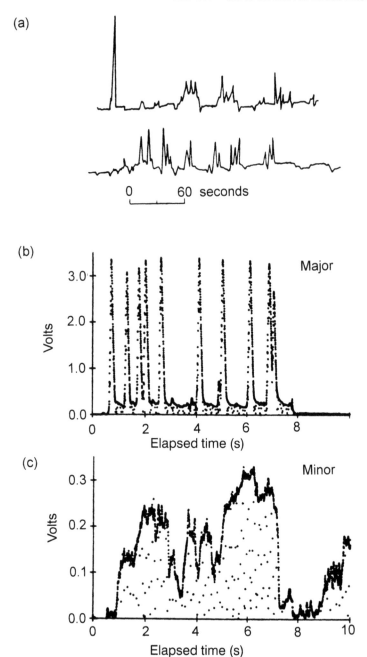

Figure 2. (a) *Time sequence of forces applied by the major claw in* Carcinus maenas. *Repeated applications are of submaximal force, as can be seen by comparison with the initial spike. Reproduced with permission from Elner (1978)* Oecologia **36**: 340, Fig. 7. Copyright Springer-Verlag 1978. (b) *Temporal pattern of force application by the major claw of* Uca pugnax. (c) *Temporal pattern of force application by the minor claw of* Uca pugnax. (b) *and* (c) *reproduced from Levinton et al (1995)* J. Exp. Mar. Biol. Ecol. **193**, pp. 153 and 155, with permission from Elsevier Science. The ordinates are arbitrary scales of applied force, from which temporal patterns, but not relative amplitude can be compared among graphs.

(fatigue loading; Sandor, 1972) inflicts damage in the form of accumulating microfractures (Boulding and LaBarbera, 1986; Elner, 1978; see also Chapter 17). When these reach a critical level the shell breaks (fatigue failure) and, although the process can take many minutes in the case of larger prey, there is little risk of damaging the claw. Apart from one exception described below, similar observations have yet to be made for the minor claw and it remains to be seen whether a different pattern of closer-muscle activity applies here.

Unlike the major claw closer muscle of molluscivorous crabs (above), the closer muscle of the minor claw in male fiddler crabs does not contract in repeated short bursts of submaximal force but shows more irregular activity with frequent, prolonged exertion of maximum force. The minor claw of male fiddler crabs and both minor and major claws of females are relatively small and delicate. Their function is to gather parcels of sediment and pass them to the mouth, involving irregular acts of prolonged, strong closure probably corresponding to the recorded activity pattern of the closer muscle (*Figure 2(c)*). The major claw of male fiddler crabs is discussed in Section 3.2.

Gripping, as opposed to squeezing, usually occurs for brief moments when manipulating food items or when grappling with opponents (Section 3.2), but sometimes occurs for prolonged periods as when a male snow crab, *Chionoecetes opilio* (O. Fabricius), carries his mate in readiness for copulation. The claw closer muscle of the male snow crab is specialized for continually gripping the base of the female's walking legs throughout periods of several weeks. The excitatory neuromuscular synapses generate slow, sustained activity (Govind *et al.*, 1992), and the muscle fibres themselves have relatively long sub-units (sarcomeres), making contraction slow (Claxton *et al.*, 1994). Moreover, the activity levels of two enzymes, ATPase regulating speed of contraction and NADH-diaphorase influencing muscle fatigability, facilitate the prolonged contraction necessary for securely gripping the female against competitors during the precopulatory period (Claxton *et al.*, 1994).

2.2 Behavioural characteristics

Behavioural applications of the claw to break into hard-shelled prey can be classified as follows (Elner, 1978; Kaiser *et al.*, 1993): *rotation*—the prey is held in the claws and turned into a new position; *crushing*—the prey is squeezed by the major claw; *sawing*—the teeth along the dactylus or pollex of the major claw are drawn across the prey; *poking*—the tip of the dactylus or pollex of the major claw is thrust at the prey.

Presenting *C. maenas* with plastic models of different shapes and sizes, each containing a standard amount of mussel tissue, reveals the development of orderly behavioural patterns as crabs gain experience of these artificial prey (Kaiser *et al.*, 1993). When first encountering the models, crabs behave in an exploratory manner, using the full repertoire of handling techniques. Consequently, there are many transitions from one technique to another and handling time is relatively long. Later, fewer transitions occur (*Figure 3(a)*) as crabs learn to favour the more effective techniques. Consequently, handling time shortens and foraging becomes more efficient (Cunningham and Hughes, 1984). Learning to apply the appropriate handling techniques will be important in most natural situations, where crabs encounter diverse types of prey (Section 2.3).

Measurements of linear size and mass of molluscan shells explain only some 65% of variance in static strength or tensile compliance, the remainder probably being due to wear or loading history (Boulding and LaBarbera, 1986; Pennington and Currey,

(a)   (b)

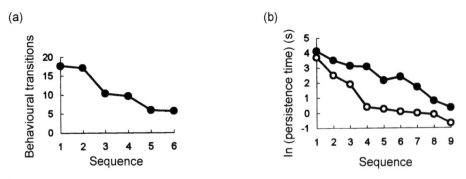

Figure 3. *(a) Number of behavioural transitions made by* Carcinus maenas *when attacking serially presented model mussels. Modified from Kaiser et al. (1993). (b) Logarithmically transformed persistence time of* Callinectes sapidus *attacking serially presented model mussels.* ●, *models presented alone;* ○, *models presented alternately with live mussels. Modified from Hughes and Seed (1995).*

1984). Resistance to fatigue loading will be even more unpredictable except in the unlikely case that minute increases in compliance can somehow be detected (Boulding and LaBarbara, 1986). Crabs therefore should persist with each new attack until the shell breaks or until a certain threshold time (persistence time) is reached. If vulnerable, energetically rewarding prey are plentiful, persistence time should be shorter, becoming longer in the converse situation. For example, when *C. sapidus* are presented with unbreakable model mussels in continuous sequence, persistence time decays exponentially as crabs habituate to the unrewarding stimuli associated with the models. When models are presented alternately with vulnerable, live mussels of similar size and shape, persistence time decays faster (*Figure 3(b)*). Persistence time hence is a dynamic variable resulting from the interaction between habituation and reinforcement, which themselves depend on the availability and quality of prey items. Opportunistic feeding based on a responsive persistence time therefore maximizes rate of energy gain while foraging in variable ecological circumstances (Hughes and Seed, 1995).

In more predictable circumstances, shell-breaking tools and associated behaviour may become more finely tuned over evolutionary time (Vermeij, 1978). The tropical crab, *Ozius verreauxii* Saussure, bears a highly developed proximal molar on the pollex of the major claw. The proximal molar (*Figure 1(c)*) is applied to the flat shelf found next to the aperture of neritid snails that make up much of the diet (Hughes, 1989). The rounded apex of the proximal molar concentrates the applied force to a small area of the shelf, strongly stressing the shell material and so promoting fatigue damage (Boulding and LaBarbera, 1986). Application of the proximal molar in this way, however, requires the shell to fit well into the angle of the claw, so excluding prey beyond a rather narrowly defined critical size. *O. verreauxii* gauges the size of each prey item by revolving the shell in the mouthparts for about 15 s, and rejects those exceeding the critical size. The biomechanical and behavioural mechanisms of prey selection are precise, in contrast to the highly opportunistic mechanisms seen, for example, in *C. sapidus* (above). In other cases, specialized feeding in molluscs may be less finely tuned than in *O. verreauxii*, but is still characterized by a robust claw, molar-iform dentition, powerful closer muscle and relatively high mechanical advantage (Behrens Yamada and Boulding, 1998).

2.3 *Application to a fundamental problem*

Knowledge of the biomechanical and behavioural principles involved when crabs use their claws as tools could generate new insight into the learning of multiple physical skills. Animals capable of learning procedural tasks, such as handling prey during an attack, will often be faced with a succession of different problems. Does learning to solve one problem interfere with or enhance learning to solve another? Theoretically, the answer should depend on the dissimilarity of the problems (Holding, 1976). If widely dissimilar, learning should be impaired if skills developed for one problem hinder performance when transferred to the other (negative transfer). For example, C. maenas presented with mussels before and after a period of training show pronounced improvement in performance (shorter shell-breaking time) when trained on mussels, no improvement when trained on fish requiring no breaking techniques and show worst performance when trained on periwinkles, *Littorina littorea* (L.), requiring different breaking techniques from mussels (*Figure 4*). The worsening performance after training on periwinkles compared with the neutral effect of training on fish is due to negative skill transfer, or interference *sensu* Cunningham and Hughes (1984). Positive skill transfer (Holding, 1976) is expected to occur when manipulative tasks become more similar; for example, learning to break open one type of mollusc might develop skills that can be applied to others of similar morphology. This still awaits experimental investigation.

The general problem of skill transfer could be studied rigorously using the crab–mollusc system. To do so, it will be necessary to relate biomechanical properties of contrasted molluscan shells to those of the crab claw. It should then be possible, for experimental purposes, to choose prey species or construct models requiring similar or dissimilar shell-breaking skills.

3. Claws as weapons

3.1 *Claws as signals of fighting potential*

Crabs use their claws defensively and offensively. Offensive usage occurs during fights over food or mates. Typical of many animal conflicts (Huntingford and Turner, 1987; but see Chapter 17), fights between crabs consist of behavioural acts in which weapons are used to signal motivational state and strength, hence fighting ability, rather than to inflict damage. A few stereotyped components account for most fighting behaviour (Huntingford et al., 1995), notably 'spread claws' in which both claws are held open and laterally outstretched while facing the adversary (*Figure 5(a)*). Physical trials of strength do sometimes occur and, indeed, to function as credible signals, the weapons must be capable of delivering the threat.

Fighting ability is determined by claw strength, which in turn is correlated with overall body size. A tighter correlation exists, however, between strength and size of the claw itself. Size of the claw rather than of the body, therefore, is the more informative signal of fighting ability, especially when opponents become more closely matched in size. This can be demonstrated by presenting crabs with model opponents constructed from empty exoskeletons. Large claws can be fastened to small bodies and vice versa, allowing body size and claw size to be varied independently (Lee and Seed, 1992). Models with large claws and small bodies suppress aggressive responses and stimulate withdrawal of live *C. maenas* more effectively than those with small claws and large bodies (*Figure 5(b)*). Not surprisingly, therefore, the winner of a fight

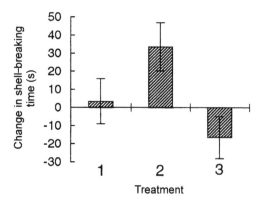

Figure 4. *Time taken by* Carcinus maenas *to break open mussels presented before and after training on similar (mussels) or dissimilar hard-shelled prey (periwinkles). Training on fish served as a control. Data are means and standard errors. ANOVA, treatment* $F_{2,18} = 4.65$, $P < 0.05$. *From Brown (unpublished MSc thesis).*

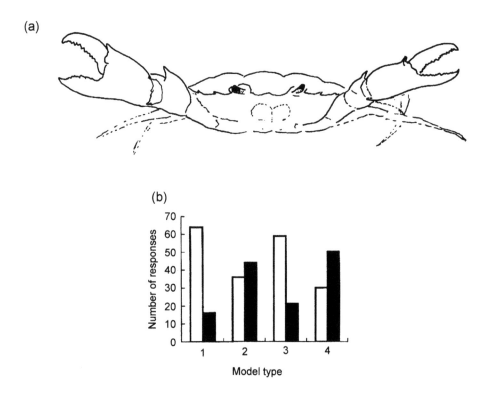

Figure 5. *(a) 'Spread claws' display, commonly used by crabs to threaten adversaries. (b) Numbers of aggressive and non-aggressive responses of* Carcinus maenas *presented with models varying in relative claw and body size. 1, large claw, large body; 2, small claw, large body; 3, large claw, small body; 4, small claw, small body. Filled bars, aggressive responses; unfilled bars, non-aggressive responses. Each of 80 crabs was presented with the four models and its response classified as aggressive or non-aggressive. Contingency* $\chi^2 = 43.6$, *d.f. = 3, P <0.001. From Hughes (unpublished data).*

between two live crabs can be predicted more accurately from its claw size than from its body size, relative to that of the opponent (Sneddon et al., 1997).

Although the correlation between claw strength and claw size is high, it is not perfect, and as the outcome of a fight is determined more by behavioural reponses to signals than by physical contact (Huntingford et al., 1995), claw strength may not always correspond to fighting success. For example, in the freshwater crab, *Potamon fluviatile* (Herbst), claw strength measured directly using a load cell was not significantly related to the outcome of a fight, whereas length of the claw was a good predictor (Gabbanini et al., 1995). In this anomalous case, the ancestral behavioural response to the signal, claw size, evidently persists even when the correlation between claw size and strength has become weakened. Further investigation of the biomechanical reasons for such weakening should explain the anomaly.

Of course, when opponents differ greatly in size, overall body dimension becomes a reliable signal of fighting ability, because smaller crabs are much weaker than larger ones. As expected, smaller C. *maenas* always avoid larger competitors (Cunningham, 1983).

3.2 Sexual dimorphism in claw size

Larger claws represent greater fighting potential, helping males to win contests over females and, once having won a female, to repel usurpers during the prolonged precopulatory period when the female becomes vulnerable before moulting (Abello et al., 1994; Sekkelsten, 1988). Large claw size therefore should be particularly advantageous in males, which could partially explain the tendency for the major claw to be relatively bigger in males than in females (Elner, 1980). Females may prefer males endowed with larger claws, the resulting process of sexual selection driving dimorphism to extremes in cases such as fiddler crabs (Christy and Salmon, 1984).

Biomechanics of the claw of male fiddler crabs. Exaggeration of the major claw in male fiddler crabs occurs through positive allometric growth (Huxley, 1924). In juveniles the major claw is the same size as the minor, but reaches half the total body mass in adults (Crane, 1975). Males attract females by waving the major claw (Oliveira and Custodio, 1998) and also use it to threaten or grapple with competitors (Levinton et al., 1995). Direct measurements of closing force in major and minor claws of *Uca pugnax* (Smith) closely match values predicted from mechanical advantage calculated from skeletal morphology and from cross-sectional area of the closer muscle (Levinton et al., 1995). Exaggeration of the major claw primarily as an organ of display therefore has not compromised its mechanical function.

Because the dactylus and pollex of the major claw become proportionately longer with increasing size in male *U. pugnax*, mechanical advantage declines but speed of movement increases. This is important during fights, as explained below.

Muscular activity of the male claw. Not only does the major claw of male *U. pugnax* share the same biomechanical relationships with normally proportioned claws, but it also shares similar patterns of muscular activity, the closer muscle exerting relatively low-amplitude pulses of force. This rhythmic activity (*Figure 2(b)*) and the predominance of slow fibres in the closer muscle are appropriate to the wrestling actions of contesting males, where grip must be taken rapidly at every opportunity (Levinton et al., 1995). Decreasing mechanical advantage accompanying elongation of the claw also contributes to the speed of gripping (above).

3.3 *Costs of weaponry*

Increased mating success of males resulting from endowment with a large major claw is gained at some cost. First, the major claw of a male fiddler crab contains about 25% of the total body musculature and so must account for a considerable share of the energy budget (Levinton and Judge, 1993). Second, the large master claw attracts predators, which consequently may preferentially attack male fiddler crabs over females (Lee and Kneib, 1994). For example, blue crabs preferentially attack male over female fiddler crabs, the larger claw of the male representing a substantial increase in prey profitability (Hughes and Seed, 1995).

4. Conclusions

Biomechanical considerations have greatly clarified our understanding of predatory and fighting behaviour of crabs, so the system has become an important model in fundamental research. Only a few types of crab have been studied in this way, however, and much further insight could be gained by exploiting the great diversity in claw morphology found among different taxa. With regard to prey-handling behaviour, biomechanical properties of the more specialized molluscivores (e.g. Hughes, 1989; Hughes and Elner, 1989) should be characterized and related to the biomechanical properties of their prey. This would inform the debate on the degree to which the shell morphology of tropical molluscs reflects a co-evolutionary response to predation pressure exerted by specialized predators (Vermeij, 1978). With regard to sexual behaviour, further attention should be paid to biomechanical allometries (Levinton *et al.*, 1995), as this will be crucial in deciding on the evolutionary significance of trade-offs between competing functions, including usage of the claw as a tool, weapon or signal of mate quality (Krebs and Davies, 1993). Finally, as discussed in Section 2.3, characterization of biomechanical properties of claws and shelled prey may produce an excellent system for investigating the general phenomenon of skill transfer, important in psychological, behavioural and ecological contexts.

References

Abello, P., Warman, C.G., Reid, D.G. and Naylor, E. (1994) Chela loss in the shore crab *Carcinus maenas* (Crustacea, Brachyura) and its effect on mating success. *Mar. Biol.* **121:** 247–252.

Alexander, R.McN. (1968) *Animal Mechanics.* Sedgwick and Jackson, London.

Behrens Yamada, S. and Boulding, E.G. (1998) Claw morphology, prey size selection and foraging efficiency in generalist and specialist shell-breaking crabs. *J. Exp. Mar. Biol. Ecol.* **220:** 191–211.

Blake, R.W. (1985) Crab carapace hydrodynamics. *J. Zool. London A* **207:** 407–423.

Blake, R.W. (1986) Hydrodynamics of swimming in the water boatman, *Cenocorixa bifida. Can. J. Zool.* **64:** 1606–1613.

Blundon, J.A. and Kennedy, V.S. (1982) Mechanical and behavioral aspects of blue crab, *Callinectes sapidus* (Rathbun), predation on Chesapeake Bay bivalves. *J. Exp. Mar. Biol. Ecol.* **65:** 47–65.

Boulding, E.G. and LaBarbera, M. (1986) Fatigue damage: repeated loading enables crabs to open larger bivalves. *Biol. Bull. Woods Hole Mass.* **171:** 538–547.

Christy, J.H. and Salmon, M. (1984) Ecology and evolution of mating systems of fiddler crabs (genus *Uca*). *Biol. Rev.* **59:** 483–509.

Claxton, W.T., Govind, C.K. and Elner, R.W. (1994) Chela function, morphometric maturity, and the mating embrace in male snow crab, *Chionoecetes opilio*. *Can. J. Fish. Aquat. Sci.* **51**: 1110–1118.

Crane, J. (1975) *Fiddler Crabs of the World*. Princeton University Press, Princeton, NJ.

Cunningham, P.N. (1983) Predatory activities of shore crab populations. Ph.D. thesis, University of Wales, Bangor.

Cunningham, P.N. and Hughes, R.N. (1984) Learning of predatory skills by shorecrabs *Carcinus maenas* feeding on mussels and dogwhelks. *Mar. Ecol. Prog. Ser.* **16**: 21–26.

Currey, J.D. and Hughes, R.N. (1982) Strength of the dogwhelk *Nucella lapillus* and the winkle *Littorina littorea* from different habitats. *J. Anim. Ecol.* **51**: 47–56.

Currey, J.D. and Kohn, A.J. (1976) Fracture in the crossed-lamellar structure of *Conus* shells. *J. Materials Sci.* **11**: 1615–1623.

Elner, R.W. (1978) The mechanics of predation by the shore crab, *Carcinus maenas* (L.), on the edible mussel, *Mytilus edulis* L. *Oecologia (Berlin)* **36**: 333–344.

Elner, R.W. (1980) The influence of temperature, sex and chela size in the foraging strategy of the shore crab, *Carcinus maenas* (L.). *Mar. Behav. Physiol.* **7**: 15–24.

Gabbanini, F., Gherardi, F. and Vannini, M. (1995) Force and dominance in the agonistic behavior of the freshwater crab *Potamon fluviatile*. *Aggr. Behav.* **21**: 451–462.

Govind, C.K., Read, A.T., Claxton, W.T. and Elner, R.W. (1992) Neuromuscular analysis of the chela closer muscle associated with precopulatory clasping in male snow crabs, *Chionoecetes opilio*. *Can. J. Zool.* **70**: 2356–2363.

Holding, D.H. (1976) An approximate transfer surface. *J. Motor Behav.* **8**: 1–9.

Hughes, R.N. (1980) Optimal foraging in the marine context. *Oceanogr. Mar. Biol. Annu. Rev.* **18**: 423–481.

Hughes, R.N. (1989) Foraging behaviour of a tropical crab: *Ozius verreauxii*. *Proc. R. Soc. Lond.* B. **237**: 201–212.

Hughes, R.N. and Elner, R.W. (1989) Foraging behaviour of a tropical crab: *Calappa ocellata* Holthuis feeding upon the mussel *Brachidontes domingensis* (Lamark). *J. Exp. Mar. Biol. Ecol.* **133**: 93–101.

Hughes, R.N. and Seed, R. (1995) Behavioural mechanisms of prey selection in crabs. *J. Exp. Mar. Biol. Ecol.* **193**: 225–238.

Huntingford, F.A. and Turner, A.K. (1987) *Animal Conflict*. Chapman and Hall, London.

Huntingford, F.A., Taylor, A.C., Smith, I.P. and Thorpe, K.E. (1995) Behavioural and physiological studies of aggression in swimming crabs. *J. Exp. Mar. Biol. Ecol.* **193**: 21–39.

Huxley, J. (1924) Constant differential growth ratios and their significance. *Nature* **114**: 895–896.

Juanes, F. (1992) Why do decapod crustaceans prefer small-sized molluscan prey? *Mar. Ecol. Prog. Ser.* **87**: 239–249.

Kaiser, M.J., Hughes, R.N. and Gibson, R.N. (1993) The effect of prey shape on the predatory behaviour of the common shore crab, *Carcinus maenas* (L.). *Mar. Behav. Physiol.* **22**: 107–117.

Krebs, J.R. and Davies, N.B. (eds) (1993) *An Introduction to Behavioural Ecology*, 3rd Edn. Blackwell Scientific, Oxford.

Lee, S.Y. (1995) Cheliped size and structure: the evolution of a multi-functional decapod organ. *J. Exp. Mar. Biol. Ecol.* **193**: 161–176.

Lee, S.Y. and Kneib, R.T. (1994) Effects of biogenic structure on prey consumption by the xanthid crabs *Eurytium limosum* and *Panopeus herbstii* in a salt marsh. *Mar. Ecol. Prog. Ser.* **104**: 39–47.

Lee, S.Y. and Seed, R. (1992) Ecological implications of cheliped size in crabs: some data from *Carcinus maenas* and *Liocarcinus holsatus*. *Mar. Ecol. Prog. Ser.* **84**: 151–160.

Levinton, J.S. and Judge, M.L. (1993) The relationship of closing force to body size for the major claw of *Uca pugnax* (Decapoda: Ocypodidae). *Functional Ecol.* **7**: 339–345.

Levinton, J.S., Judge, M.L. and Kurdziel, J.P. (1995) Functional differences between the major and minor claws of fiddler crabs (*Uca*, family Ocypodidae, Order Decapoda, Subphylum Crustacea): a result of selection or developmental constraint? *J. Exp. Mar. Biol. Ecol.* **193**: 147–160.

Oliveira, R.F. and Custodio, M.R. (1998) Claw size, waving display and female choice in the European fiddler crab, *Uca tangeri*. *Ethol. Ecol. Evol.* **10**: 241–251.

Pennington, B.J. and Currey, J.D. (1984) A mathematical model for the mechanical properties of scallop shells. *J. Zool.* **202**: 239–263.

Sandor, B.I. (1972) *Fundamentals of Cyclical Stress and Strain.* University of Wisconsin Press, Madison, WI.

Seed, R. and Hughes, R.N. (1995) Criteria for prey-size selection in molluscivorous crabs with contrasting claw morphologies. *J. Exp. Mar. Biol. Ecol.* **193**: 177–195.

Seed, R. and Hughes, R.N. (1997) Chelal characteristics and foraging behaviour of the blue crab *Callinectes sapidus* Rathbun. *Estuarine Coastal Shelf Sci.* **44**: 221–229.

Sekkelsten, G.I. (1988) Effect of handicap on mating success in male shore crabs *Carcinus maenas*. *Oikos* **51**: 131–134.

Sneddon, L.U., Huntingford, F.A. and Taylor, A.C. (1997) Weapon size versus body size as a predictor of winning in fights between shore crabs, *Carcinus maenas* (L.). *Behav. Ecol. Sociobiol.* **41**: 237–242.

Vermeij, G.J. (1978) *Biogeography and Adaptation: Patterns of Marine Life.* Harvard University Press, Cambridge, MA.

12

Linking feeding behaviour and jaw mechanics in fishes

Peter C. Wainwright, Mark W. Westneat and
David R. Bellwood

1. Introduction

Research on the functional morphology of feeding in fishes has produced a large body
of work that characterizes general features of the musculoskeletal and mechanical
mechanisms used during feeding behaviours (e.g. Alexander, 1967; Bellwood and
Choat, 1990; Gillis and Lauder, 1995; Lauder, 1985; Motta, 1984; Muller *et al.*, 1982;
Osse and Muller, 1980; Wainwright and Turingan, 1993; Westneat, 1990). Workers
have identified elements of skull structure, neuromuscular patterns and key hydrodynamic features that are widespread among fishes and these insights have provided the
basis for our general understanding of the complex functional system that fishes
employ during feeding behaviour (Barel, 1983; Lauder, 1985; Liem, 1979). However,
among the roughly 25 000 living fish species there is great diversity in patterns of prey
use, foraging behaviour, prey capture behaviour, prey capture abilities and design of
the feeding mechanism (Motta, 1984; Sanderson and Wassersug, 1993; Wainwright and
Richard, 1995a; Westneat, 1994) and efforts have been made to identify principles of
how these aspects of fish feeding biology are interrelated.

In this chapter we focus on the link between mechanics of the lower jaw in bony
fishes and feeding behaviour. We summarize recent research that provides a
framework for interpreting the consequences of variation in jaw design for prey
capture kinematics and ultimately for feeding patterns in the wild. We begin with a
description of the fish jaw as a mechanical system and relate jaw design to the three
major feeding modes that have been identified in fishes. We show in a phylogenetically broad sample of fish species that a strong association exists between the
mechanics of the jaw, feeding mode and feeding habits. We then focus on research
that has asked specifically whether jaw linkage mechanics can be used to predict variation among species in kinematics of prey capture. We discuss two studies that have
attempted to link jaw mechanics to prey capture kinematics of fishes. The first study
tests predictions of prey capture speed, based on differences among three species of
centrarchid sunfishes in the scaling of jaw mechanics. The second study examines

---

Biomechanics in Animal Behaviour, edited by P. Domenici and R.W. Blake.
© 2000 Taylor & Francis, Oxford.

how well jaw mechanics accounts for differences in prey capture kinematics among 20 species of labrid fishes.

One aim of this chapter is to advocate the use of quantitative mechanical models in comparative studies of animal behaviour. Certain animal behaviours (e.g. locomotion, feeding, vocalization) are generated by complex neuromuscular systems that frequently lend themselves to being abstracted as mechanical networks of stiff skeletal links, soft connective tissues, and contracting muscles. One goal of a quantitative mechanical model is to account for the dominant physical properties of the system under consideration. By accomplishing this the model helps identify the key elements of design of a system that determine the details of forces and motion associated with the behaviour. Often these key elements of design are accessible morphological traits, such as various physical dimensions, that are easily measured. The great utility of mechanical models in comparative biology is that to the extent that a model is applicable across taxa, it provides a causal link between morphological diversity and behavioural diversity. We hope to illustrate this potential power of mechanical models by reviewing some of their applications to the feeding behaviour of fishes.

2. Jaw mechanics, feeding mode, and patterns of prey use

2.1 *The lower jaw as a mechanical system*

The linkage mechanics of the fish jaw have been recognized as an important source of functionally significant structural variation with implications for fish feeding behaviour. Included in this category would be the four-bar linkages (*Figure 1*) that have been proposed to govern the movements of lower jaw depression (Aerts and Verraes, 1974; Anker, 1974), upper jaw protrusion (Westneat, 1990), and hyoid depression (Muller, 1987, 1989, 1996). Attempts to test the last two mechanisms have met with considerable success, although the four-bar linkage proposed to mediate jaw depression does not appear to function during feeding in some groups (Westneat, 1990) and may have limited applicability. Four-bar linkages are complex lever systems and as models they provide a framework for understanding the consequences that

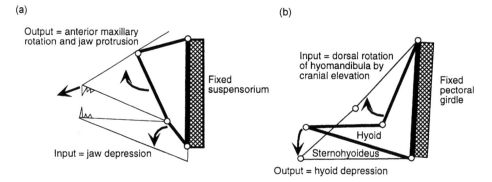

Figure 1. *Diagrams of the four-bar linkage systems that have been proposed to underlie upper jaw protrusion (a) and hyoid depression (b) in the teleost feeding mechanism. (a) As proposed by Westneat (1990), jaw depression results in rotation of the maxilla, which protrudes the upper jaw. (b) The hyoid four-bar linkage (Muller, 1987) uses head elevation to produce hyoid depression. This linkage is shown before and after 20° of hyomandibular rotation.*

specific morphological changes will have for skull motion during feeding behaviour. Hyoid depression is one of the principal mechanisms of buccal expansion and the generation of subambient pressures during suction feeding in fishes. Thus, the hyoid four-bar linkage is a promising model for interpreting the significance of fish structural diversity for feeding performance (Muller, 1996; Westneat, 1994).

The mechanics of the mandible, illustrated in *Figure 2*, have also received some attention, and it is here that we focus our attention. The mandible of teleosts is a single stiff skeletal link that rotates about the joint between the articular and quadrate bones. Jaw depression, as during mouth opening, is effected by the application of a posterior tension on the interoperculo-mandibular ligament, which attaches some distance ventral to the jaw joint, providing a moment arm for the jaw depression forces. Jaw adduction is provided by a complex of adductor mandibulae muscles that originate from the suspensorium and attach onto the mandible, at some distance dorsal to the jaw joint, thus providing the adductor muscle with a measurable moment arm for the jaw closing action.

The mechanical advantages (see also Chapter 11) of the jaw opening ligament and the adductor mandibulae muscle have been shown to vary considerably among species (Barel, 1983; Wainwright and Richard, 1995a; Westneat, 1994). This has implications for the strength and speed of jaw motions. In a simple lever system, such as the depression system of the mandible, the force that is realized at the outlever, in this case the tip of the toothed jaw, is given by the equation

$$F_o = F_i \, (L_i/L_o) \tag{1}$$

where F_o is force at the outlever, F_i is force at the inlever, L_i is the length of the inlever, and L_o is the length of the outlever. The important point that follows from this equation is that, assuming a constant value of force at the inlever, increasing the ratio of the inlever to the outlever will result in a net increase in the force that is realized at the outlever.

In the same lever system the velocity of movement at the jaw tip is given by

$$V_o = V_i \, (L_o/L_i) \tag{2}$$

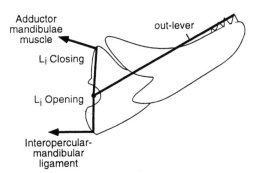

Figure 2. *Diagram of the mechanical model of the mandible referred to in this paper. Posterior tension on the interoperculo-mandibular ligament causes the mandible to rotate clockwise during mouth opening. The moment arm of this ligament is equal to the distance between the rotational centre of the jaw joint and the attachment site of the ligament. During mouth closing the adductor mandibulae rotates the mandible in the reverse direction and has a moment arm equal to the distance from attachment of the muscle on the mandible to the centre of the joint.*

where V_o and V_i are velocity at the outlever and inlever, respectively. It should be noted that velocity at the outlever is a function of the ratio of the outlever to the inlever, whereas force at the outlever is a function of the inverse of this ratio. The result is that force and velocity are simultaneously altered by changes in the ratio of inlever and outlever.

This model of the mandible has clear implications for the effect of variation in jaw shape on the relative speed and strength of mouth opening and closing, which are critical elements of fish feeding behaviour. The major implication of this model is the presence of a direct trade-off between transmission of force and velocity in fish feeding mechanisms.

2.2 *Three modes of prey capture in fishes*

Liem (1980) made the useful observation that fishes use three basic methods to capture prey: *suction feeding*, *ram feeding* and *manipulation*. These are methods that fishes use to trap prey in their jaws or buccal cavity. In many cases, additional behaviours and methods are subsequently used to separate prey from the water (i.e. suspension feeding), from their protective shell (i.e. crushing), or debris that is captured with the prey (i.e. winnowing). These processing methods typically involve branchial structures and take place after the prey has been captured (see discussion in Chapter 14). Here we focus on methods of prey capture in relation to jaw movements.

Suction feeding is associated with a rapid expansion of the oral cavity that generates a pressure gradient between the inside of the buccal cavity and the ambient water, causing water to rush into the opened mouth. Prey are entrained in this current and pulled into the mouth (Lauder, 1985). Suction feeding is believed to be the primitive mode of feeding in bony fishes (Lauder, 1980).

In *ram feeding*, the fish overtakes the prey with forward movement of the body or jaw. The key distinction between ram and suction feeding is whether the predator's mouth is thrust over the prey (ram feeding) or the predator generates a water flow that pulls the prey into the jaws (suction feeding). Ram may be accomplished by whole body movements or by rapid jaw protrusion (Motta, 1984). Classic ram feeders include animals such as the paddlefish, *Polyodon*, and the manta ray, *Manta*, that filter relatively tiny prey suspended in the water column (Sanderson and Wassersug, 1993) as well as predators like *Esox*, that feed on large elusive prey. In reality, ram and suction feeding appear to exist as the theoretical extremes of a continuous spectrum of the two modes (see the excellent discussion by Norton and Brainerd (1993)). Thus, the majority of species employing suction also incorporate forward body or jaw movements in their strike to help traverse the last few centimetres between predator and prey.

Manipulation is a broad category that incorporates a variety of behaviours involving the use of the oral jaws to directly grasp prey. Examples are many of the benthic-scraping coral reef taxa including some butterflyfishes (Motta, 1988), surgeonfishes (Winterbottom and McLennan, 1993), parrotfishes and wrasses (Randall, 1967), file-fishes and other tetraodontiform taxa (Randall, 1967). The benthic scraping cichlids fall into this category (Fryer and Iles, 1972). In this feeding mode the oral jaws are applied to the substrate (or directly to the prey) and a biting action is used to remove the prey from the substrate or break off pieces of a larger prey.

Virtually all fish species known use one of these three feeding modes or, more typically, a combination of at least two of them (Liem, 1980). This paradigm of feeding modes in fishes provides a useful framework to consider the relationship between the structural design of the jaw as discussed in the previous section and the feeding mode used by different taxa. There is a basic distinction between prey capture by manipulation and capture by either ram or suction. In manipulating taxa one might predict that the forcefulness of jaw adduction would be of greater relevance to feeding performance than the speed of jaw motion. In contrast, during ram and suction feeding speed of mouth opening and closing are likely to be more important elements of prey capture behaviour than the strength of these motions. Thus, one might expect to see larger ratios of inlever to outlever in the jaw closing system of manipulating taxa than in suction feeders or ram feeders.

Several studies have found a strong correspondence between jaw mechanics, feeding mode and patterns of prey use (Barel, 1983; de Visser and Barel, 1996; Wainwright and Richard, 1995a). Evolutionary changes in the mechanical advantage of jaw opening and closing were found to be significantly correlated with changes in diet within cheiline labrids (Westneat, 1995). Even across a broad phylogenetic sample of taxa a clear distinction exists between manipulating taxa and ram or suction feeders in the mechanical advantage of the adductor mandibulae muscle (*Figure 3*).

The observations of a strong ecomorphological correlation between linkage mechanics and diet in fishes suggests that variation in linkage mechanics should have

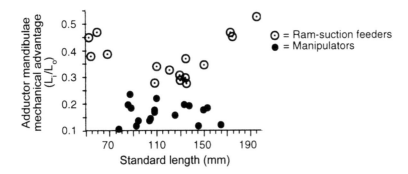

Figure 3. Mechanical advantage of the adductor mandibulae muscle in representatives of 34 species of teleost fishes illustrating the tendency for taxa that using a biting or manipulating behaviour during prey capture to have a higher force transmission of the adductor muscle than species that capture prey by ram or suction feeding. Species with the lower mechanical advantage gain more efficient velocity transfer of the adductor muscle. The species included in this analysis, listed in order of increasing jaw closing ratio are as follows. Rapid-strike suction feeders: Pempheris schomburgki (94 mm), Chromis cyanea (78.5 mm), Cynoscion nebulosus (142 mm), Hypoplectrus indigo (95.5 mm), Ocyurus chrysurus (102 mm), Cephalopholis fulva (103 mm), Mycteroperca bonaci (163 mm), Sphyraena barracuda (148 mm), Paranthias furcifer (126 mm), Holocentrus ascensionis (108 mm), Lutjanus grisseus (110 mm), Lutjanus analis (153 mm), Clepticus parra (92 mm), Epinephalus guttatus (88 mm), Haemulon sciurus (138 mm), Inermia vittata (135 mm), Sargocentron vexillarium (90 mm), Halichoeres pictus (107 mm). Biting and crushing species: Halichoeres garnoti (135 mm), Halichoeres maculipinna (105 mm), Halichoeres cyanocephalus (132 mm), Lachnolaimus maximus (132 mm), Halichoeres bivittatus (130 mm), Bodianus rufus (111 mm), Halichoeres radiatus (135 mm), Lagodon rhomboides (70 mm), Stegastes partitus (50 mm), Stegastes planifrons (63 mm), Melichthys niger (195 mm), Archosargus probatocephalus (123 mm), Sparisoma rubripinne (151 mm), Chaetodon ocellatus (56 mm), Balistes vetula (170 mm), Diodon hystrix (195 mm). Data from Wainwright and Richard (1995a).

predictable consequences for prey capture kinematics, but few studies have directly tested these predictions. The four-bar linkages of upper jaw protrusion and hyoid depression (*Figure 1*) have proven to be useful models of these elements of feeding kinematics (Muller, 1996; Westneat, 1990, 1991, 1994, 1995). In the following two sections we review results of our research, which explores the consequences of jaw mechanical design (*Figure 2*) for movements of the mandible during prey capture behaviour (Wainwright and Shaw, 1999).

3. Scaling of jaw mechanics in sunfishes

3.1 *Rationale and predictions*

Wainwright and Shaw (1999) asked if differences among species of centrarchid sunfishes in the lever systems for jaw depression and jaw adduction could be used to predict differences in feeding kinematics. They focused on the speed of two key movements that occur during prey capture—the time taken to open the mouth during the strike and the time taken to close the mouth. Observations were made on the large-mouth bass (*Micropterus salmoides*), the spotted sunfish (*Lepomis punctatus*), and the bluegill sunfish (*Lepomis macrochirus*). These species vary in external morphology as well as trophic biology, and the functional morphology of feeding in the largemouth bass and bluegill has been well studied (Gillis and Lauder, 1995; Grubich and Wainwright, 1997; Lauder, 1983; Nyberg, 1971; Richard and Wainwright, 1995; Wainwright and Lauder, 1986; Wainwright and Richard, 1995b).

Wainwright and Shaw (1999) assessed the influence of species differences in lever lengths by assuming that all other elements of the jaw opening and closing systems scale similarly in the three study species. Contractile properties of muscles and the scaling of forces that resist jaw movement were held constant, to generate predictions of interspecific differences in jaw kinematics based purely on measured differences in the scaling of the jaw levers. Given that these other features of the jaw opening and closing systems may be free to change during ontogeny and evolution, the central question of the study was to ask if differences between species in the scaling patterns of jaw opening and closing moment arms explain major differences between species in kinematics.

Let us consider the case of movement of the jaw during mouth opening for an individual fish (*Figure 2*). An input velocity (V_i) applied to the mandible through the interoperculo-mandibular ligament moves the jaw through an arc proportional to the distance of the inlever (L_i). As L_i increases, the input muscle must shorten a longer distance to cause the same angular rotation of the jaw. If we assume that V_i is a constant, then the time taken to fully depress the jaw (T_o) is directly proportional to L_i. Under these simplifying assumptions, T_o depends on L_i and is independent of jaw length, or the outlever of the jaw depression system:

$$T_o \propto L_i/V_i \tag{3}$$

As a fish grows, the scaling exponents of V_i and L_i will determine the scaling exponent of T_o. If V_i increases in direct proportion to the number of sarcomeres acting in series and the muscle grows isometrically, then V_i will scale directly with body or muscle length (i.e. the slope of a log–log plot of V_i on body length would be 1.0). If the shortening distance of the muscle (D_s) were to remain constant during growth, then T_o

would scale with an exponent of −1.0 (its scaling would be determined directly by the scaling of V_i). However, if L_i also scales isometrically, then D_s increases in proportion with body length. Thus, under this set of assumptions, if V_i and L_i both scale isometrically, the net effect will be that they cancel each other out and T_o will not change as the fish grows.

Following Equation (3), if we hold the scaling of V_i at isometry (exponent = 1.0), the scaling of T_o will increase from zero directly as the scaling of L_i increases above 1.0 and will decrease below zero as the scaling of L_i decreases below 1.0. In other words, the scaling exponent for T_o = (exponent for L_i −1.0). For example, if L_i scales to body length with a slope of 1.3, then T_o will scale with a slope of 0.3. If L_i scales with a slope of 0.8, then T_o will scale with a slope of −0.2. Similarly, changes in the scaling of V_i will have an inverse effect on the scaling of T_o. An increase in the slope of V_i will result in a decrease in the slope of T_o of the same magnitude.

To simplify predictions, the scaling of V_i can initially be assumed to be the same across species (in fact, as we shall see, this is a hazardous assumption) and quantitative estimates made of the consequences of variation between species in the scaling of L_i for the scaling of T_o. The scaling of T_o should change in the same direction and magnitude as the scaling of L_i. Wainwright and Shaw (1999) expressed predictions of the scaling of prey capture kinematics in the two sunfish species relative to patterns observed in the largemouth bass. A distinctive feature of the ontogeny of the largemouth bass is that growth is generally isometric and the scaling of prey capture morphology, kinematics and motor pattern are all well known (Richard and Wainwright, 1995; Wainwright and Richard, 1995a,b). Thus, for each of the two sunfish species the prediction was calculated as follows:

$$\text{scaling exponent of } T_o = \text{scaling exponent of } T_o \text{ in largemouth bass} +$$
$$(\text{scaling of } L_i \text{ in target species} - \text{scaling of } L_i \text{ in bass}). \quad (4)$$

In other words, the exponent that describes scaling of T_o was predicted to differ from that observed in largemouth bass by the difference in the scaling of L_i in the two species. The scaling of time to close the mouth (T_c) was calculated in a similar fashion. The predicted scaling relationships for time to open and close the mouth in the two sunfish species are given in Table 1.

3.2 Testing predictions of kinematic patterns

Feeding events were video recorded at 200 images s^{-1} and the times taken for the fish to completely open and close the mouth were measured for each successful strike (the reader is referred to Wainwright and Shaw (1999) for full methodological details). Data were collected from 1015 prey capture sequences from 10 largemouth bass, 22 spotted sunfish and 14 bluegill. T_o and T_c were taken from the fastest feeding sequence for each individual fish. Least-squares regression and analysis of covariance revealed that scaling of the lever systems varied among species (Table 1). Largemouth bass showed isometric relationships for both the jaw opening inlever and the mouth closing inlever. Scaling of both inlever measurements showed positive allometry in the two sunfishes, with the bluegill showing the steepest slopes in both systems (Table 1).

Although jaw morphology of the largemouth bass showed isometry in all anatomical measurements, the scaling of T_o (slope = 0.592) and T_c (slope = 0.572)

Table 1. Summary of predicted and observed scaling exponents for time to open the mouth (T_o) and time to close the mouth (T_c) based on the differences between species in scaling of jaw levers in centrarchid sunfishes

Variable	Slope for largemouth bass	Slope for spotted sunfish	Slope for bluegill sunfish
Opening inlever of lower jaw	1.00	1.183	1.248
Time to open mouth	0.592	0.745 (0.775 predicted) t (d.f. = 31) = 0.43, $P = 0.8$	0.854 (0.84 predicted) t (11) = 0.23, $P = 0.8$
Closing inlever of lower jaw	1.01	1.094	1.402
Time to close mouth	0.572	0.671 (0.656 predicted) t (31) = 0.17, $P = 0.9$	0.878 (0.964 predicted) t (11) = 0.86, $P = 0.4$

Predicted and observed slopes were compared with one sample Student's t-tests. Scaling exponents are based on least-squares regressions of $\log_{10}$ transformed data.

indicates a relative decrease in V_i in larger fish for both systems (*Figures 4* and *5*). The scaling exponent of T_o in the spotted sunfish and the bluegill differed from that seen in the bass (*Figure 4*). These differences were in agreement with those predicted by the scaling of the inlevers (*Table 1*). Scaling of T_c also showed a steeper slope in the two sunfishes than the bass, again closely matching the predictions based on scaling of jaw levers (*Figure 5, Table 1*).

The jaw model (Equation (3)) predicts that when V_i scales similarly in two species and the angular excursion of the jaw is constant, the difference in scaling of T_o will be equal to the difference in scaling of L_i. This prediction was borne out by the observed kinematics in these three species. Indeed, for the jaw closing system, interspecific differences in lever morphology account for almost all of the variation among species in T_c (lower panel of *Figure 5*). The species effect on the data in *Figure 5* is not significant. It is important to emphasize that this does not mean that the length of the jaw closing inlever is the only factor determining the time to close the mouth. Rather it suggests that at any given inlever length the other factors determining the rate of jaw movement are either the same across the three species, or they cancel each other in such a way as to yield a common value of V_i.

Variation in T_o is also largely accounted for by inlever length (top panel, *Figure 4*). However, in this case the largemouth bass shows a higher intercept value (species effect from an analysis of covariance $P < 0.002$). The implication of this result is that at all body sizes, the bass has a slower V_i for the jaw depression system than the two species of *Lepomis*.

This analysis reveals two ways in which the centrarchid feeding mechanism has been modified during evolution to produce the existing variation among species in prey capture behaviour. First, ontogenetic growth patterns of the mandible differ among species, resulting in species-specific relationships between the mechanical advantage of the adductor mandibulae and the interoperculo-mandibular ligament. These differences in the ontogeny of jaw levers result in differences in the ontogeny

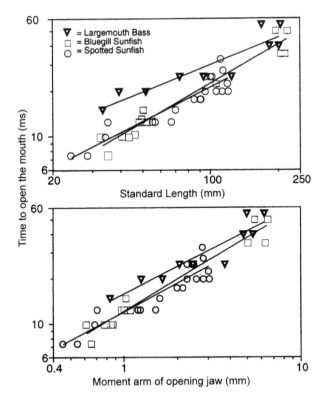

Figure 4. *Plots of the fastest times to open the mouth during prey capture in three species of centrarchid sunfishes. Each point in the graphs represents the fastest opening time for one fish. Differences between species in the scaling of time to open the mouth are eliminated when the kinematic variable is plotted against the moment arm of the interopercular-mandibular ligament (see* Figure 2). *Although a common scaling relationship is seen in the three species when mouth opening time is viewed against the moment arm of the jaw, at all sizes largemouth bass are slower than the other two species. Data from Wainwright and Shaw (1999).*

of the times to open and close the mouth during prey capture strikes. In addition, largemouth bass have a slower input velocity for opening the mouth at all body sizes than either of the sunfishes. This difference could have several sources, including the relative size of the mandible, the linkage involved in delivering jaw opening input velocity and the shortening speed of the jaw depression muscles (principally the sternohyoideus, but also possibly the epaxialis).

4. Diversity of prey-capture kinematics in labrid fishes

As a broader test of the applicability of the lever model of lower jaw mechanics we review data from 20 species in the family Labridae (wrasses and parrotfishes). Labrids exhibit exceptional diversity in trophic morphology, prey capture kinematics and patterns of prey use (Bellwood, 1994; Bellwood and Choat, 1990; Wainwright, 1988; Westneat, 1995), and provide a powerful test of the effect of jaw mechanics on the diversity of feeding behaviour.

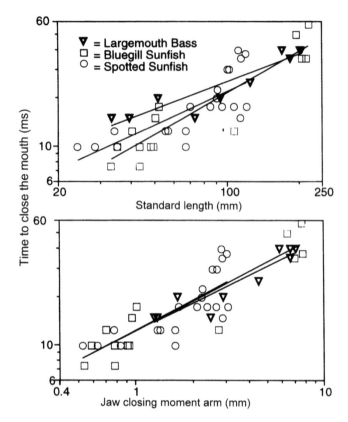

Figure 5. *Plots of the fastest times to close the mouth during prey capture in three species of centrarchid sunfishes. Each point in the two graphs represents the fastest mouth closing time for an individual fish. The scaling relationship of this kinematic variable with standard length differs among the three species. However, when scaled against the moment arm of the adductor mandibulae muscle a common scaling relationship and y-intercept is seen in the three species. Data from Wainwright and Shaw (1999).*

4.1 Predictions and tests

From Equation (2) and with the assumptions that the angular excursion of the jaw is a constant across species and V_i has a constant relationship across species, we predict a general relationship between T_o or T_c and the moment arm of the input muscle (L_i). In other words, we expect L_i to explain a large amount of the variation in mouth opening and closing times across a broad range of species.

Labrids were videotaped (300–1000 images s^{-1}) feeding on pieces of penaeid shrimp, and T_o and T_c were measured from the sequences. Data are shown here for one individual from each of 20 species of the Labridae, including four parrotfish and 16 wrasses, all native to the Great Barrier Reef of Australia. The fastest prey capture cycle recorded for each fish is reported from a data set of over 10 sequences per specimen.

Among the 16 wrasse species, the moment arms of the jaw depression and jaw adduction system explained about 60% of the variance in the times to open and close the mouth during prey capture (*Figure 6*). In neither case can this be explained as a simple consequence of increasing body size, as standard length explained less than

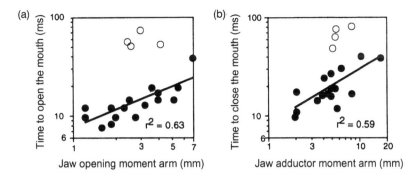

Figure 6. Plots of the fastest time to open (a) and close (b) the mouth for representatives of 16 species of wrasses (closed symbols) and four species of parrotfishes (open symbols) against the moment arms of the interoperculo-mandibular ligament and the adductor mandibulae muscle. Points represent the fastest mouth opening or closing time for an individual fish. Each species is represented by one individual. Least-squares regression lines and r^2 values are based on the wrasse data. Wrasse species shown in order of increasing jaw opening moment arm (rank in ascending order of jaw adductor moment arm is shown in parentheses): Coris dorsomacula *(1)*, Cirrhilabrus exquisitus *(has a longer opening time than previous species) (3)*, Cheilinus oxycephalus *(4)*, Coris aygula *(2)*, Thalassoma lunare *(5)*, Bodianus axillaris *(6)*, Hologymnosus doliatus *(7)*, Coris gaimard *(11)*, Bodianus mesothorax *(10)*, Gomphosus varius *(12)*, Bodianus loxozonus *(8)*, Thalassoma leutescens *(9)*, Oxycheilinus digrammus *(14)*, Novaculichthys taeniourus *(13)*, Cheilinus fasciatus *(15)*, Choerodon anchorago *(16)*; *Parrotfishes:* Cetoscarus bicolor *(1)*, Scarus schlegeli *(4)*, Scarus chameleon *(3)*, Scarus niger *(2)*.

30% of the variation in either variable (the regressions of fish standard length versus T_o and T_c gave $r^2 = 0.29$ and 0.26, respectively). The unexplained variation in the plots of *Figure 6* is due to two major sources. First, individual variation can be expected to be reasonably high in these data (see, for example the variation among individual centrarchids in *Figures 4* and *5*). Second, just as L_i may vary among species, changes in V_i would also lead to diversity of mouth movement times. Let us consider the jaw adductor muscle that shortens to provide V_i during mouth closing. Changes in the velocity of contraction of this muscle could be brought about by modifications to the muscle at several levels of design, including the angle of fibre pinnation, the isoform make-up of the actin–myosin complex, and the activity of various metabolic enzymes. This possible interspecific variation in the input velocities of fish jaw movements and their basis is a virtually unexplored area and yet clearly represents a potential focus of evolution. Nevertheless, the strong relationships between movement times and L_i for the two systems support the prediction that mechanical consequences of jaw design play an important role in shaping prey capture kinematics.

The jaw motions of the four parrotfishes clearly do not conform to the patterns observed in the 16 wrasse species. All four parrotfish species were much slower than expected based on the length of the input levers of their adductor muscle and jaw depression system (*Figure 6*). Recent studies show that parrotfishes have highly variable feeding motor patterns (Alfaro and Westneat, 1999) and that many taxa have an additional rotation point in the jaw as a result of a novel intramandibular joint (Bellwood, 1994). These factors, in addition to any modifications of the contractile properties of jaw muscles, probably influence the timing and motion of the jaws during the parrotfish bite. Our sample of four parrotfishes is inadequate to address whether a common relationship exists within parrotfishes, but it is clear that this group has diverged considerably from the pattern found in wrasses.

5. Discussion

5.1 *Linking morphological diversity with behavioural diversity*

Fish feeding mechanisms are complex mechanical networks of actuating muscles and skeletal elements linked together in such a fashion as to permit the explosive three-dimensional expansion of the mouth and buccal cavity that occurs during feeding behaviour. Mechanical models of hyoid motion (Muller, 1987, 1989, 1996; Westneat, 1990), upper jaw protrusion (Westneat, 1990), and lower jaw motion (this chapter) have been used as frameworks for interpreting the consequences of morphological variation among fish species for kinematic patterns during feeding. To the extent that these models provide accurate predictions of ontogenetic and interspecific differences in kinematics and allow us to mechanistically tie anatomy to feeding performance, they provide a valuable link between morphological and behavioural diversity.

Our studies of the relationship between jaw lever mechanics and jaw movement times in sunfishes and labrids suggest two summary points. First, within fairly closely related groups of species variation in jaw lever mechanics is a powerful predictor of variation within and among species in T_o and T_c during prey capture. Furthermore, the model provides a basis for interpreting variation in intraspecific allometry of kinematics as seen in the sunfish. Even among the morphologically diverse wrasses, input levers accounted for about 60% of the kinematic variation. Within these two groups of fishes it appears that the mechanical advantage of the jaw opening and closing muscles could be used as a rough indicator of the relative speed of mouth movements during prey capture.

However, the second major point to be drawn from these observations is that diversity in L_i is apparently not the only source of diversity in prey capture kinematics. Considerable variation can be inferred in the input velocity of jaw movements, most prominently in the difference between wrasses and parrotfishes. This pattern was also seen in the mouth opening time results for the sunfishes, which indicated a slower V_i in *Micropterus* at all body sizes, relative to the two *Lepomis* species. Although much of the variation among species in V_i may have its basis in linkage mechanics of more proximal systems, for example in the jaw depression mechanism, our observations of considerable diversity in V_i suggests an important limitation on inferring variation in kinematic parameters based only on the length of the jaw input levers. Although V_i and L_i are likely to covary in many instances, as we might expect that taxa with relatively rapidly shortening muscles may also have short L_i, it is nevertheless possible for these two determinants of the time course of jaw opening and adduction to be independently modified. L_i does account for 60% of the variation among species in the time course of jaw motion but the strikingly poor fit of parrotfish to the wrasse pattern highlights the potential for changes in V_i to contribute to kinematic diversity. Differences between parrotfish and wrasses substantially underestimate the differences between these groups in jaw movement times.

5.2 *Conclusions*

Teleost fishes are a morphologically and trophically variable group that represent about half of all living vertebrate species. In spite of the remarkable diversity of the group the feeding apparatus is largely based on a common mechanical design. The simplified linkage models that have been identified to function in fish skulls provide a framework for assessing the consequences that morphological evolution has for several key aspects of feeding behaviour, including the speed of jaw movement, the strength of the bite and the rate of suction generation. The key result that emerges from the research reviewed in this chapter is that within phylogenetic groups that have a common general design of the feeding mechanism, jaw lever mechanics can be used as a morphological index of differences in speed of the prey capture strike. Thus, moment arm variables could prove useful in a variety of ecomorphological analyses and other attempts to gain insight into the diversity of fish feeding behaviour. Mechanical models of fish skull motion show promise as revealing indicators of feeding behaviour.

5.3 *Issues for the future*

Research on the mechanical mechanisms that underlie feeding behaviour in fishes has provided a simplified way of evaluating key elements of the design of specific prey capture systems. By summarizing the major mechanical properties of the feeding mechanism these models offer a common scale along which many species can be evaluated. We highlight two general areas of future research aimed at understanding how fish feeding mechanisms are modified during evolution to produce trophic diversity.

First, comprehensive studies are still lacking in which all major determinants of jaw motion are evaluated in groups of closely related species. For example, the studies reported in this chapter are focused on the mechanical advantage of third-order lever systems of the jaw, and ignore the physiological properties of the muscles that provide the input motion for jaw movement. Synthetic analyses that incorporate all major levels of design will identify which aspects of feeding mechanisms are most frequently modified during evolution and which show patterns of conservation.

Second, by repeating in several fish lineages analyses of how feeding mechanisms are modified to produce trophic diversity it will be possible to identify generalities in how feeding mechanisms evolve. Some of the issues that could be addressed include the following. (i) Do some elements of design (e.g. jaw levers) repeatedly show themselves to be hotspots of modification in trophic radiation, whereas other levels are relatively conserved in spite of potential contributions to trophic diversity? (ii) Are there repeated sequences in trophic radiations that involve initial divergence in specific elements of the feeding mechanism, followed by secondary patterns of radiation? For example, do we see major shifts in mechanical properties of the jaw initially, followed by non-mechanical behavioural divergence in the more recent stages of the radiation? Recent progress in understanding the mechanical basis of fish feeding behaviour sets the stage for future research on their evolution.

References

Aerts, P. and Verraes, W. (1974) Theoretical analysis of a planar four-bar linkage in the teleostean skull. The use of mathematics in biomechanics. *Ann. Soc. R. Zool. Belg.* **114**: 273–290.

Alexander, R.McN. (1967) *Functional Design in Fishes.* Hutchinson, London.

Alfaro, M. and Westneat, M.W. (1999) Motor patterns of herbivorous feeding: electromyographic analysis of biting in the parrotfishes *Cetoscarus bicolor* and *Scarus iserti*. *Brain Behav. Evol.* **54**: 205–222.

Anker, G.C. (1974) Morphology and kinetics in the stickleback, *Gasterosteus aculeatus. Trans. Zool. Soc.* **32**: 311–416.

Barel, C.D.N. (1983) Towards a constructional morphology of the cichlid fishes (Teleostei, Perciformes). *Neth. J. Zool.* **33**: 357–424.

Bellwood, D.R. (1994) A phylogenetic study of the parrotfishes family Scaridae (Pisces: Labroidei), with a revision of genera. *Rec. Aust. Mus. Suppl.* **20**: 1–84.

Bellwood, D.R. and Choat, J.H. (1990) A functional analysis of grazing in parrotfishes (family Scaridae): the ecological implications. *Environ. Biol. Fishes* **28**: 189–214.

de Visser, J. and Barel, C.D.N. (1996) Architectonic constraints on the hyoid's optimal starting position for suction feeding of fish. *J. Morphol.* **228**: 1–18.

Fryer, G. and Iles, T.D. (1972) *The Cichlid Fishes of the Great Lakes of Africa*. TFH, Neptune City, NJ.

Gillis, G.B. and Lauder, G.V. (1995) Kinematics of feeding in bluegill sunfish: is there a general distinction between aquatic capture and transport behaviors? *J. Exp. Biol.* **198**: 709–720.

Grubich, J. and Wainwright, P.C. (1997) Motor basis of suction feeding performance in large-mouth bass (*Micropterus salmoides*). *J. Exp. Zool.* **277**: 1–13.

Lauder, G.V. (1980) Evolution of the feeding mechanism in primitive actinopterygian fishes: a functional analysis of *Polypterus, Lepisosteus*, and *Amia. J. Morphol.* **163**: 283–317.

Lauder, G.V. (1983) Prey capture hydrodynamics in fishes: experimental test of two models. *J. Exp. Biol.* **104**: 1–13.

Lauder, G.V. (1985) Aquatic feeding in lower vertebrates. In: *Functional Vertebrate Morphology* (eds M. Hildebrand, D. Bramble, K. Liem and D. Wake), Harvard University Press, Cambridge, MA, pp. 210–229.

Liem, K.F. (1979) Modulatory multiplicity in the feeding mechanism of cichlid fishes, as exemplified by the invertebrate pickers of Lake Tanganyika. *J. Zool.* **189**: 93–125.

Liem, K.F. (1980) Acquisition of energy by teleosts: adaptive mechanisms and evolutionary patterns. In: *Environmental Physiology of Fishes* (ed. M.A. Ali), Plenum, New York, pp. 299–334.

Motta, P.J. (1984) Mechanics and function of jaw protrusion in teleost fishes: a review. *Copeia* **1984**: 1–18.

Motta, P.J. (1988) Functional morphology of the feeding apparatus of ten species of Pacific butterflyfishes (Perciformes, Chaetodontidae): an ecomorphological approach. *Environ. Biol. Fishes* **22**: 39–67.

Muller, M. (1987) Optimization principles applied to the mechanism of neurocranium levation and mouth bottom depression in bony fishes (Halecostomi). *J. Theor. Biol.* **126**: 343–368.

Muller, M. (1989) A quantitative theory of expected volume changes of the mouth during feeding in teleost fishes. *J. Zool.* **217**: 639–661.

Muller, M. (1996) A novel classification of planar four-bar linkages and its application to the mechanical analysis of animal systems. *Philos. Trans. R. Soc. Lond. B* **351**: 689–720.

Muller, M., Osse, J.W.M. and Verhagen, J.H.G. (1982) A quantitative hydrodynamical model of suction feeding in fish. *J. Theor. Biol.* **95**: 49–79.

Norton, S.F. and Brainerd, E.L. (1993) Convergence in the feeding mechanism of ecomorphologically similar species. *J. Exp. Biol.* **176**: 11–29.

Nyberg, D.W. (1971) Prey capture in the largemouth bass. *Am. Midl. Nat.* **86**: 128–144.

Osse, J.W.M. and Muller, M. (1980) A model of suction feeding in teleostean fishes with some implications for ventilation. In: *Environmental Physiology of Fishes* (ed. M.A. Ali). Plenum, New York, pp. 335–351.

Randall, J.E. (1967) Food habits of reef fishes of the West Indies. *Stud. Trop. Oceanogr.* **5**: 665–847.

Richard, B.A. and Wainwright, P.C. (1995) Scaling the feeding mechanism of largemouth bass (*Micropterus salmoides*): I. Kinematics of prey capture. *J. Exp. Biol.* **198**: 419–433.

Sanderson, S.L. and Wassersug, R. (1993) Convergent and alternative designs for vertebrate suspension feeding. In: *The Skull, Vol. III. Functional and Evolutionary Mechanisms* (eds. J. Hanken, and B.K. Hall). University of Chicago Press, Chicago, IL, pp. 37–112.

Wainwright, P.C. (1988) Morphology and ecology: the functional basis of feeding constraints in Caribbean labrid fishes. *Ecology* **69**: 635–645.

Wainwright, P.C. and Lauder, G.V. (1986) Feeding biology of sunfishes: patterns of variation in the feeding mechanism. *Zool. J. Linnean Soc.* **88**: 217–228.

Wainwright, P.C. and Richard, B.A. (1995a) Predicting patterns of prey use from morphology with fishes. *Environ. Biol. Fishes* **44**: 97–113.

Wainwright, P.C. and Richard, B.A. (1995b) Scaling the feeding mechanism of the largemouth bass (*Micropterus salmoides*): motor patterns. *J. Exp. Biol.* **198**: 1161–1171.

Wainwright, P.C. and Shaw, S.S. (1999) Morphological basis of kinematic diversity in feeding sunfishes. *J. Exp. Biol.* **202**: 3101–3110.

Wainwright, P.C. and Turingan, R.G. (1993) Coupled vs uncoupled functional systems: motor plasticity in the queen triggerfish, *Balistes vetula* (Teleostei, Balistidae). *J. Exp. Biol.* **180**: 209–227.

Westneat, M.W. (1990) Feeding mechanics of teleost fishes (Labridae: Perciformes): a test of four-bar linkage models. *J. Morphol.* **205**: 269–295.

Westneat, M.W. (1991) Linkage biomechanics and evolution of the jaw protrusion mechanism of the sling-jaw wrasse, *Epibulus insidiator*. *J. Exp. Biol.* **159**: 165–184.

Westneat, M.W. (1994) Transmission force and velocity in the feeding mechanisms of labrid fishes (Teleostei, Perciformes). *Zoomorphology* **114**: 103–118.

Westneat, M.W. (1995) Feeding function and phylogeny: analyses of historical biomechanics in labrid fishes using comparative methods. *Syst. Biol.* **44**: 361–383.

Winterbottom, R. and McLennan, D.A. (1993) Cladogram versatility: evolution and biogeography of acanthuroid fishes. *Evolution* **47**: 1557–1571.

13

Biomechanics and feeding behaviour in carnivores: comparative and ontogenetic studies

Blaire Van Valkenburgh and Wendy J. Binder

1. Introduction

The predatory and feeding activities of carnivorous mammals tend to be relatively dramatic and consequently have been the focus of numerous biomechanical studies. The act of killing prey often involves considerable risk to the predator, as jaws, teeth and limbs can be loaded heavily in unpredictable directions. Most analyses of the biomechanics of carnivore feeding behaviour have not included direct behavioural observations. Instead they have been comparative osteological studies that have relied on the published literature for information on diet and/or predatory techniques. Frequently, the analysis of skull and dental dimensions has led to predictions about feeding behaviour in the wild, but these predictions have rarely been tested. For example, Biknevicius *et al.* (1996) suggested that the parabolic shape of the incisor arcade in canids as opposed to its more linear shape in extant felids might allow canids to use their incisors more selectively, but this has not been tested by observation. Of course, many of the predictions involve the relative magnitude and direction of the loads experienced by teeth, jaws, and skull, and these are impossible as yet to measure in the field.

In the last 5 years, there have been several new approaches to the study of the biomechanics of carnivore feeding behaviour. The first is the analysis of the ontogeny of adult feeding morphology (e.g. Biknevicius, 1996; Biknevicius and Leigh, 1997). As will be discussed below, these studies have highlighted several possible constraints on juvenile feeding performance, as well as documenting some compensatory adaptations exhibited by juveniles. Second, tooth blade sharpness has been quantified for the first time and examined across a variety of both herbivorous and faunivorous (consumers of invertebrates and vertebrates) mammals (Binder, 1998; Freeman and Wiens, 1997;

Biomechanics in Animal Behaviour, edited by P. Domenici and R.W. Blake.
© 2000 Taylor & Francis, Oxford.

Popowics and Fortelius 1997). Third, there have been two observational studies of carnivore feeding behaviour, one in the wild (Van Valkenburgh, 1996) and the other in captivity (Binder, 1998), with the latter focusing on the development of bite strength in growing juvenile spotted hyenas. Observation improves our understanding of function in part because animals do not always behave as expected, thereby forcing us to rethink our paradigms.

In this paper, we will first review briefly the past literature on the relationship between diet and/or killing behaviour and cranial and dental shape in carnivores, noting where the results from behavioural studies are relevant. Second, we will examine the more recent comparative work on the ontogeny of feeding adaptations in carnivores, including some new data on tooth sharpness. Third, we will describe a case study in which feeding biomechanics and behaviour are used to understand the life history characteristics of a species, the spotted hyena, *Crocuta crocuta*.

2. Cranial and dental morphology

Explorations of craniodental adaptations of carnivorans (members of the order Carnivora), have emphasized two aspects of feeding behaviour, diet and killing technique (Biknevicius and Van Valkenburgh, 1996). Not all carnivorans are highly carnivorous; diets vary from extreme herbivory (e.g. giant panda), to varying degrees of omnivory (e.g. foxes, bears), to complete carnivory (e.g. cats, some canids and hyenids). This variation is reflected in the dentition and much effort has gone towards the development of quantitative dietary indicators that can also be applied to extinct species. Killing technique has been studied similarly. Among those carnivorans that kill regularly, there are two fairly distinct killing techniques, each of which is expected to load skulls and teeth differently. Felids and mustelids tend to use a single strong killing bite, whereas canids and hyaenids are more prone to using multiple bites or biting and shaking of the prey (Ewer, 1973). These behavioural differences have led to predictions concerning the shape and strength characteristics of teeth, skulls, and jaws that have been tested by morphometric analysis.

2.1 *Diet*

The tooth row of carnivorans is typically composed of teeth that vary in size and shape, reflecting their different functions as piercers, grinders or slicers. Using the lower jaw of a canid as a model, there are incisors and canines anteriorly that can be used to threaten, groom, grasp prey or pick fruit (*Figure 1*). Directly posterior are the premolars, which function as piercers in most species but as bone crackers in hyenas (*Figure 1*). These are followed by the primary cutting tooth in the lower jaw, the first molar or carnassial. The carnassial of canids and most carnivorans is a large, multi-purpose tooth; the anterior portion acts as a cutting blade for slicing, and the posterior more basin-like portion functions as a grinding surface in combination with the rearmost molars, all of which tend to bear low rounded cusps (*Figure 1*, shaded areas). The grinding surface is assumed to be used for breaking up small bones, and invertebrate and plant foods, whereas the blade is assumed to function in slicing skin and muscle. Meat specialists such as the lion tend to have reduced or no grinding areas on their molars. More omnivorous species, such as foxes, exhibit relatively larger molars behind the carnassial. Thus, the relative importance of meat as opposed to plant foods

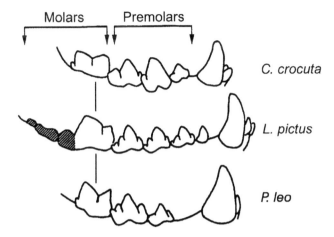

Figure 1. *Lateral view of the lower dentitions of three carnivorans: (top) spotted hyena (Crocuta crocuta), (middle) African wild dog (Lycaon pictus); (bottom) lion (Panthera leo). The vertical line indicates the position of the first lower molar or carnassial. The portion of the tooth row devoted to grinding function is shaded.*

in the diet of a carnivore can be estimated by the proportion of the lower molar row devoted to grinding as opposed to slicing function (Jaslow, 1987; Van Valkenburgh, 1989; Van Valkenburgh and Koepfli, 1993). These and other quantitative measures of teeth have been used successfully to examine resource partitioning in past and present suites of co-existing carnivorans (e.g. Dayan and Simberloff, 1996; Van Valkenburgh, 1989, 1991; Van Valkenburgh and Wayne, 1994; Werdelin, 1996a), as well as carnivorous marsupials (Jones, 1995, 1997).

Special attention has been paid to the habit of bone cracking because bone is an exceptionally tough food that is expected to require strong skulls and teeth. The spotted hyena and, to a lesser extent, the brown and striped hyenas are the living examples of habitual consumers of fairly large bones. Tooth fracture surveys of large predatory mammals revealed a relatively high incidence of fracture among hyenas, suggesting they may be working at the limits of tooth strength (Van Valkenburgh, 1988; Van Valkenburgh and Hertel, 1993). Indeed, studies of tooth microstructure have revealed that the enamel of hyena teeth has a very complex, zigzag organization that probably evolved to resist crack propagation (Rensberger, 1995; Stefan, 1997). In addition to hyenids, the mammalian fossil record provides us with several other examples of bone crackers within canids, creodonts, and marsupials (Werdelin, 1989, 1996b). These extinct species have been recognized as probable bone crackers because of their similarity to extant hyenas in having an enhanced mechanical advantage of the jaw closing muscles, relatively robust premolars, and deep mandibles that resist dorsoventral bending under load (Biknevicius and Ruff, 1992; Radinsky, 1981; Van Valkenburgh, 1989).

Most of the above analyses were based on largely untested assumptions about how carnivores feed. One possible exception is that by Jaslow (1987), who documented the digestive efficiency of two canids, grey fox and red fox, that differ in the slicing and grinding proportion of their tooth rows. Although she did not report on their feeding behaviour, she found that the species with the larger grinding area, the grey fox, digested fruit more thoroughly than did the red fox. However, it is not clear whether the increased efficiency is due to differences in oral processing and/or digestive enzymes.

Recently, Van Valkenburgh (1996) published a study of feeding behaviour in free-ranging large African carnivorans. The purpose was to test several assumptions concerning carnivore teeth. For example, it was assumed that spotted hyenas used their enlarged premolars for bone-cracking and that all carnivorous species used their blade-like carnassials to cut flesh. In general, these assumptions appeared correct but there were some surprises. In all four species that were observed (cheetah, wild dog, lion, spotted hyena), the carnassials were used primarily to cut skin and rarely for muscle or other softer tissues. Rather, these were swallowed with little or no chewing after being pulled from the carcass with incisors, or a combination of incisors and canines. As for bone-cracking, wild dogs behaved as expected, relying almost entirely on their post-carnassial molars. However, results on hyenas did not match the predictions. Although they did use their premolars predominantly, there were numerous observations of other teeth being used, including canines, incisors and especially carnassials. This lack of specificity of tooth use might explain why teeth other than the premolars exhibit complex, 'crack-stopping' enamel in spotted hyenas (Rensberger, 1995; Stefan, 1997). Thus the behavioural data provided some new insights into the function of previously defined biomechanical adaptations. The data suggest that carnassial design and evolution should be examined in view of cutting skin, a tissue with very different material properties than muscle, and that incisors deserve more attention given their heavy use (Van Valkenburgh, 1996).

2.2 Killing technique

The relationship between craniodental form and killing technique has been best studied in large placental carnivores (but see Jones (1995, 1998) on marsupials). Large canids, such as wolves, typically use multiple, slashing bites to subdue their prey. Felids, on the other hand, use a single strong killing bite in concert with grappling fore-limbs to hold prey immobile. The placement of the bite varies according to relative predator and prey size. Cats often use a dorsal neck bite for prey much smaller than themselves; when taking down much larger prey, they will use either a ventral throat hold or muzzle hold (Ewer, 1973; Kitchener, 1991). In either case, felid bite forces and the loads experienced by their jaws and canine teeth are expected to be larger and perhaps more unpredictable in direction than those of canids.

Although there have been no comparative studies of observed bite force in felids and canids, biomechanical analyses of their skulls and jaws suggest that felids can produce relatively more force for a bite with the canines. This is largely a consequence of the improved leverage of their jaw adductor muscles as a result of their shortened snouts (Maynard Smith and Savage, 1959; Radinsky, 1981; Thomasen 1991; Turnbull, 1970). Further biomechanical explorations of the canine teeth and lower jaws of carnivorans have used beam theory to test predictions concerning stress magnitude and direction. Consistent with expectation, the upper canine teeth of felids are rounder in cross-section, and thus more resistant to fracture from loads directed both mediolaterally and anteroposteriorly than are those of canids. Canids have much narrower, more knife-like upper canine teeth that are more vulnerable to mediolaterally directed loads, but more effective at producing slashing wounds (Van Valkenburgh and Ruff, 1987).

Interestingly, hyaenids have intermediate shaped upper canines despite the fact that they kill prey with multiple bites. As noted above, the robustness of all their teeth may reflect their bone-cracking habit, which is not confined to premolars alone.

When mammals bite with their canines, their lower jaw is loaded dorsoventrally, and the mandible is potentially deformed by bending. To resist these loads, dentary bones have an oval cross-section in which the dorsoventral dimension is greatest (Hildebrand, 1988). Bending rigidity can also be enhanced by thickening the cortical bone in the regions of greatest stress. Felids and hyaenids, both of which load their mandibles heavily, display enhanced cortical thickness of their dentaries relative to canids (Biknevicius and Ruff, 1992). Moreover, the jaws of felids are better buttressed for mediolateral stresses than are those of canids, presumably because the single strong killing bite of felids often produces torsional loads as the prey struggles (Biknevicius and Ruff, 1992). However, actual strain measurements during killing have not been made for any carnivoran. It would be very difficult and highly invasive to obtain such data on wild animals. Captive animals (e.g. cats, ferrets) might be induced to kill domestic mice and rats, but monitoring strain in the dentary during the act of killing would be difficult. At present we must infer function from anatomy, and our functional arguments remain as predictions.

3. Ontogeny and feeding adaptations

In most predatory mammals for which data are available, juveniles are weaned well before they acquire their adult dentition and body size (Binder and Van Valkenburgh, 2000; Hillson, 1986). Most often, they then feed on prey provided to them by their mothers until they are capable of hunting on their own. During this interval, they have to consume muscle, skin, organs and perhaps bones with smaller jaws and teeth than those of adults. In some social species, they may also have to compete with other juveniles and adults for food. Even in solitary species, there can be competition among littermates. Consequently, the question arises of how juveniles feed successfully with the apparently limited feeding equipment they possess. Moreover, how is function maintained during the changeover from juvenile to adult dentition and skull morphology? Do juvenile predators exhibit compensatory adaptations that mitigate the shortcomings of small size?

The ontogeny of craniodental biomechanics in carnivorans was first explored in a comparative analysis of two species, the cougar (*Puma concolor*) and the spotted hyena (*Crocuta crocuta*) (Biknevicius, 1996). On the basis of museum specimens of juveniles and adults, she quantified many of the features described above such as tooth shape, proportion of the carnassial tooth devoted to slicing function, and the location of maximum potential bite force. The deciduous cheek teeth of juvenile hyenas and cougars were more similar to one another in morphology than either was to the adult teeth that replace them. Moreover, they were less specialized for carnivory, suggesting that they are not as efficient at processing a carcass as those of adults. Using a model developed by Greaves (1983, 1985, 1988), Biknevicius further demonstrated that the region of maximum bite force production shifts anteriorly as juveniles of both species grow, and that the process of replacement of juvenile by adult dentition is likely to entail transient reductions in feeding efficiency. As will be discussed below, her results led to testable predictions about feeding behaviour in juveniles.

In a companion study, Biknevicius and Leigh (1997) documented dramatic changes in mandibular cross-sectional geometry with growth in the same two species. In both, juveniles were characterized by mandibles (dentary bones) that appeared to be as strong as or stronger in torsion than those of adults. This was accomplished by having dentaries that were much rounder in cross-section than those of adults, but with thinner cortical walls. Housed within these somewhat cylindrical, hollow dentaries of juveniles are of course the developing teeth, and this may explain dentary shape in part. Nevertheless, in species in which it has been examined, juvenile bone is less mineralized and thus less stiff than mature bone (Brear *et al.*, 1990; Currey and Pond, 1989). By expanding their dentaries externally, juveniles may maintain sufficient torsional strength despite the weaker bone.

The search for compensatory adaptations among juvenile carnivores was continued by Binder (1998), who undertook a broader comparative analysis of two aspects of feeding morphology, mechanical advantage of the jaw muscles and tooth sharpness. In her study of 20 species of canids, felids, hyenids and ursids, she found no evidence that juveniles have improved leverage of their jaw muscles for bites placed at either the carnassials or canines. In all but two species juveniles were at a disadvantage relative to adults. The two exceptions were the red fox (*Vulpes vulpes*) and the bush dog (*Speothos venaticus*) in which juvenile and adult values were similar for some measures. However, tooth sharpness showed a different pattern. Here, at last, juveniles appear to have a possible advantage.

Sharper teeth concentrate sites of contact and increase force per unit area. In a comparative study of bat canine teeth, Freeman and Wiens (1997) found that tooth sharpness declined with increasing body size. This same result was found by Popowics and Fortelius (1997) in a study of blade sharpness in a broad taxonomic sample of herbivorous and faunivorous mammals. As they pointed out, the sharper blades of smaller taxa allow them to employ lower bite forces in feeding on the same foods as larger species. Thus a hyrax and a rhinoceros are capable of feeding efficiently on the same tough plant foods.

Binder (1998) compared tooth sharpness between juvenile and adult carnivorans for three lower teeth, canines, premolars (adult fourth to juvenile third), and carnassial (adult first molar to juvenile fourth). These represent analogous pairs of teeth (*Figure 2*). Tooth sharpness was estimated by two measures, tip sharpness and whole tooth sharpness (*Figure 3*). Her results showed that in almost all cases,

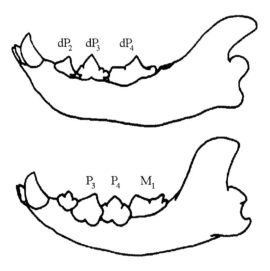

Figure 2. Lateral views of the lower jaws of a juvenile (top) and adult (bottom) spotted hyena, scaled to the same length.

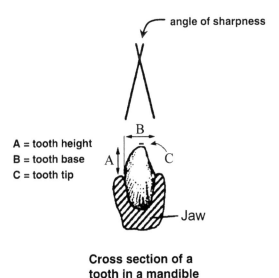

Figure 3. Two measures of tooth sharpness used by Binder (1998). Whole tooth sharpness was estimated using tooth height (A) and bucco-lingual breadth (B) at the tooth base to calculate a triangle. The angle of the apex of the triangle was used as the index of whole tooth sharpness. Tip sharpness was measured as the bucco-lingual distance across the flat occlusal portion of the tooth tip (C).

juveniles have significantly sharper teeth than do adults. Because tooth wear in juveniles had often blunted the tips slightly, the difference was more apparent in the measure of whole tooth sharpness. In addition to having sharper teeth, juveniles also tended to have more cusps and accessory cusps on their teeth. Bite force is concentrated at cusp tips, and thus additional sharp cusps should favour crack propagation in tough, fibrous foods such as skin and muscle. It seems that the sharp, pointed teeth of juvenile carnivorans may compensate somewhat for their reduced jaw musculature and lesser mechanical advantage. The question of how much they improve feeding performance might be approached by mechanically forcing tooth models of varying sharpness through muscle and skin, and quantifying the loads necessary.

Despite improved tooth sharpness, juvenile carnivorans are unlikely to be capable of feeding as efficiently as adults. Their jaw muscles are smaller and have less leverage, their teeth are less specialized than those of adults, and the bones that make up the skull and jaws are not as strong as those of adults. These factors are likely to contribute to slower overall rates of feeding, because of limits on the rate at which they can ingest food and limits on the size and type of foods they can rapidly consume. To discover how much less efficient juveniles might be and the rate at which they achieve adult morphology, we initiated the first observational study of the development of feeding behaviour in a carnivoran.

4. Development of bite strength and feeding performance in the spotted hyena, *Crocuta crocuta*

To understand the relationship between the developing skull and the acquisition of functional aptitude in feeding, ontogeny of the jaw and characteristics associated with feeding were closely examined in a single large carnivore, the spotted hyena (*Crocuta crocuta*), an extreme example of an animal adapted to a carnivorous lifestyle. They are social animals who hunt and kill like members of the family Canidae, yet whose cheek teeth and jaw characteristics more closely resemble those of members of the family Felidae (Biknevicius and Ruff, 1992; Ewer, 1973; Gittleman and Van Valkenburgh, 1997). In addition, they are distinguished by their ability to consume large bones regularly, an activity that places large strains upon their jaws (Kruuk, 1972; Mills, 1990). The development of adult levels of bite strength and feeding proficiency is important in young spotted hyenas. Weaned at around 12 months of age, they must compete side by side with adults for part of a carcass. Access to the carcass is largely determined by social rank within the clan (Frank, 1986; Kruuk, 1972), and a cub's rank depends on that of its mother (Holekamp and Smale, 1993; Smale *et al.*, 1993). This suggests that newly weaned sub-adults are not capable of obtaining sufficient food on their own, perhaps because their jaws and skulls are not fully developed. To investigate this, we wanted to measure performance changes with age, along with morphological measures that would give more detailed information about these performance measures.

To examine the relationship between feeding performance and morphology in growing hyenas, we tracked the development of nine young spotted hyenas in captivity for as much as 3.3 years. The hyenas were part of a colony of captive individuals housed at the Field Station for Biological Research at the University of California at Berkeley, California. All nine juveniles had been weaned at about 6 months, and their development was tracked for periods of 18–40 months, with data being collected at 8–12 week intervals. Morphological measurements were made in parallel with the behavioural observations. In addition to the nine juveniles, several adults of varying ages were tested for several measures. Bite force was measured with a force transducer similar to those used by previous workers (e.g. DeChow and Carlson, 1983; Oyen and Tsay, 1991; Thomasen *et al.*, 1990). Bite forces were recorded for unrestrained individuals who voluntarily bit the transducer device. Individuals that chose to bite the device were typically actively engaged in their efforts, and maximum forces obtained in a session were used. Feeding performance was assessed by bone-cracking behaviour in a series of 'bone-crunching tests'. Bone-crunching tests consisted of sessions during which a food deprived, isolated hyena was given a single, defleshed sheep femur of known weight and then its behaviour observed for 15 min. Feeding behaviour was quantified as per cent time spent using incisors, front premolars, or rear premolars, which was relatively easy to determine based on a consistent relationship between eye and tooth position. Feeding performance was estimated in two ways: first, the total weight of bone consumed in the 15 min bone test and, second, the time taken to finish a bone. The second measure of feeding performance was possible only for those cases in which the bone was completely consumed.

Our results are more fully described elsewhere (Binder and Van Valkenburgh, 2000), but relevant aspects are summarized here. Maximum bite force increased throughout the juvenile to sub-adult transition, and continued to increase until at least 5 years (60 months) of age, seeming to level off at about 6.5 years (80 months) (*Figure 4(a)*). All individuals showed this increase, but their individual trajectories could not be compared statistically because of missing data as individuals were not always co-operative. Notably, this is far beyond the age at which the adult dentition is fully erupted (12–14 months), and well beyond the age (20 months) at which our measures of skull size had peaked (*Figure 4(b)*).

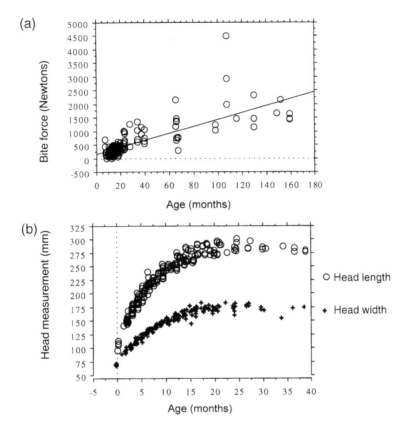

Figure 4. (a) Maximum measured bite force of spotted hyenas against age (months). Some individuals were sampled repeatedly as they grew; thus they are represented by more than one point in this graph. The equation for the least-squares regression line is y = 165.952 + 12.683 x, r² = 0.62, P < 0.0001. (b) Head length and width plotted against age (months) for nine growing spotted hyenas.

Feeding behaviour changed demonstrably with age. In our sample, the use of anterior teeth dropped dramatically after the adult teeth were fully in (~ 12–14 months), suggesting that their arrival altered feeding behaviour (*Figure 5*). Beyond 14 months of age, individuals exhibited much less variance in tooth use, as though they had learned how to efficiently break bones. The rearward shift in tooth use might be explained by alterations in jaw muscle mechanics with growth that favoured parallel changes in behaviour. As the jaw grows, the deciduous carnassials and premolars are replaced by permanent ones in slightly more posterior positions relative to the jaw joint (Biknevicius, 1996). Because the mechanical advantage of the jaw adductor muscles tends to be greater for bites placed nearer the jaw joint rather than farther away (Hildebrand, 1988; Turnbull, 1970), the relative rearward movement of the bone-crunching teeth (premolars) shifts the teeth into a more advantageous position, and the observed change in feeding behaviour reflects this.

Given that Biknevicius (1996) determined that the location of maximal bite force potential in the juvenile hyena jaw was at the most posterior premolars (the deciduous

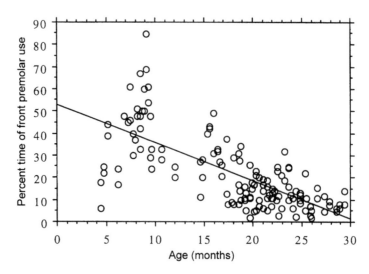

Figure 5. *Percentage of time young hyenas spent chewing with the anterior premolars in the 15 min bone-crunching tests. (Note the decrease in both magnitude and variance with age.) Regression line:* $y = 52.829 - 1.707\,x$, $r^2 = 0.5$, $P < 0.0001$.

carnassials) it is surprising that young juveniles did not always use these teeth for forceful biting. Although her conclusions are theoretically correct, the data presented here show an opposing trend, where juveniles initially preferred to use their front premolars, shifting to their rear premolars only as they age. Here is an example where behavioural observation can lead to a revised view of biomechanics. Juveniles may be unwilling or not able to use their rearmost teeth for bone cracking because of: (i) the need to protect the deciduous carnassials from fracture; (ii) limited gape size; (iii) the tendency to overstretch their jaw muscles. Bone crunching with the blade-like carnassials is likely to blunt them at least and fracture them at worst. Gape size is simply a function of jaw length; the longer the jaws, the greater the size of the object that can be placed between upper and lower teeth. The small jaws of juvenile hyenas must be opened very wide to accommodate most ungulate bones, especially if they wish to position the bone closer to the jaw joint. In opening their jaws so wide, the short jaw adductor muscles of juveniles will probably be overstretched, resulting in a significant loss of force-producing capability (Herring, 1975; Hildebrand, 1988). As their jaws and jaw adductor muscles lengthen with age, gape potential is increased and the problem of overstretching is lessened. Because the captive hyenas were fed similar-sized bones (sheep femora) throughout the study, they gradually positioned the bones more posteriorly in the region of near-maximum bite force as the constraints on gape size lessened with growth. Although the problem of gape limitation could have been realized based on skulls alone, it was not considered until contrary behavioural observations forced us to reconsider our explanations.

    In association with their greater bite strength and altered feeding behaviour, older individuals tended to consume more bone and consume it more rapidly than younger individuals. Although performance measures seemed to increase gradually with age, some individuals show a marked jump in speed of processing around 20–22 months of age, the time when skull growth reached a plateau. This follows by several months the eruption of the adult teeth and subsequent emphasis of rear premolars in feeding, and might imply that additional bite strength was needed to achieve higher performance levels. However, it is also possible that feeding behaviour may change as a result of learning, potentially indicated by a decrease in the variance of anterior premolar use after about 15 months of age (*Figure 5*), suggesting more directed feeding behaviour.

Our study of development in captive hyenas provides quantitative support for the presumed difficulties faced by young sub-adults in the wild. Before the eruption of the adult dentition, feeding behaviour was highly variable and bone-crunching ability was limited. A cub between the ages of 7 and 12 months would have difficulty dealing with the tougher parts of a carcass (skin, bones), and probably would be restricted to eating the more favoured parts, flesh and organs. Unless prey are exceptionally abundant, access to these is probably possible only if the female can dominate the carcass, preventing other adults from feeding. Low-ranked females cannot do this, and thus it is critical that their cubs be better equipped physically when they are weaned. Our data from the captive hyenas indicate that significant improvements in bone-crunching performance were apparent some 5–8 months after the typical weaning age of 12 months, probably as a result of both learning and growth. Given this, a newly weaned 12–13-month-old offspring of a low-ranking female would seem to be especially vulnerable to starvation in the wild. Unfortunately, there are not yet sufficient field data on mortality, age and social rank to test this idea.

5. Concluding remarks

Our understanding of the biomechanics of feeding behaviour in mammals can be improved by behavioural observation, as shown by this study as well as those of others (see German and Crompton 1996; Herring, 1985; Hiiemae and Crompton, 1985). It is certainly a two-way street; biomechanical analyses of jaw muscle leverage, craniodental strength, and tooth strength can yield predictions about behaviour, and behavioural observations can in return provide biomechanical hypotheses to be tested. The next level of integration is to place the biomechanics and behavioural combination into an ecological and evolutionary context. This was possible in the case of spotted hyenas because of the substantial number of superb field studies that have been published (see Frank, 1986; Holekamp *et al.*, 1997; Kruuk, 1972; Mills, 1990). There certainly are other mammal species, many of them primates, that have been similarly well studied, and they are excellent candidates for the integration of biomechanics, behaviour, ecology and evolution.

References

Biknevicius, A.R. (1996) Functional discrimination in the masticatory apparatus of juvenile and adult cougars (*Puma concolor*) and spotted hyenas (*Crocuta crocuta*). *Can. J. Zool.* **74**: 1934–1942.

Biknevicius, A.R. and Leigh, S.R. (1997) Patterns of growth of the mandibular corpus in spotted hyenas (*Crocuta crocuta*) and cougars (*Puma concolor*). *Zool. J. Linnean Soc.* **120**: 139–161.

Biknevicius, A.R. and Ruff, C.B. (1992) The structure of the mandibular corpus and its relationship to feeding behaviours in extant carnivorans. *J. Zool.* **228**: 479–507.

Biknevicius, A. and Van Valkenburgh, B. (1996) Design for killing: craniodental adaptations of predators. In: *Carnivore Behavior, Ecology, and Evolution, Vol. II* (ed. J.L. Gittleman), Cornell University Press, Ithaca, NY, pp. 393–428.

Biknevicius, A., Van Valkenburgh, B. and Walker, J. (1996) Incisor size and shape: implications for feeding behaviors in sabertoothed 'cats'. *J. Vertebr. Paleontol.* **16**: 510–521.

Binder, W.J. (1998) Functional aspects of tooth and jaw development in large carnivores. Ph.D. dissertation, University of California, Los Angeles.

Binder, W.J. and Van Valkenburgh, B. (2000) Development of bite strength and feeding behaviour in juvenile spotted hyaenas (*Crocuta crocuta*). *J. Zool.* (in press).

Brear, K., Currey, J.D. and Pond, C.M. (1990) Ontogenetic changes in the mechanical properties of the femur of the polar bear *Ursus maritimus*. *J. Zool.* **222**: 49–58.

Currey, J.D. and Pond, C.M. (1989) Mechanical properties of very young bone in the axis deer (*Axix axis*) and humans. *J. Zool.* **218**: 59–67.

Dayan, T. and Simberloff, D. (1996) Patterns of size separation in carnivore communities. In: *Carnivore Behavior, Ecology, and Evolution, Vol. II* (ed. J.L. Gittleman). Cornell University Press, Ithaca, NY, pp. 243–266.

DeChow, P.C. and Carlson, D.S. (1983) A method of bite force measurement in primates. *J. Biomech.* **16**: 797–802.

Ewer, R.F. (1973) *The Carnivores*. Cornell University Press, Ithaca, NY.

Frank, L.G. (1986) Social organization of the spotted hyena *Crocuta crocuta*. II. Dominance and reproduction. *Anim. Behav.* **34**: 1510–1527.

Freeman, P.W. and Wiens, W.N. (1997) Puncturing ability of bat canine teeth: the tip. In: *Life Among the Muses: Papers in Honor of James S. Findley* (eds T.L. Yates, W.L. Gannon, and D.E. Wilson). University of New Mexico, Albuquerque, NM, pp. 225–232.

German, R.Z. and Crompton, A.W. (1996) Ontogeny of suckling mechanisms in opossums. *Brain Behav. Evol.* **48**: 157–164.

Gittleman, J.L. and Van Valkenburgh, B. (1997) Sexual dimorphism in the canines and skulls of carnivores: effects of size, phylogeny, and behavioural ecology. *J. Zool.* **242**: 97–117.

Greaves, W.S. (1983) A functional analysis of carnassial biting. *Biol. J. Linnean Soc.* **20**: 353–363.

Greaves, W.S. (1985) The generalized carnivore jaw. *Zool. J. Linnean Soc.* **85**: 267–274.

Greaves, W.S. (1988) The maximum average bite force for a given jaw length. *J. Zool.* **214**: 295–306.

Herring, S.W. (1975) Adaptations for gape in the hippopotamus and its relatives. *Forma Functio* **8**: 85–100.

Herring, S.W. (1985) The ontogeny of mammalian mastication. *Am. Zool.* **25**: 339–350.

Hiiemae, K.M. and Crompton, A.W. (1985) Mastication, food transport and swallowing. In: *Functional Vertebrate Morphology* (eds M. Hildebrand, D.M. Bramble, K.F. Liem and D.B. Wake). Harvard University Press, Cambridge, MA, pp. 262–290.

Hildebrand, M. (1988) *Analysis of Vertebrate Structure*, 3rd Edn. John Wiley, New York.

Hillson, S. (1986) *Teeth*. Cambridge University Press, Cambridge.

Holekamp, K.E and Smale, L. (1993) Ontogeny of dominance in free-living spotted hyaenas juvenile rank relations with other immature individuals. *Anim. Behav.* **46**: 451–466.

Holekamp, K.E., Smale, L., Berg, R. and Cooper, S.M. (1997) Hunting rates and hunting success in the spotted hyena (*Crocuta crocuta*). *J. Zool.* **242**: 1–15.

Jaslow, C.R. (1987) Morphology and digestive efficiency of red foxes (*Vulpes vulpes*) and grey foxes (*Urocyon cinereoargenteus*) in relation to diet. *Can. J. Zool.* **65**: 72–79.

Jones, M. (1995) Guild structure of the large marsupial carnivores in Tasmania. Ph.D. dissertation, University of Tasmania, Hobart.

Jones, M. (1997) Character displacement in Australian dasyurid carnivores: size relationships and prey size patterns. *Ecology* **78**: 2569–2587.

Jones, M. (1998) Reconstruction of the predatory behaviour of the extinct marsupial thylacine (*Thylacinus cynocephalus*). *J. Zool.* **246**: 239–246.

Kitchener, A. (1991) *The Natural History of the Wild Cats*. Comstock, Ithaca, NY.

Kruuk, H. (1972) *The Spotted Hyena: A Study of Predation and Social Behavior*. University of Chicago Press, Chicago, IL.

Maynard Smith, J. and Savage, R.J.G. (1959) The mechanics of mammalian jaws. *School Sci. Rev.* **141**: 289–301.

Mills, M.G.L. (1990) *Kalahari Hyaenas: Comparative Behavioural Ecology of Two Species.* Unwin Hyman, London.

Oyen, O.J. and Tsay, T.P. (1991) A biomechanical analysis of craniofacial form and bite force. *Am. J. Orthod. Dentofac. Orthop.* **99**: 298–309.

Popowics, T.E. and Fortelius, M. (1997) On the cutting edge: tooth blade sharpness in herbivorous and faunivorous mammals. *Ann. Zool. Fenn.* **34**: 73–88.

Radinsky, L.B. (1981) Evolution of skull shape in carnivores. 1. Representative modern carnivores. *Biol. J. Linnean Soc.* **15**: 369–388.

Rensberger, J.M. (1995) Determination of stresses in mammalian dental enamel and their relevance to the interpretation of feeding behaviors in extinct taxa. In: *Functional Morphology in Vertebrate Paleontology* (ed. J.J. Thomason). Cambridge University Press, Cambridge, pp. 151–172.

Smale, L., Frank, L.G. and Holekamp, K.E. (1993) Ontogeny of dominance in free-living spotted hyaenas juvenile rank relations with adult females and immigrant males. *Anim. Behav.* **46**: 467–477.

Stefan, C. (1997) Differentiations in Hunter–Schreger bands of carnivores. In: *Tooth Enamel Microstructure* (eds W.V. Koenigswald and P.M. Sander). Balkema, Rotterdam, pp. 123–135.

Thomasen, J.J. (1991) Cranial strength in relation to estimated biting forces in some mammals. *Can. J. Zool.* **69**: 2326–2333.

Thomasen, J.J., Russell, A.P. and Morgeli, M. (1990) Forces of biting, body size, masticatory muscle tension in the opossum, *Didelphis virginiana. Can. J. Zool.* **68**: 318–324.

Turnbull, W.D. (1970) Mammalian masticatory apparatus. *Fieldiana (Zool.)* **18**: 147–356.

Van Valkenburgh, B. (1988) Incidence of tooth breakage among large, predatory mammals. *Am. Nat.* **131**: 291–300.

Van Valkenburgh, B. (1989) Carnivore dental adaptations and diet: a study of trophic diversity within guilds. In: *Carnivore Behavior, Ecology, and Evolution, Vol. I* (ed. J.L. Gittleman). Cornell University Press, Ithaca, NY, pp. 410–436.

Van Valkenburgh, B. (1991) Iterative evolution of hypercarnivory in canids (Mammalia: Canidae): evolutionary interactions among sympatric predators. *Paleobiology* **17**: 340–362.

Van Valkenburgh, B. (1996) Feeding behavior in free-ranging, large African carnivores. *J. Mammal.* **77**: 240–254.

Van Valkenburgh, B. and Hertel, F. (1993) Tough times at La Brea: tooth breakage in large carnivores of the late Pleistocene. *Science* **261**: 456–459.

Van Valkenburgh, B. and Koepfli, K. (1993) Cranial and dental adaptations for predation in canids. In: *Mammals as Predators, Symposia of the Zoological Society of London, 65* (eds N. Dunstone and M.L. Gorman). Oxford University Press, Oxford, pp. 15–37.

Van Valkenburgh, B. and Ruff, C.B. (1987) Canine tooth strength and killing behaviour in large carnivores. *J. Zool.* **212**: 379–397.

Van Valkenburgh, B. and Wayne, R.K. (1994) Shape divergence associated with size convergence in sympatric East African jackals. *Ecology* **75**: 1567–1581.

Werdelin, L. (1989) Constraint and adaptation in the bone-cracking canid *Osteoborus* (Mammalia: Canidae). *Paleobiology* **15**: 387–401.

Werdelin, L. (1996a) Carnivoran ecomorphology: a phylogenetic perspective. In: *Carnivore Behavior, Ecology, and Evolution, Vol. I* (ed. J.L. Gittleman). Cornell University Press, Ithaca, NY, pp. 582–624.

Werdelin, L. (1996b) Community-wide character displacement in Miocene hyaenas. *Lethaia* **29**: 97–106.

14

Predator–prey relationships in fish and other aquatic vertebrates: kinematics and behaviour

Robert S. Batty and Paolo Domenici

1. Introduction

The kinematic analysis of aquatic vertebrate behaviour is a powerful tool for under-standing predator–prey interactions. In this context, behaviour encompasses a number of levels from encounter rate and patterns of foraging to the processes of capture and escape. At its most simple, the capture and escape processes the attack of one indi-vidual upon another, while complex interactions can involve attacks by individuals or groups on aggregations such as fish schools. Studies of predator attack and prey escape in vertebrates have ranged in scale from fish larvae to whales.

Interaction between predators and their prey can be divided into a series of discrete stages (*Figure 1*) starting with an encounter between the two parties. We define encounter as the prey being within the range at which it can possibly be detected by the predator or vice versa. For this encounter to be relevant to either predator or prey detection must occur. Having detected a prey item, identification of the item as suitable may result in an attack, which may be followed by an attempt to escape – an escape response – by the prey. Finally, the process culminates in capture and ingestion or escape and survival. Not all prey detections will result in attack; the prey may be too large or too distant. More interestingly, the predator may adjust its attack according to prey type, behaviour or distance. Attack behaviour of a species can be variable; some fish predators can distinguish prey types and tailor the dynamics (velocity or acceler-ation) of their attack according to escape capability or distance to their prey. Encounter may also be altered following identification of prey by a predator; the predator may adopt an alternative foraging strategy and this may be on the basis of biomechanical efficiency. For example, the acceleration and manoeuvrability of verte-brates decreases with size and may explain why some large vertebrates have developed alternative feeding strategies that do not involve 'whole body' accelerations when

Biomechanics in Animal Behaviour, edited by P. Domenici and R.W. Blake.

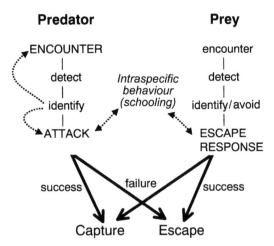

Figure 1. Processes involved in predator–prey interactions. Those stages at which behaviour and biomechanics interact are highlighted in upper case. The influence of intraspecific behaviour and prey identification on predator–prey interactions are also indicated.

capturing small and manoeuvrable prey. Killer whales 'tail-slapping' to stun fish while feeding on herring schools is a good example.

Temporal and spatial aspects of escape behaviour can be related to sensory capability, social context (schooling or solitary) and scale effects of both predator and prey. Spatial characteristics of responses vary depending upon the relative positions of predator and prey but the resulting escape trajectories are, in general, directed away from the predator. Intra-specific behaviour can also have a bearing upon the kinematics of attack and escape. Schooling fish must communicate and coordinate their escape responses both to avoid collisions and, if possible, to confuse the predator. To acheive this, the escape behaviour of fish within schools is modified temporally and spatially to allow co-ordinated responses.

We examine the kinematics and behaviour of predator–prey interactions in fish and other vertebrates by focusing on either the predator or the prey and therefore treat the process as either attack or escape response. The two are interrelated but for clarity we deal with them separately.

2. Attack

Fish and other aquatic vertebrates use a variety of methods for capturing their prey. A feeding method is selected by the predator to maximize profitability: minimizing energy expenditure and maximizing energy intake. The most profitable feeding method will depend upon the interacting effects of size, behaviour and density of the prey, together with the size and feeding apparatus of the predator. With a biomechanical approach, we will consider the three main classes of feeding method; filtering, particulate feeding (divided into ram and suction feeding), and other methods, including co-operative feeding and the use of weapons.

These categories correspond to different 'styles' of predation, rather than different feeding movements of the jaw, as presented in Chapter 12.

2.1 *Filtering*

This feeding method is used by many species of fish and also by whales. Kinematically it is relatively simple but it does rely on complex anatomical structures. Filtering depends upon a large size difference between predator and prey so that prey escape attempts do not result in a significant number of prey avoiding the trajectory of the predator's mouth. Filtering can indeed be defined by the predator being very much larger than the prey; prey organisms are not identified and captured individually (Weihs and Webb, 1983). The predator also needs a filtering apparatus; structures adapted for effective filtering may preclude other feeding modes. Although many fish species are either obligate filter feeders or biters some are facultative filter feeders in that they filter or bite depending upon prey density, type and/or light intensity. The herring (*Clupea harengus*), together with some other clupeids, is a facultative filter feeder that has been extensively studied (Batty *et al.*, 1986, 1990; Gibson and Ezzi, 1985, 1992; James and Findlay, 1989). Although feeding profitability of herring and other clupeid species has been studied and the size and density of prey at which filtering becomes profitable considered (Gibson and Ezzi, 1992), the energetic cost of filtering has not been measured. Metabolic rates while filtering in Cape anchovies (James and Probyn, 1989) and Atlantic menhaden (Durbin *et al.*, 1981) have been measured and compared with routine swimming but the actual cost of filtering has not been calculated. The proportion of time with the mouth open and volume filtered was not measured in these metabolic studies and therefore the cost cannot be estimated. Gibson and Ezzi (1992) demonstrated that if relative profitability can be measured, the prey concentration at which filtering becomes more profitable could be predicted (*Figure 2*). Yowell and Vinyard (1993) have gone some way towards this by measuring and comparing the increased metabolic costs of filtering and biting over routine swimming. To put real figures on this relationship further experimental work is required. In particular, an integrated study of filtering behaviour, combining energetics with kinematic analysis, would be required.

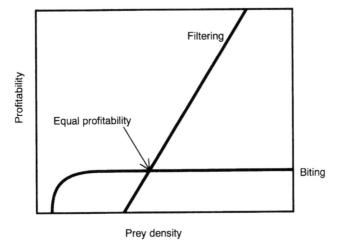

Figure 2. *The dependence of relative profitability of filter feeding and biting on prey density (adapted from Gibson and Ezzi, 1992).*

The profitability of filtering will not be simply the ratio of the cost of swimming while filtering to the energetic value of prey in the volume of water filtered. The probability of escape by the prey should also be considered. Likelihood of capture will increase as predator velocity and gape increase. It will be a function of the product of these two factors. Probability of escape by the prey will be a function of its escape speed and reaction latency. A detailed study using kinematic analysis and modelling would help to reveal the relationship between predator size, prey size, prey density and profitability.

The filtering apparatus within the buccal cavity, the gill rakers, and their structure development and function have been extensively studied. Gibson (1988) studied the particle retention capability of herring gill rakers, and van den Berg et al. (1992, 1993, 1994a–c) made extensive studies of the filtering mechanisms of bream (*Abramis brama*), white bream (*Blicca bjoerkna*) and roach (*Rutilus rutilus*). These studies culminated in a kinematic analysis of branchial and gill raker movements using X-ray cinematography to reveal the mechanism by which particles are retained during the transition from filtering to ingestion (Hoogenboezem et al., 1990; van den Berg et al., 1994b).

2.2 Particulate feeding

This is the method most commonly studied with a biomechanical or kinematic approach. Two case studies of the biomechanics of particulate feeding in vertebrates from an evolutionary perspective are illustrated in Chapter 2, and the jaw mechanics of feeding behaviour in fishes are discussed in Chapter 12. We will not consider the foraging process before identification and attack but will concentrate on the attack and capture process. As with filtering, profitability must be maximized to optimize fitness. In this case there is a trade-off between capture probability and cost of attack. Particulate feeding directed at individual prey organisms includes both suction and ram or lunge feeding. In suction feeding, water and the prey are drawn rapidly into the buccal cavity by abduction of the jaw and, in fishes, opercula. The alternative feeding method is to lunge forward rapidly with the mouth opened at the appropriate time to engulf the prey. This type of feeding can also be called ram feeding and is often referred to as biting. Ram feeding often includes suction and can vary considerably in terms of forward velocity and kinematics.

Ram feeding The energy expended during an attack largely depends upon acceleration of the predator's body in the lunge forward to engulf the prey. A number of experimental studies of the kinematics of attack have revealed that attacks can be modulated to minimize energy expenditure. Harper and Blake (1991) showed that pike (*Esox lucius*) employ four varieties of S-strike, which are expected to have a different energetic cost and also attack velocity. Harper and Blake interpreted their results as indicating that the choice of S-strike type by the predator depended upon the apparent size of the prey. Their observations, however, show that attack velocity was independent of actual prey size. Their data can also be interpreted as resulting from judgement of prey distance by the predator (see *Figure 3*).

Attack by the fifteen-spined stickleback (*Spinachia spinachia*) (Kaiser et al., 1992) is also modulated following prey identification, but in this case in response to prey type rather than distance. Kaiser et al. did not consider attack distance and therefore did not

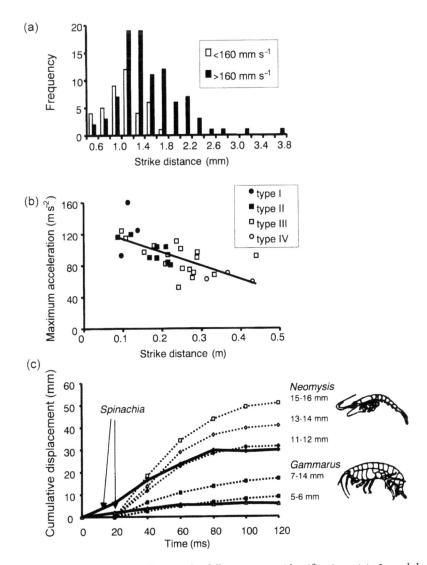

Figure 3. *Modulation of S-strike attacks following prey identification. (a) Larval herring attacking* Artemia nauplii *vary velocity according to prey distance (from Morley and Batty, 1996). (b) Pike vary acceleration according to prey distance (Harper and Blake, 1991). (c) Fifteen-spined stickleback displacement (solid lines) during attack on two different prey types compared with displacement of those prey types,* Gammarus sp. *or neomysis sp. (broken lines). Stickleback use two different attack velocities matched to prey performance, but attacks did not vary according to prey size (Kaiser et al., 1992).*

show an absence of distance judgement by the predator. In their experiments two types of prey (*Gammarus* sp. and/or *Neomysis* sp.) were offered separately and *Spinachia* modulated their attack velocity according to the type of prey encountered and the prey species' known escape performance. The results are replotted in *Figure 3* and also reveal that attack velocity does not vary with prey size. Clearly, identification of prey type leads to either of two stereotyped responses and prey size is of lesser importance. Interestingly, attack velocities closely match prey escape speeds, indicating that energy expenditure is minimized and the predator relies on the latency of the prey's response to effect capture. It is likely that the probability of capture success is less than one with this mechanism. Vinyard (1982) in similar experiments with Sacramento perch (*Archoplites interuptus*) also revealed that fish will learn very rapidly to increase their attack velocity when presented with a novel prey with much greater escape capabilities that those previously experienced. Further experiments might reveal the theoretical optimum attack velocity for a particular prey type depending upon the trade-off between energetic cost of attack, the product of capture success probability and the energetic value of the prey item.

Fish larvae, although less advanced and sophisticated (neurologically and behaviourally) than adult fish, also have the ability to modulate attack kinematics. The mechanism of velocity control in herring (*Clupea harengus*) larvae (Morley and Batty, 1996) works in two stages: the coiling of the body into the S-strike followed by the acceleration phase. Both stages can be modulated. *Figure 3* shows how strike velocity of herring larvae feeding on *Artemia* sp. nauplii is dependent upon the attack distance. Morley and Batty revealed that herring were able to adjust the S-strike coil (measured as the ratio of the coiled length to straight length) to vary velocity during the strike. Furthermore, the dependence of velocity on body coiling allows velocity to be linked to distance to prey and results in a constant time to capture regardless of prey distance.

The snapping turtle (*Chelydra serpentina*) thrusts its head forward in feeding strikes (Lauder and Prendergast, 1992) and the kinematics of this variation on ram feeding also vary according to the type of prey being attacked. When feeding on earthworms relatively low head extension velocities of 54 cm s^{-1} are employed but when feeding on fish velocity increases to 152.5 cm s^{-1}.

In all the above examples attack performance modulation following prey identification serves to minimize expenditure on prey capture. The energetic cost of attacks can be estimated as the cost of accelerating the predator's body to the attack velocity but in one study Yowell and Vinyard (1993) measured oxygen consumption during particulate feeding by blue tilapia (*Tilapia aurea*). The work done during attacks was estimated from kinematic analyses of video recordings taken during feeding and showed that the energetic cost of particulate feeding increased both with predator weight and with attack distance.

Suction feeding We have discussed particulate feeding involving lunges at individual prey items but predation on discrete prey items is often mediated by suction. This method of particulate feeding can be used by stationary fish or combined with forward motion (Weihs, 1980), and has been extensively studied in teleost fish. There is a considerable body of work on detailed kinematic analysis of suction mechanisms, and hydrodynamic models have been generated. This work has covered both larval (Drost et al., 1986, 1988a,b) and adult fish (Lauder, 1986; Lauder and Prendergast, 1992; Muller, 1989; Muller and Osse, 1984; Muller et al., 1982, 1985; Osse et al., 1985; van Leeuwen and Muller, 1984). The debate over models of the mechanism of suction generation centres on the role of the opercula bones in generating negative pressure and the gill arches in determining the pattern of pressure change (Lauder, 1986). A full discussion on this issue is beyond the scope of the present paper and we will therefore present only the main issues. It does, however, provide a basis for studies of the variability of feeding kinematics within species to cope with different prey types.

Transition from suction to ram feeding Behaviour experiments with kinematic analyses have shown how the different particulate feeding methods, ram and suction, are adaptations to feeding on prey with different escape abilities. Some recent work by Nemeth (1997a,b), however, extends the theme of varying strike velocity according to prey type. She found that kelp greenling *Hexagrammos decagarmmus* used increased negative buccal pressure as prey became more difficult to capture (Nemeth, 1997a) and also switched to ram feeding for the most elusive prey (Nemeth, 1997b). Norton (1991) compared the feeding kinematics and success on different prey types with the mouth size of a range of marine cottid species. In general, small-mouthed cottids tended to use suction feeding and were more successful preying on less elusive prey such as isopods, whereas larger-mouthed species used ram feeding and were more successful tackling elusive prey such as shrimps. All species, however, were able to modulate their attack kinematics according to prey type.

2.3 *Other methods, including cooperative feeding and use of 'weapons'*

Larger predators feeding on manoeuvrable prey may be disadvantaged by their relatively poor acceleration performance and turning ability. Webb and de Buffrénil (1990) considered this problem and using theoretical models equating thrust with resistance during acceleration, predicted that the relationship between body length and acceleration should be n-shaped. Animals longer than 0.4 m would be expected to fall on the descending portion of the curve where acceleration should decrease in relation to L^{-1}. Some large species such as tuna (*Thunnus albacares*) and yellowtail (*Seriola dorsalis*), which are specialized pelagic cruisers, have increased body stiffness, which helps in reducing drag during continuous swimming but is detrimental for their turning radius performance (Blake et al., 1995; Webb and Keyes, 1981). It has been suggested that the consequent impairment of manoeuvrability in these predator species may be mitigated by group foraging behaviours (Blake et al., 1995; Webb and de Buffrénil, 1990).

Very large vertebrates such as cetaceans will, therefore, find it difficult, unrewarding (in terms of success rate), energetically expensive and unprofitable to feed using whole-body attacks on prey within the predicted size range where acceleration and manoeuvrability are maximum (0.15–0.4 m). An example of this type of feeding that has been studied in detail is the use of tail fluke slaps by killer whales (*Orcinus orca*) while 'carousel feeding' on herring (Domenici et al., 2000; Similä, 1997; Similä and Ugarte, 1993). The whales did use whole-body lunges but were never observed to capture herring during these lunges, which appeared to be used to herd the herring, maintaining them in a very dense school. The killer whales slapped the herring with their tails, stunning the herring so they could then be engulfed in large numbers without need for high-speed attacks. Velocities of the lunges and tail slaps were measured during detailed kinematic analysis (Domenici et al., 2000) of underwater video recording taken in a fjord in northern Norway. In *Figure 4*, killer whale performance is compared with the predicted performance (Batty and Blaxter, 1992) of the herring that they were attacking. Only the largest whales employed lunges that exceeded the herring's escape performance but tail slaps were at least four times herring escape speed.

An interesting feeding method is employed by baleen whales and could be described as a combination of lunge, suction and filtering. The fin whale (*Balaenoptera physalis*) feeds by engulfing 70 l of water by lowering its jaw, which includes an elastic throat

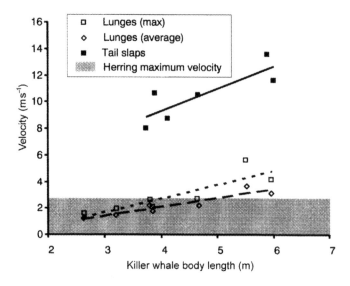

Figure 4. *Killer whale attacking herring schools use tail fluke slaps rather than whole-body attacks. Lunging speed and fluke slap velocity (Domenici et al., 2000) are compared with the predicted burst speed of 35 cm herring at 7°C (Batty and Blaxter, 1992).*

pouch, before filtering. While examining fresh carcasses, Brodie (1993) observed a noise, a synovial joint-crack, produced during realignment of the tips of the mandibles when the jaw is at full gap. Brodie proposed that this sound could induce startle responses by piscine prey that will tend to turn away from the sound stimulus (Blaxter and Batty, 1987; Domenici and Batty, 1994, 1997) further into the mouth and pouch to be captured, rather than escape.

3. Escape response

Many factors determine success or failure of prey escape responses: sensory capability, reaction distance, latency, turning ability, locomotor performance. As the aim of this paper is to review the relevance of kinematic data for behaviour, we will focus on those aspects that are particularly relevant within the context of escape behaviour and the animals' biomechanics and functional morphology. For this reason sensory capability and the sensory input (mechanical, auditory, visual, etc.) will not be considered except where sensory factors have an influence on latency or reaction distance, for example.

3.1 *Latencies*

Escape latencies in fish, defined as the interval between stimulus onset to the first detectable movement of the animal, are of the order of 10–100 ms (Domenici and Batty, 1994; Eaton and Hackett, 1984). For any given species, variability in escape latency is due to three main factors: the environmental temperature, the sensory system for detection and the neural pathway leading to the escape response. The effect of temperature is predicted from the positive effect on both nerve conduction time and muscle contraction time. Batty and Blaxter (1992) found that a rise in temperature did indeed reduce the latency of herring larvae to tactile stimuli.

Prey may detect attack by mechano-acoustic (sound pressure, displacement or touch), visual and chemical stimulation, separately or in combination. Mechano-acoustic, sound pressure or displacement stimulation, results in the shortest latency, of the order of 10–30 ms depending on prey size, species and temperature. The very short delay of this reflex response results from the direct neural connections of the octavolateralis nerve and the lateral line nerve to the giant Mauthner neurons in the retinospinal system (Eaton and Hackett, 1984). Visual stimulation results in longer latencies, of the order of 100 ms, as a result of a longer neural pathway between the optic nerve and the Mauthner cell, including an intermediate processing by the optic tectum (Zottoli et al., 1987). Herring larvae (Batty, 1989) have much longer latencies (120 ms or longer) to visual flash stimuli than to sound stimuli (25 ms). Chemical signals give rise to escape responses with latencies of the order of 2–120 s (Pfeiffer et al., 1986). Such long latencies were measured as the time between the addition of a chemical alarm substance to the water and the fish's reaction. Therefore, the time course of such latencies also depends on the slow diffusion of the chemical alarm substance in the water to reach the olfactory organs, as well as the processes of detection and neural command.

For any given sensory channel, it has been suggested that fish may use different neural pathways, resulting in different latencies (Eaton et al., 1984). Domenici and Batty (1994, 1997) have found that herring responding to a sound stimulus responded with one of two latencies, that is, around 30 and around 100 ms. Domenici and Batty (1994) showed that the occurrence of one or the other latency was related to stimulus distance, and therefore intensity. This is in agreement with Blaxter et al. (1981), who found shorter latencies for responses to stimuli at threshold than above threshold.

Domenici and Batty (1997) found that schooling modulates such a relationship by raising the sensory threshold for fast escape responses, because for any given stimulus distance solitary herring are more likely to use a short latency than are schooling herring. Domenici and Batty (1997) and Domenici and Blake (1997) suggested that these two latencies (30 or 100 ms), which are found in escapes with different kinematics, correspond to two escape response types. They result in different angular velocities during stage 1 of the c-start (*Figure 5*) and are likely to be driven by different neural pathways; the Mauthner system for the fast escape response, and possibly an alternative neural circuit for the slow escape response. In addition, in schooling fish stimulated by a transient sound (Domenici and Batty, 1997), some of the fish responding with longer latencies may depend on visual signals from other members of the school rather than the original sound stimulus.

3.2 *Reaction distance*

Timing is one of the most important components in predator–prey encounters. Timing is dictated not only by escape latency. Latency, as defined above, is the time between the stimulus onset and the fish's reaction. This definition is particularly appropriate for laboratory experiments in which a sudden stimulus such as a flashlight (Batty, 1989) or a transient sound or mechanical disturbance (see review by Eaton and Hackett (1984)) is produced to provoke an escape response. However, in natural situations, fish may not necessarily receive 'sudden' stimulation, but stimuli with increasing strength related to a

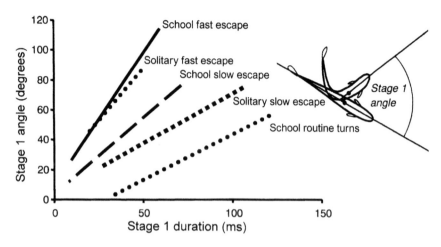

Figure 5. *A comparison of C-start stage 1 angle and duration for adult herring. The slope of this relationship gives the angular velocity in these turns. Different classes of responses can be identified dependent on social interaction and the latency (delay between stimulus and response). Based on Domenici and Batty (1994, 1997).*

predator's approach. A fish does not escape as soon as it perceives a visual or mechanical signal produced by a predator; such a strategy would not be an efficient way of avoiding predation. Fish respond when the rate of change of such a signal exceeds a certain threshold (i.e. when the danger is real). In this respect, vision is the most studied sense (e.g. Anderson, 1988; Dill, 1974). Dill (1974) showed that the prey's reaction distance increases with the speed and the depth of the body profile of a predator. This phenomenon has important consequences for predator strategies in terms of approaching speed and for predator's functional morphology. Batty (1989) found that looming targets with vertical edges expanding in the horizontal plane were most effective for eliciting an escape response from herring larvae. Webb (1982, 1984) in experiments on the responses of fathead minnows to predators of different cross-sectional shape found that predators presenting a lateral ellipse were more likely to elicit a response than ones that are rounded in cross-section. These findings are consistent with the hypothesis that a rapid approach by a predator with a deep body profile (laterally compressed) may cause the prey to flee too early for the predator to capture them. A similar phenomenon of dependence of reaction distance on speed and size of the approaching object is likely to occur for the mechano-sensory system as well, as observed by Dijgraaf (1963).

3.3 C-start turning angle

Turning angles are an important factor in a prey's ability to escape from predators. The 'choice' of turning angle (if it is a 'choice') determines a fixed preferred escape trajectory (Domenici and Blake, 1993a). Particular trajectories may maximize parameters related to escape success such as sensory awareness (Domenici and Blake, 1993a), or the distance maintained between the prey and the threat (Domenici and Blake, 1993a; Weihs and Webb, 1984). Recent studies (Domenici and Blake, 1991; Foreman and Eaton, 1993) show that the turning angles of escape responses can span more than about 180° on either side of the fish.

Escape responses may consist of a single C-start bend (followed by displacement as the fish straightens) or a double bend: a C-start immediately followed by a bend in the opposite direction. The latter type of response is often followed by a burst of swimming at or near maximum velocity. Double-bend escape responses can be divided into two stages: stage 1 is the initial turn during the formation of the 'C' in a C-start; stage 2 results from the subsequent contralateral contraction of myotomal muscle causing a rotation of the head in the opposite direction. Various reserchers have studied and compared stage 1 and stage 2 turning angles (Domenici and Blake, 1991, 1993a,b; Eaton and Emberley, 1991; Foreman and Eaton, 1993; Kasapi et al., 1993). Stage 1 is the main turning component of an escape response, and therefore stage 1 angles are generally larger than stage 2 angles (Domenici and Blake, 1991; Eaton et al., 1988). Stage 1 angles were found to be correlated with electromyograph duration (Eaton et al., 1988) and with stage 1 duration (Domenici and Blake, 1991). Turning rate (angular velocity) and angle of turn vary with the type of stimulus evoking the response and also differ in schooling fish when compared with solitary individuals. Responses evoked by sense organ input have a higher average turning rate than responses resulting from direct electrical stimulation of the M-cell (Nissanov et al., 1990). When schooling herring (Clupea harengus) made C-starts in response to acoustic stimuli (Domenici and Batty, 1994) two distinct response latencies were observed, each associated with a different stage 1 angular velocity (observed as the relationship between stage 1 duration and turning angle) (Figure 5). The angle versus time curve of an individual escaping fish, however, is sinusoidal (Kasapi et al., 1993). Differences in average turning rates could, therefore, be due to maximum turning rate, stage 1 duration or to a difference in the shape of this curve.

As would be expected, Domenici and Blake (1991) found that single-bend escape responses (without a stage 2) have larger turn angles than double-bend responses. Rotation of the head in stage 2 of a double-bend response is to some extent correlated with stage 1 angle; both Eaton et al. (1988) and Domenici and Blake (1991) found a weak inverse relationship between stage 1 and stage 2 turning angles. Large stage 1 turns are associated with small stage 2 opposite turns, and vice versa. In consequence, Eaton et al. (1988) suggested that neural commands for the escape trajectory could be organized before the end of stage 1.

3.4 Escape trajectories

So far we have considered the angle turned during a C-start with respect to the orientation of the body at the time that the stimulus was presented. We will now consider the actual trajectory followed by an animal with respect to the orientation of the direction to the stimulus. Classifying responses as away from or towards the stimulus direction, response directionality (Blaxter and Batty, 1987), although simple, can reveal a great deal about interactions between biomechanics and behaviour. Response rates to sound stimuli have been found to vary between periods of swimming and gliding during intermittent locomotion. Blaxter and Batty (1987) found that herring were more likely to respond when they were in the glide phase and that when actively swimming responses were not directional: stage 1 trajectories away from or towards the stimulus occurred with equal frequency. A more detailed analysis showed that during active swimming responses were orientated according to tail-beat phase. C-starts were initiated by contraction of myotomal muscle on the side that was already contracting during swimming; the fish turned in a direction away from the orientation of the caudal fin. In electrophysiological experiments on *Xenopus* tadpole larvae, Sillar and Roberts (1988) found that Mauthner neurons were gated during swimming to prevent simultaneous contraction of muscle on both sides of the body.

In more detailed analyses of trajectories, both Eaton and Emberley (1991) and Domenici and Blake (1993a) showed a weak linear relationship between escape angles and the angle at which the stimulus is presented. Domenici and Blake (1993a) found that the distribution of the escape trajectories (measured as angles relative to the stimulus direction) in angelfish showed a bimodal pattern with trajectories at 130° and 180°. Such a bimodal pattern is also found in escape responses of solitary herring separated from the school (Domenici and Batty, 1997). Domenici and Blake (1993a) hypothesised that escaping at 180° and 130° may correspond to maximizing the distance away from the stimulus and escaping at the furthest limits of the fish's discrimination zone (defined as the angular sector of orientation relative to a stimulus, which resulted in non-random distribution of escape trajectories), respectively. Such patterns were not present in schooling herring, which have a unimodal pattern of trajectory distribution with a peak around 150° (*Figure 6*). This may be attributed to interactions between neighbouring fish (Domenici and Batty, 1994).

Directionality can also vary depending on the orientation of the prey to the stimulus or predator attack direction. Predators tend to attack prey from the side more often than towards the head or caudally (Moody *et al.*, 1983; Webb, 1986; Webb and Skadsen, 1980) and the probability of successful capture is even more frequent when the attack is directed toward the centre of the body (Webb, 1986). It is, therefore, interesting that Domenici and Blake (1993a) observed that escapes can be directed away from the stimulus more reliably when the stimulus is from the side of the prey, between 60 and 120°. Stimuli presented frontally or caudally, however, produce randomly directed escape trajectories. It is likely that left–right discrimination of the orientation of sensory stimuli will decrease when the stimulus is more in line with the longitudinal axis of the prey. A random frequency of 'left or right' trajectories may be of advantage to a prey animal when attacked towards the head and close to the body axis. Both left and right responses will take the prey away from the path of the predator and unpredictability will disadvantage the predator. Boothby and Roberts (1995) found that *Xenopus* embryos responded to tactile stimulation of the head with random escape trajectories but trunk stimulation resulted in more away responses. Has the predator tactic of attacking from the side, which Webb (1986) reported as being more frequent and more successful, evolved because prey trajectory is more predictable when attacking from this position? Or do predators attack more often from this direction because the prey presents a larger visual target and is more likely to be detected?

In schooling fish such as herring, the trajectories of escape responses in which the initial turn (during stage 1) is towards the stimulus are corrected to the extent that the final angle of escape (at the end of stage 2) is directed away from the stimulus (Domenici and Batty, 1994). Domenici and Batty (1994, 1997) suggested that this behaviour may be due to the influence of schooling neighbours, most of which show stage 1 angles away from the stimulus.

The type of stimulus can also determine the directionality of the response. Data from three studies of escape behaviour by herring larvae indicate that, for this species at least, visual stimuli (Batty, 1989), with 91.4% of escapes away from the stimulus, are the most reliable. Tactile stimuli (Yin and Blaxter, 1987) are less likely to result in an away response than visual stimuli, and sound stimuli produce only

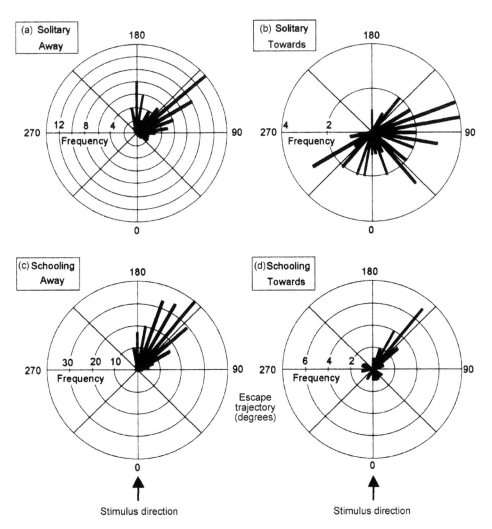

Figure 6. *Herring escape trajectories. Stage 1 turns may be directed away from (a, c) or towards the stimulus (b, d). Stage 2 trajectories are very different in schooling (c, d) and solitary fish (a, b). Solitary fish have bimodal trajectories similar to those of other species. Unimodal trajectories of schooling fish probably result from interactions with other members of the school. From Domenici and Batty (1997).*

73.8% away responses (Blaxter and Batty, 1985). Visual stimuli may be received by both eyes but more often by only one eye. Because the visual field is mapped in the optic tectum, it is unlikely that the 'wrong' Mauthner cell could be activated. In their study of the ontogeny of escape behaviour to sound stimuli in herring (Blaxter and Batty, 1985), the ability to correct trajectories following initial responses towards the stimulus direction was related to increasing sophistication of mechanoreception. In herring, the enclosed lateral line system and its connection to the inner ear via the lateral recess develops during and after metamorphosis, and the observed improvement in trajectory correction was related to the development of this integrated mechanosensory system.

3.5 *Turning radius*

Species that live in complex environments such as coral reefs and weedy rivers (angelfish, *Pterophyllum eimekei*; pike, *Esox lucius*; knifefish, *Xenomystus nigri*) have smaller turning radii than pelagic species (tuna, *Thunnus albacares*; yellowtail, *Seriola dorsalis*). A short turning radius is also beneficial during escape from a predator and has been shown to vary between species (Blake *et al.*, 1995; Domenici and Blake, 1991; Howland, 1974; Webb, 1976; Webb and Keyes, 1981; Weihs and Webb, 1984). Turning radius is independent of velocity but within a species is proportional to length (Domenici and Blake, 1991; Howland, 1974; Webb, 1976).

3.6 *Propulsive performance*

Both high speed and rapid acceleration are essential for successful escape from a predator. Prey with high maximum burst speeds will not necessarily be the most successful in escaping a predator unless acceleration is also rapid. Various distance–time parameters (average velocity, maximum velocity, and both average and maximum acceleration) have been measured in many studies (Beddow *et al.*, 1995; Domenici and Blake, 1991, 1993b; Frith and Blake, 1995; Gamperl *et al.*, 1991; Harper and Blake, 1990, 1991; Kasapi *et al.*, 1993; Webb, 1976, 1978; Weihs, 1973). In most studies the analysis of distance-related parameters is based on the centre of mass of the fish when stretched straight. This approximates the instantaneous centre of mass, the point about which propulsive forces act (Webb, 1978).

    Most studies of fast-start performance are based on fish starting from a still position. Speed starts from zero and increases during the two stages of a fast start, and may further increase after the end of stage 2 if the fish continues active swimming. The pattern of the velocity curve is related to the timing of the two kinematic stages (Domenici and Blake, 1991). Although velocity increases throughout the fast start, the rate of increase drops around the end of stage 1 and the beginning of stage 2. There are, therefore, two peaks of acceleration: one per stage. The timing of propulsive power is therefore related to the kinematics of the tail beats. When a full return flip of the tail is not present (single-bend response; Domenici and Blake, 1991), velocity does not increase beyond stage 1, and the high acceleration peak in stage 1 is followed by a much lower peak. There is evidence that in such escape responses a contralateral contraction during stage 2 is absent (Foreman and Eaton, 1993). Acceleration after stage 1 in single-bend responses may be due to thrust produced by passive elastic elements, such as skin and collagen fibres, and elasticity in the muscle itself (Kasapi *et al.*, 1993).

    The locomotor performance of fish striking at prey is often lower than when escaping from predators (Harper and Blake, 1991; Kaiser *et al.*, 1992; Morley and Batty, 1996) possibly because of the employment of sub-maximal performance by attacking predators. Initially, it seems counter-intuitive that escaping prey should not always use maximum burst speeds, but Webb (1986) found that when attacks were not followed by chases prey escape speed was often sub-maximal. Habituation of responses in goldfish has been reported (Meuller, 1978) and it seems likely that habituation and/or identification of predator type may also modulate escape performance.

    There may, in some circumstances, be a trade-off between locomotor performance and manoeuvrability. Domenici and Blake (1993b) found that the escape speed of large angelfish is compromised when large escape angles are employed.

3.7 *Scale effects*

How escape responses are affected by prey size is a major issue; it could theoretically have an important influence on prey size preference of predators as well as size-differential survival from predator attacks. Of the factors discussed above, scaling has mainly been investigated in linear and angular variables. Although, on the basis of theoretical considerations, both latency and reaction distance may be related to prey size, there is no experimental evidence of such a relationship.

The effects of body size on locomotor performance have been extensively studied. Wardle (1975) showed that the limit to fish swimming speed is the minimum muscle contraction time, which increases with fish length and sets the limit to tail beat frequency. In a later paper, Wardle and Videler (1980) established that swimming speed is the product of tail beat frequency and stride length. The relationship between burst swimming performance and body length is, therefore, non-linear and smaller fish attain higher specific speeds (body lengths per second; L s^{-1}) than larger fish (Batty and Blaxter, 1992). Archer *et al.* (1990) measured the fast muscle contraction times of cod and found that they scaled in proportion to $L^{0.29}$. Kinematic analyses of larval fish burst swimming have shown that their tail beat frequencies are very much higher than those of juvenile or adult fish (Batty and Blaxter, 1992). Batty and Blaxter (1992) used their own data together with data from the literature for many species (Archer *et al.*, 1990; Wardle, 1977, 1980) with a wide range of body size to fit a model predicting maximum tail beat frequency from body length and temperature. Tail beat frequency declined with body length in proportion to $L^{-0.266}$.

Stride length (the distance covered per tail beat), however, is reduced at low Reynolds numbers (Batty and Blaxter, 1992; Fuiman and Batty, 1997) but resulting specific swimming speeds, measured in L s^{-1}, will still show an inverse relationship to body length. Batty and Blaxter (1992) used their equation for predicting tail beat frequency together with the relationship between stride length and Reynolds number to predict the maximum burst speeds of various species of fish larvae (anchovy, Hunter, 1972; Webb and Corolla, 1981; zebra danios, Fuiman, 1986; *Harpagifer antarcticus*, Johnston *et al.*, 1991; herring and plaice, Batty and Blaxter, 1992) and to compare the predictions with experimental measurements. *Harpagifer antarcticus* and the two clupeid species (herring and anchovy) have similar performance to the prediction but plaice has relatively poor performance. In contrast, zebra danios swim at twice the expected maximum velocity. As Fuiman (1986) did not carry out detailed kinematic analyses of his films, it is not known how this feat was achieved. Wardle and Videler (1980) suggested that very high burst speeds could be attained if stride length is increased by including less than one complete propulsive wave on the body; the fish uses a wavelength λ_b greater than body length L. No kinematic evidence of fish employing wavelengths greater than L has yet been presented.

Very small aquatic vertebrates such as fish larvae are particularly disadvantaged in predator–prey interactions by their confinement in the viscous or intermediate flow regimes at low Reynolds numbers where resistive forces are more important than inertial forces. Burst swimming performance is restricted as evinced by reduced stride length occurring at low Reynolds numbers (Batty and Blaxter, 1992). Both Muller and Videler (1996) and Webb and Weihs (1986) have suggested that fish larvae should increase their length to escape viscous drag. Müller and Videler (1996) described inertia as 'safe harbour' where susceptibility to predation is reduced. Batty (1984) and

Webb and Weihs (1986) pointed out the importance of the timing of fin development with transition between the viscous and inertial regimes.

In contrast to swimming speed, fast-start acceleration is size-independent in fish species investigated thus far (Domenici and Blake, 1993b; Webb, 1976). Theoretical models equating thrust with resistance during acceleration (Daniel and Webb, 1987; Webb and de Buffrénil, 1990) predict that the relationship between body length and acceleration should be n-shaped. Animals longer than 0.4 m would be expected to fall on the descending portion of the curve where acceleration should decrease in relation to L^{-1}. However, Webb and Johnsrude (1988) suggested that, because of summation of muscle twitches, this may not be so in some cases.

Because, for many adult fish, maximum acceleration is size-independent and fast-start duration increases with size, speed should be measured within a given time interval (Webb, 1976). Both Webb (1976) and Domenici and Blake (1993b) found that speed within a given time is size independent. Domenici and Blake (1993b) found that the occurrence of low-performance fast starts with large turning angles increased with size. They suggested that larger fish may not be able to employ high-performance fast starts with large turning angles, as a result of morphological constraints such as limited flexibility at the centre of mass. The maximum angle of single-bend escapes is inversely proportional to body length in angelfish (Domenici and Blake, 1993b), possibly because of a decrease in flexibility around the centre of mass in larger fish. Single-bend responses allow large fish to achieve large turning angles through a coasting phase, despite their limited flexibility around the centre of mass, and Domenici and Blake (1993b) found that the proportion of single-bend responses increased with angel fish body size. Turning radius is also linearly related to size (Webb, 1976). The relationship between manoeuvrability and size awaits further investigation.

4. Discussion

Many predators use the most energetically efficient foraging and prey capture methods that environmental conditions and their sensory capabilities will allow. It is clear that predators aim to minimize energy expenditure while maximizing energy intake in their feeding activities. To achieve this, predators are able to recognize prey and adjust their capture method or S-strike performance accordingly. There is also evidence that predators, even small fish larvae, can either estimate the distance to the prey or use apparent prey size to optimize capture performance by using the minimum necessary capture velocity. There is a need for more work to more fully understand both the methods used by predators and the ecological implications of these processes in terms of prey selectivity and community structure.

The orientation of predator and prey is an important factor determining successful capture or escape. The unpredictability of head-on attack orientations benefits the prey and it is not surprising that more attacks from the side have a successful outcome for the predator. A number of papers report a greater frequency of attacks from the side. What is not clear, however, is if this is a tactic on behalf of the predator to maximize its success, or if it is simply a result of increased prey conspicuousness when viewed from this orientation.

Kinematic analyses of the behaviour of fish schools during escape responses have revealed greater complexity in prey escape performance. It is clear that certain rules are followed by members of schools, leading to communication and co-ordination of the school. There is a requirement for further work to elucidate the means of communication, visual, sound or both, for the propagation of responses across a school. There is evidence that fish emit sounds during escape responses (Gray and Denton, 1991) and that mirror scales may produce meaningful signals (Denton and Rowe, 1994).

During escape responses after the C-start has been initiated sensory information could still be received but it is not known if prey can still detect and use it to modify their response: in stage 2 for example. The particularly violent displacement of the head in a C-start may saturate mechanoreceptors (inner ear and neuromasts) and movement of the head itself is so rapid that visual information may not be interpretable. Lighthill (1993) developed a mathematical model that predicts optimum rotation and lateral displacement of the head during swimming for the minimization of both drag and stimulation of the sense organs (Denton and Gray, 1993). Comparison of predicted motion with kinematic analyses of routine swimming of adult herring (Rowe et al., 1993) showed a close correlation. Further similar work is necessary to ascertain sensory capability during C-starts.

Although laboratory experiments, if appropriately interpreted, can provide an understanding of the mechanisms and processes underlying predator–prey interactions, they cannot substitute for observations of interactions in the natural environment. There is a challenging need for new techniques to be developed for recording actual encounters between predators and prey avoiding disturbance that may influence their behaviour. Recordings will need to be of adequate quality to allow kinematic analysis and sufficiently practical to record many interactions and permit statistically valid interpretations.

References

Anderson, J.J. (1988) A neural model for visual activation of startle behavior in fish. *J. Theor. Biol.* **131**: 279–288.

Archer, S.D., Altringham, J.D. and Johnston, I.A. (1990) Scaling effects on the neuromuscular system, twitch kinetics and morphometrics of the cod, *Gadus morhua. Mar. Behav. Physiol.* **17**: 137–146.

Batty, R.S. (1984) Development of swimming movements and musculature of larval herring. *J. Exp. Biol.* **110**: 217–229.

Batty, R.S. (1989) Escape responses of herring larvae to visual stimuli. *J. Mar. Biol. Assoc. UK* **69**: 647–654.

Batty, R.S. and Blaxter, J.H.S. (1992) The effect of temperature on the burst swimming performance of fish larvae. *J. Exp. Biol.* **170**: 187–201.

Batty, R.S., Blaxter, J.H.S. and Libby, D.A. (1986) Herring (*Clupea harengus*) filter feeding in the dark. *Mar. Biol.* **91**: 371–375.

Batty, R.S., Blaxter, J.H.S. and Richard, J.M. (1990) Light intensity and the feeding behaviour of herring, *Clupea harengus. Mar. Biol.* **107**: 383–388.

Beddow, T.A., van Leeuwen, J.L. and Johnston, I.A. (1995) Swimming kinematics of fast starts are altered by temperature acclimation in the marine fish *Myoxocephalus scorpius. J. Exp. Biol.* **198**: 203–208.

Blake, R.W., Chatters, L.M. and Domenici, P. (1995) Turning radius of yellowfin tuna (*Thunnus albacares*) in unsteady swimming manoeuvres. *J. Fish Biol.* **46**: 536–538.

Blaxter, J.H.S. and Batty, R.S. (1985) The development of startle responses in herring larvae. *J. Mar. Biol. Assoc. UK* **65**: 737–750.

Blaxter, J.H.S. and Batty, R.S. (1987) Comparisons of herring behaviour in the light and dark: changes in activity and responses to sound. *J. Mar. Biol. Assoc. UK* **67**: 849–860.

Blaxter, J.H.S., Denton, E.J. and Gray, J.A.B. (1981) The auditory bullae–swimbladder system in late stage herring larvae. *J. Mar. Biol. Assoc. UK* **61**: 315–326.

Boothby, K.M. and Roberts, A. (1995) Effects of site of tactile stimulation on the escape swimming responses of hatchling *Xenopus laevis* embryos. *J. Zool. Lond.* **235**: 113–125.

Brodie, P.F. (1993) Noise generated by jaw actions of feeding whales. *Can. J. Zool.* **71**: 2546–2550.

Daniel, T.L. and Webb, P.W. (1987) Physical determinants of locomotion. In: *Comparative Physiology: Life in Water and on Land* (eds P. de Jours, L. Bolis, C.R. Taylor and E.R. Weibel). Liviana Press, New York, pp. 343–369.

Denton, E.J. and Gray, J.A.B. (1993) Stimulation of the acostico-lateralis of clupeid fish by external sources and their own movements. *Philos. Trans. R. Soc. Lond., Ser. B* **341**: 113–127.

Denton, E.J. and Rowe, D.M. (1994) Reflective communication between fish, with special reference to the greater sand eel, *Hyperoplus lanceolatus*. *Philos. Trans. R. Soc. Lond.* **344**: 221–237.

Dijgraaf, S. (1963) The functioning and significance of the lateral line organs. *Biol. Rev.* **38**: 51–105.

Dill, L.M. (1974) The escape response of the zebra danio (*Brachydanio rerio*). I. The stimulus for escape. *Anim. Behav.* **22**: 710–721.

Domenici, P. and Batty, R.S. (1994) Escape manoeuvres of schooling *Clupea harengus*. *J. Fish Biol.* **45** (Suppl A): 97–110.

Domenici, P. and Batty, R.S. (1997) The escape behaviour of solitary herring (*Clupea harengus* L.) and comparisons with schooling individuals. *Mar. Biol.* **128**: 29–38.

Domenici, P. and Blake, R.W. (1991) The kinematics and performance of the escape response in the angelfish (*Pterophyllum eimekei*). *J. Exp. Biol.* **156**: 187–205.

Domenici, P. and Blake, R.W. (1993a) Escape trajectories in angelfish (*Pterophyllum eimekei*). *J. Exp. Biol.* **177**: 253–272.

Domenici, P. and Blake, R.W. (1993b) The effect of size on the kinematics and performance of angelfish (*Pterophyllum eimekei*) escape responses. *Can. J. Zool.* **71**: 2319–2326.

Domenici, P. and Blake, R.W. (1997) The kinematics and performance of fish fast-start swimming. *J. Exp. Biol.* **200**: 1165–1178.

Domenici, P., Batty, R.S., Similä, T. and Ogam, E. (2000) Killer whales (*Orcinus orca*) feeding on schooling herring (*Clupea harengus*) using underwater tail slaps: kinematic analyses of field observations. *J. Exp. Biol.* **202**: 283–294.

Drost, M.R. and van den Boogaart, J.G.M. (1986) The energetics of feeding strikes in larval carp, *Cyprinus carpio*. *J. Fish Biol.* **29**: 371–379.

Drost, M.R., Muller, M. and Osse, J.W.M. (1988a) A quantitative hydrodynamical model of suction feeding in larval fishes—the role of frictional forces. *Proc. R. Soc. Lond., Ser. B* **234**: 263–281.

Drost, M.R., Osse, J.W.M. and Muller, M. (1988b) Prey capture by fish larvae, water-flow patterns and the effect of escape movements of prey. *Neth. J. Zool.* **38**: 23–45.

Durbin, A.G., Durbin, E.G., Verity, P.G. and Smayda, T.J. (1981) Voluntary swimming speeds and respiration rates of a filter-feeding planktivore, the Atlantic menhaden, *Brevoortia tyrannus* (Pisces: Clupeidae). *Fish. Bull. US* **78**: 877–886.

Eaton, R.C. and Emberley, D.S. (1991) How stimulus direction determines the trajectory of the Mauthner-initiated escape response in a teleost fish. *J. Exp. Biol.* **161**: 469–487.

Eaton, R.C. and Hackett, J.T. (1984) The role of Mauthner cells in fast-starts involving escape in teleost fish. In: *Neural Mechanisms of Startle Behavior* (ed. R.C. Eaton). Plenum, New York, pp. 213–266.

Eaton, R.C., Nissanov, J. and Wieland, C.M. (1984) Differential activation of Mauthner and non-Mauthner startle circuits in the zebrafish—implications for functional substitution. *J. Comp. Physiol.* **155**: 813–820.

Eaton, R.C., DiDomenico, R. and Nissanov, J. (1988) Flexible body dynamics of the goldfish C-start: implications for reticulospinal command mechanisms. *J. Neurosci.* **8**: 2758–2768.

Foreman, M.B. and Eaton, R.C. (1993) The direction change concept for reticulospinal control of goldfish escape. *J. Neurosci.* **13**: 4101–4133.

Frith, H.R. and Blake, R.W. (1995) The mechanical power output and hydromechanical efficiency of northern pike (*Esox lucius*) fast-starts. *J. Exp. Biol.* **198**: 1863–1873.

Fuiman, L.A. (1986) Burst-swimming performance of larval Zebra danios and the effects of diel temperature fluctuations. *Trans. Am. Fish. Soc.* **115**: 143–148.

Fuiman, L.A. and Batty, R.S. (1997) What a drag it is getting cold: partitioning the physical and physiological effects of temperature on fish swimming. *J. Exp. Biol.* **200**: 1745–1755.

Gamperl, A.K., Schnurr, D.L. and Stevens, E.D. (1991) Effect of a sprint-training protocol on acceleration performance in rainbow trout (*Salmo gairdneri*). *Can. J. Zool.* **69**: 578–582.

Gibson, R.N. (1988) Development, morphometry and particle retention capability of gill rakers in the herring, *Clupea harengus* L. *J. Fish Biol.* **32**: 949–962.

Gibson, R.N. and Ezzi, I.A. (1985) Effect of particle concentration on filter- and particulate-feeding in the herring *Clupea harengus*. *Mar. Biol.* **88**: 109–116.

Gibson, R.N. and Ezzi, I.A. (1992) The relative profitability of particulate- and filter-feeding in the herring *Clupea harengus*. *J. Fish Biol.* **40**: 577–590.

Gray, J.A.B. and Denton, E.J. (1991) Fast pressure pulses and communication between fish. *J. Mar. Biol. Assoc. UK* **71**: 83–106.

Harper, D.G. and Blake, R.W. (1990) Fast-start performance of rainbow trout *Salmo gairdneri* and northern pike *Esox lucius*. *J. Exp. Biol.* **150**: 321–342.

Harper, D.G. and Blake, R.W. (1991) Prey capture and the fast-start performance of northern pike *Esox lucius*. *J. Exp. Biol.* **155**: 175–192.

Hoogenboezem, W., Sibbing, F.A., Osse, J.W.M., van den Boogaart, J.G.M., Lammens, E.H.R.R. and Terlouw, A. (1990) X-ray measurements of gill-arch movements in filter-feeding bream, *Abramis brama* (Cyprinidae). *J. Fish Biol.* **36**: 47–58.

Howland, H. (1974) Optimal strategies for predator avoidance: the relative importance of speed and manoeuverability. *J. Theor. Biol.* **47**: 333–350.

Hunter, J.R. (1972) Swimming and feeding behaviour of larval anchovy, *Engraulis mordax*. *Fish. Bull.* **70**: 821–838.

James, A.G. and Findlay, K.P. (1989) Effect of particle size and concentration on feeding behaviour, selectivity and rates of food ingestion by the Cape anchovy *Engraulis capensis*. *Mar. Ecol. Prog. Ser.* **50**: 275–294.

James, A.G. and Probyn, T. (1989) The relationship between respiration rate, swimming speed and feeding behaviour in the Cape anchovy *Engraulis capensis* Gilchrist. *J. Exp. Mar. Behav. Ecol.* **131**: 81–100.

Johnston, I.A., Johnson, T.P. and Battram, J.C. (1991) Low temperature limits swimming performance in Antarctic fish. In: *Biology of Antarctic Fishes* (eds G. di Prisco, B. Maresca and B. Tota). Springer-Verlag, Heidelberg, pp. 179–190.

Kaiser, M.J., Gibson, R.N. and Hughes, R.N. (1992) The effect of prey type on the predatory behaviour of the fifteen-spined stickleback, *Spinachia spinachia* (L.). *Anim. Behav.* **43**; 147–156.

Kasapi, M.A., Domenici, P., Blake, R.W. and Harper, D. (1993) The kinematics and performance of escape responses of the knifefish (*Xenomystus nigri*). *Can. J. Zool.* **71**: 189–195.

Lauder, G.V. (1986) Aquatic prey capture in fishes—experimental and theoretical approaches. *J. Exp. Biol.* **125**: 411–416.

Lauder, G.V. and Prendergast, T. (1992) Kinematics of aquatic prey capture in the snapping turtle *Chelydra serpentina*. *J. Exp. Biol.* **164**: 55–78.

Lighthill, M.J. (1993) Estimates of pressure differences across the head of a swimming clupeid fish. *Philos. Trans. R. Soc. Lond., Ser. B* **341**: 129–140.

Meuller, T.J. (1978) Factors modulating the occurrence and direction of the startle response in goldfish. *Neurosci. Abstr.* **4**: 363.

Moody, R.C., Helland, J.M. and Stein, R.A. (1983) Escape tactics used by bluegills and fathead minnows to avoid predation by tiger muskellunge. *Envir. Biol. Fish.* **8**, 61–65.

Morley, S.A. and Batty, R.S. (1996) The effects of temperature on S-strike feeding of larval herring, *Clupea harengus* L. *J. Mar. Physiol. Behav.* **28**: 123–136.

Muller, M. (1989) A quantitative theory of expected volume changes of the mouth during feeding in teleost fishes. *J. Zool.* **217**: 639–661.

Muller, M. and Osse, J.W.M. (1984) Hydrodynamics of suction feeding in fish. *Trans. Zool. Soc. Lond.* **37**: 51–135.

Muller, M., Osse, J.W.M. and Verhagen, J.H.G. (1982) A quantitative hydrodynamical model of suction feeding in fish. *J. Theor. Biol.* **95**: 49–79.

Muller, M., van Leeuwen, J.L., Osse, J.W.M. and Drost, M.R. (1985) Prey capture hydrodynamics in fishes: two approaches. *J. Exp. Biol.* **119**: 389–394.

Müller, U.K. and Videler, J.J. (1996) Inertia as a 'safe harbour': do fish larvae increase length growth to escape viscous drag? *Rev. Fish Biol. Fish.* **6**: 353–360.

Nemeth, D.H. (1997a) Modulation of buccal pressure during prey capture in *Hexagrammos decagarmmus* (Teleostei: Hexagrammidae). *J. Exp. Biol.* **200**: 2145–2154.

Nemeth, D.H. (1997b) Modulation of attack behaviour and its effect on feeding performance in a trophic generalist fish, *Hexagrammos decagarmmus*. *J. Exp. Biol.* **200**: 2155–2164.

Nissanov, J., Eaton, R.C. and DiDomenico, R. (1990) The motor output of the Mauthner cell, a reticulospinal command. *Neuron Brain Res.* **517**: 88–98.

Norton, S.F. (1991) Capture success and diet of cottid fishes: the role of predator mophology and attack kinematics. *Ecology* **72**: 1807–1819.

Osse, J.W.M., Muller, M. and van Leeuwen, J.L. (1985) The analysis of suction feeding in fish. *Fortschr. Zool.*, **30**: 217–221.

Pfeiffer, W., Denoix, M., Wehr, R., Gnass, D., Zachert, I. and Breisch, M. (1986) Slow motion analysis of the fright reaction of ostariophysean fish and the significance of the Mauthner reflex. *Zool. Jahrb. Abt. Allg. Zool. Physiol. Tiere* **90**: 115–165.

Rowe, D.M., Denton, E.J. and Batty, R.S. (1993) Head turning in herring and some other fishes. *Philos. Trans. R. Soc. Lond., Ser. B* **341**; 141–148.

Sillar, K.T. and Roberts, A. (1988) A neuronal mechanism for sensory gating during locomotion in a vertebrate. *Nature* **331**: 262–265.

Similä, T. (1997) Sonar observations of killer whales (*Orcinus orca*) feeding on herring schools. *Aquat. Mammals* **23**: 119–126.

Similä, T. and Ugarte, F. (1993) Surface and underwater observation of cooperatively feeding killer whales in Northern Norway. *Can. J. Zool.* **71**: 1494–1499.

van den Berg, C., Sibbing, F.A., Osse, J.W.M. and Hoogenboezem, W. (1992) Structure, development and function of the branchial sieve of bream (*Abramis brama*), white bream (*Blicca bjorkna*) and roach (*Rutilus rutilus*). *Envir. Biol. Fish.* **33**: 105–124.

van den Berg, C., van den Boogaart, J.G.M., Sibbing, F.A., Lammens, E.H.R.R. and Osse, J.W.M. (1993) Shape of zooplankton and retention in filter-feeding. A quantitative comparison between industrial sieves and the branchial sieves of common bream (*Abramis brama*) and white bream (*Blicca bjoerkna*). *Can. J. Fish. Aquat. Sci.* **50**: 49–62.

van den Berg, C., van den Boogaart, J.G.M., Sibbing, F.A. and Osse, J.W.M. (1994a) Zooplankton feeding in common bream (*Abramis brama*), white bream (*Blicca bjoerkna*) and roach (*Rutilus rutilus*), experiments, models and energy-intake. *Neth. J. Zool.* **44**: 15–42.

van den Berg, C., van den Boogaart, J.G.M., Sibbing, F.A. and Osse, J.W.M. (1994b) Implications of gill arch movements for filter-feeding—an X-ray cinematographical study of filter-feeding white bream (*Blicca bjoerkna*) and common bream (*Abramis brama*). *J. Exp. Biol.* **191**: 257–282.

van den Berg, C., van Snik, G.J.M., van den Boogaart, J.G.M., Sibbing, F.A. and Osse, J.W.M. (1994c) Comparative microanatomy of the branchial sieve in 3 sympatric cyprinid species, related to filter-feeding mechanisms. *J. Morphol.* **219**: 73–87.

van Leeuwen, J.L. and Muller, M. (1984) Optimum sucking techniques for predatory fish. *Trans. Zool. Soc. Lond.* **37**: 137–169.

Vinyard, G.L. (1982) Variable kinematics of Sacramento perch (*Archoplites interruptus*) capturing evasive and nonevasive prey. *Can. J. Fish. Aquat. Sci.* **39**: 208–211.

Wardle, C.S. (1975) Limit of fish swimming speed. *Nature* **255**: 725–727.

Wardle, C.S. (1977) Effects of size on the swimming speeds of fish. In: *Scale Effects in Animal Locomotion* (ed. T.J. Pedley), pp. 299–313. Academic Press, London.

Wardle, C.S. (1980) Effects of temperature on the maximum swimming speed of fishes. In: *Environmental Physiology of Fishes* (ed. M.A. Ali) Plenum, New York, pp. 519–531.

Wardle, C.S. and Videler, J.J. (1980) How do fish break the speed limit? *Nature* **284**: 445–447.

Webb, P.W. (1976) The effect of size on the fast-start performance of rainbow trout, *Salmo gairdneri*, and a consideration of piscivorous predator–prey interactions. *J. Exp. Biol.* **65**: 157–177.

Webb, P.W. (1978) Temperature effects on acceleration of rainbow trout, *Salmo gairdneri*. *J. Fish. Res. Board Can.* **35**: 1417–1422.

Webb, P.W. (1982) Avoidance responses of fathead minnow to strikes by four teleost predators. *J. Comp. Physiol.* **147**: 371–378.

Webb, P.W. (1984) Body and fin form and strike tactics of four teleost predators attacking fathead minnow (*Pimephales promelas*) prey. *Can. J. Fish. Aquat. Sci.* **41**: 57–165.

Webb, P.W. (1986) Effect of body form and response threshold on the vulnerability of four species of teleost prey attacked by largemouth bass (*Micropterus salmoides*). *Can. J. Fish. Aquat. Sci.* **43**: 763–771.

Webb, P.W. and Corolla, R.T. (1981) Burst swimming performance of northern anchovy, *Engraulis mordax*, larvae. *Fish. Bull.* **79**: 143–150.

Webb, P.W. and de Buffrénil, V. (1990) Locomotion in the biology of large aquatic vertebrates. *Trans. Am. Fish. Soc.* **119**: 629–641.

Webb, P.W. and Johnsrude, C.L. (1988) The effect of size on the mechanical properties of the myotomal–skeletal system of rainbow trout (*Salmo gairdneri*). *Fish Physiol. Biochem.* **5**: 163–171.

Webb, P.W. and Keyes, R.S. (1981) Division of labour between median fins in swimming dolphin (Pisces: Coryphaenidea). *Copeia* **1981**: 901–904.

Webb, P.W. and Keyes, R.S. (1982) Swimming kinematics of sharks. *Fish. Bull.* **80**: 803–812.

Webb, P.W. and Skadsen, J.M. (1980) Strike tactics of *Esox*. *Can. J. Zool.* **58**: 1462–1469.

Webb, P.W. and Weihs, D. (1986) Functional locomotor morphology of early life history stages of fish. *Trans. Am. Fish. Soc.* **115**: 115–127.

Weihs, D. (1973) The mechanism of rapid starting of slender fish. *Biorheology* **10**: 343–350.

Weihs, D. (1980) Hydrodynamics of suction feeding of fish in motion. *J. Fish Biol.* **16**: 425–433.

Weihs D. and Webb P.W. (1984) Optimal avoidance and evasion tactics in predator–prey interactions. *J. Theor. Biol.* **106**: 189–206.

Yin, M.C. and Blaxter, J.H.S. (1987) Escape speeds of marine fish larvae during early development and starvation. *Mar. Biol.* **96**: 459–468.

Yowell, D.W. and Vinyard, G.L. (1993) An energy-based analysis of particulate feeding and filter-feeding by blue tilapia, *Tilapia aurea*. *Environ. Biol. Fish.* **36**: 65–72.

Zottoli, S.J., Hordes, A.R. and Faber, D.S. (1987) Localization of optic tectal input to the ventral dendrite of the goldfish Mauthner cell. *Brain Res.* **401**: 113–121.

15

Biomechanics of display behaviour in tetrapods: throat display in squamates

Vincent L. Bels

1. Introduction

All organisms express a full range of behaviours to interact with their environment. These behaviours have been classified into two main categories following the context in which they occur. The 'everyday' behaviours deal with any kind of 'environmental necessities' (i.e. feeding, drinking and moving), and the 'display' behaviours are involved in inter- and intra-specific communicative contexts (Lorenz, 1966). Since the early stages of ethology, research on the causal mechanisms that underlie the display behaviours and determination of these mechanisms have immediately posed general questions on the origin and phylogenetic changes of these display behaviours (Huntingford and Turner, 1987). Valuable information on the course of the evolution of these behaviours comes from descriptions of behavioural traits among members of a taxonomic group. This approach permits deduction of the behaviour of a hypothetical common ancestor and their origin (Huxley, 1966).

Display behaviours in all animals are the result of movements of skeletal elements under a specific sequence of muscle activities driven by specific neuronal command. The skeletal elements are structures used in other key behaviours (e.g. feeding, locomotion, escape, and ventilation) or are structures that are used or exhibited only within the context of communication. For example, wings play a major role in courtship of birds (Tinbergen, 1952), but the major change for display purposes seems to relate to the plumage and not the structure of the wing itself. In contrast, horns and antlers of Bovidae and Cervidae are strongly correlated with their fighting and reproductive strategy (Kitchener, 1991). Subsequently, the display behaviour of an animal is subject to a large number of constraints directly related to the context of communication and/or constraints imposed by other environmental demands.

To understand the encoding, transmission, detection and decoding of a communicative signal between two or more individuals, ethologists respond to questions such as why animals respond to environmental or internal stimuli in a particular way, or why

Biomechanics in Animal Behaviour, edited by P. Domenici and R.W. Blake.

some animals respond in one way and others in another way in the same situation (MacFarland, 1993). Results from a large number of studies have provided data on the cause, the development, the adaptive significance and the phylogeny of display behaviours. More recently, a vast literature in neuroethology provides valuable data on the sensorimotor control of several behaviours (Alcock, 1998; Guthrie, 1987).

Questions about 'why' immediately raise questions about 'how' these behaviours are generated and controlled in a particular context of communication. Could their neuromotor origin be determined? What are the biomechanical and functional implications of the display behaviour on the complexity of the structures? How do we relate evolution of display behaviour and the structures involved in the communicative context?

The questions 'why' and 'how' cannot be posed separately in a global understanding of complex animal behaviours such as ritualized display behaviours. Integration of biomechanical and behavioural studies by using functional, ethological and neuroethological concepts and methods is probably one of the most valuable ways to achieve this understanding. Such integration needs to explore the display behaviours with diverse approaches including anatomy, histology, kinematics and electromyography, classical ethology and neurobiology.

1.1 Display behaviour: definitions

Before I provide a case study that documents the key importance of the integrative study of biomechanics and behaviour, it is necessary to look briefly to the concept of display behaviour used in classical ethology and to provide a basic picture of their evolution and origin. Records of 'instinctive behaviours' in birds (Heinroth, 1911; Huxley, 1914) probably offer one of the first complete detailed descriptions of display behaviour in tetrapods. To date, display behaviour can be defined as follows: 'Displays are stereotyped motor patterns involved in animal communication. Displays are largely genetically determined and specific to each species ... The function of a display is of considerable interest in understanding animal communication' (MacFarland, 1981, pp. 133–134); 'Behavior having a communication function; expressive behavior; signaling behavior; advertisement behavior ... Display thus includes all behavior patterns that are specially differentiated to serve intra-species (and sometimes inter-species) communication, such as courtship, threat and appeasement postures, and begging by young' (Immelmann and Beer, 1989, pp. 75–76).

The display behaviour consists of a series of discrete elements determined by a limited number of actions and postures. Actions are either organized in sequences of movements of one or several morphological structures or the repetition of movements of a given structure. Postures are the results of movements of structures 'frozen' in a typical position. These actions and postures have been described as 'instinctive action', 'hereditary coordination', 'fixed-action-pattern' or 'typical form' (Baerends, 1958; Lorenz, 1966; Schleidt, 1974; Tinbergen, 1952). Barlow (1977) proposed the alternative term of 'modal-action-pattern' (MAP) defined as a recognizable spatiotemporal pattern of movement widely distributed in similar form throughout an interbreeding population. Carpenter (1978) suggested that movements of the body of lizards in social context should be called display-action-patterns (DAPs). All this terminology was used to define ritualized behaviours used in social interactions within different clades of animals.

During the social display behaviour, a message is transmitted from one individual to another. By this message, one individual may indicate position and/or state of arousal (Payne and Pagel, 1997), or convey information such as size, strength, or intention (Colgan, 1989). To improve this message, display behaviour has a 'typical intensity' corresponding to a constancy of the form of the actions of animals independent of frequency (Morris, 1957). Colgan (1989) indicated that animals may produce a consistent signal in a particular context or, alternatively, exhibit behavioural flexibility according to the social context, and Moynihan (1982) reported that the significance of the communicative message may relate to ritualized and unritualized display behaviours.

1.2 *Origin of display behaviour and ritualization*

MacFarland (1981) defined ritualization as an evolutionary process by which any kind of behaviour patterns can become modified in display behaviour to serve a communication function, and Immelman and Beer (1989) defined it as the evolutionary transformation of non-display behaviour into display behaviour. Display behaviours have several possible origins. They can have their own genetic basis as 'everyday behaviour' (i.e. feeding, drinking, locomotion) or derive from another behaviour (Mayr, 1975). The derived nature of the display behaviour (Tinbergen, 1952) has been emphasized by many workers (Blest, 1961; Daanje, 1950; Hinde, 1966; Manning, 1972; Zahavi, 1979). Display behaviours are supposed to have derived from a wide variety of other behavioural activities such as 'everyday activities' and 'motivational conflict' (e.g. intention movements, displacement activities, and defensive behaviour) through the process of ritualization (Huxley, 1914, 1966).

Lorenz (1966) suggested that the term ritualization can be defined only by the enumeration of properties of the ritualized behaviours 'which constitute the essence of the concept only by summation' (Lorenz, 1966, p. 276). Three properties are of particular importance to explain the evolutionary changes of behaviour within the process of ritualization by which signals evolve from non-signals (Colgan, 1989). First, the behaviour originally used for responding to some environmental constraints acquires a new biological role, that of communication. Second, the display behaviour involves a change of form in the service of its communicative role. Third, the display behaviour must transmit an unambiguous communicative message (Lorenz, 1966), and the display behaviours also change during the evolution within a clade (Cullen, 1966). To emphasize this evolution of display behaviour in lizards, Carpenter (1978) used the term 'greatest degree of ritualism' in his description of agonistic and sexual behaviours in lizards. In their review of animal interactions, Huntingford and Turner (1987) proposed five plausible behaviours at the origin of the ritualized display behaviour: locomotion, exploration, physiological arousal, protective movements, and attack–flight conflicts.

2. Case study of throat display in lizards

The ritualized display behaviours of lizards provide an interesting model to address the general problem of understanding the evolution of display behaviour and its origin by means of the integration of biomechanics and behaviour. Here, the discussion is limited to the ritualized movements of the throat called *throat display* in lizards.

The aim of this paper is to show the role of this integration by documenting the complex relationships between the structure and the display behaviour. First, throat displays are defined on the basis of behavioural and kinematic studies. Second, the anatomy of the hyoid apparatus producing the throat displays is briefly described to investigate the possible relationships between its structural properties and throat displays. Third, the motor output of the throat muscles is explained to test the hypothesis of a similar mechanism for all throat displays, although their neuromotor controls remain to be investigated in detail. On the basis of recorded data, a possible (but not exclusive) scenario of the evolution of throat displays is proposed in a discussion of the process of ritualization and the origin of these displays.

2.1 Display-action-pattern (DAP)

The social behaviour of lizards, classified into challenge, assertion, threat and sexual contexts (*Table 1*), consists of complex actions and postural changes of various parts of the body, resulting in species-specific DAPs (Carpenter, 1978; Carpenter and Ferguson, 1977; Chiszar, 1978). Throat display is one of the major elements of the DAP. The size of the display repertoire in lizards varies from two DAPs in *Anolis aeneus* (Stamps and Barlow, 1972) to five in *Anolis limifrons* (Jenssen, 1977, 1978) and *Anolis chlorocyanus* (Bels, 1986). However, Jenssen (1978) suggested that the size of display repertoire in a large number of species has not yet been determined. Within this display repertoire, Jenssen (1978) defined the signature display of one species as the DAP performed in a context in which there is little or no social interaction ('the assertion' context, *Table 1*).

Table 1. *Definitions of the social display contexts*

Display	Definition	Reference
Threat	'Perhaps the simplest type of agonistic behaviour, as a type of aggression, is threat display'	Carpenter (1978)
	'A form of communication that usually occurs in situation involving mild aggression, or conflict between aggression and fear ... Threat sometimes takes the form of intention movements of attack'	MacFarland (1981)
	'...When an observer (or predator) approached the resident male suddenly and quickly or in challenge behaviors'	Bels (1986)
Assertion	'The signature display would be defined as that display type performed in a context in which there is little or no social interaction (the 'assertion' context)'	Jenssen (1978)
Challenge	'The greatest degree of ritualism, stereotypy or rigidity of behavior appears in the display'	Carpenter (1978)
Courtship	'Courtship may be defined as those actions, performed primarily by the male and, to a lesser extent, by the female, that bring the two sexes together to permit copulation'	Carpenter (1978)

Each DAP is divided into discrete units divided into *display type* and *display modifiers* (*Figure 1(a)* and (*b*)). The display type relates to (1) movements of the body (i.e. inflation, pushups on the forelimbs), the head (headbob) and/or the throat, and (2) any consistently associated movements of other body part (e.g. tail). Display modifiers are either dynamic or static (Jenssen, 1978, 1979). Roughly, dynamic modifiers are optional added movements of any parts of the body and static modifiers are optionally added postural changes (i.e. nuchal crest, open mouth, colour change).

As an example of DAP in lizards, we briefly describe one typical display type within the display repertoire of *Anolis carolinensis* reported by Bels and Goosse (1987) and extensively analysed by DeCourcy and Jenssen (1994), who recognized three display types in the repertoire of this species. Ventral view of displaying males shows that vertical displacements of the body and the head are always preceded by positioning of the fore limbs on a para-sagittal plane. *A. carolinensis* then exhibits a series of pushups.

The fore body is elevated by increase of the knee angle and angle of the shoulder (*Figure 1(a)* and (*b*)). The fore digits remain strongly in contact with the substratum, and the angle of the ankle strongly decreases during body elevation. The head rotates during each elevation of the fore body, and dewlap is always fully extended at the end of series. Throat movements to different extents during the series of pushup movements or at any other moment of the interaction are called throat displays (*Figure 1(c)*).

2.2 *Definition of throat display*

Six main modes of throat display have been recognized among lizards: throat inflation, dewlap, gular flap, frill erection, neck-frill extension, and lateral cheek flap extension. *Figure 2* summarizes the distribution of behavioural characters of throat displays of lizards. Throat extension is recorded in all families of lizards. Dewlap and gular flapping are exhibited by iguanids, and frill erection (*Pogona barbata*) and frillneck erection (*Chlamydosaurus kingii*) have been reported only in agamids.

The throat display of *Varanus griseus* in threat context provides a unique example of association of ventilatory cycle and throat movements. *V. griseus* performs two types of throat display in threat and challenge contexts called bucco-pharyngeal breathing pump (BPBP) and ventilatory bucco-pharyngeal breathing pump (VBPBP) (Bels *et al.*, 1995). The salient difference between VBPBP in *V. griseus* and other throat displays (e.g. throat extension, dewlap, frill erection) is the integration of ventilatory exhalation and inhalation and throat cyclic depression–elevation (*Figure 3(a)* and (*b*)). A first throat cycle is associated with exhalation of the air within the lungs and a second with inhalation of new air towards the lungs. This association results in complex displacements of air flow from the lungs through the throat as revealed by X-ray films during the threat response of the animal (Bels *et al.*, 1995).

2.3 *Throat display and anatomy of the hyoid apparatus*

All throat displays relate to movement of the hyoid apparatus incorporated in the floor of the oro-pharyngeal cavity. The hyoid apparatus is composed of cartilaginous and ossified elements with variable anatomical properties (*Figure 4(a)*). Differences in the structural properties of the hyoid elements (i.e. shape, length, articulation) have been observed (see, e.g. Avery and Tanner, 1971; Gnanamuthu, 1937; Jollie, 1960; Sondhi, 1958; Tanner and Avery, 1982). Quantitative investigation of morphology of the hyoid apparatus shows

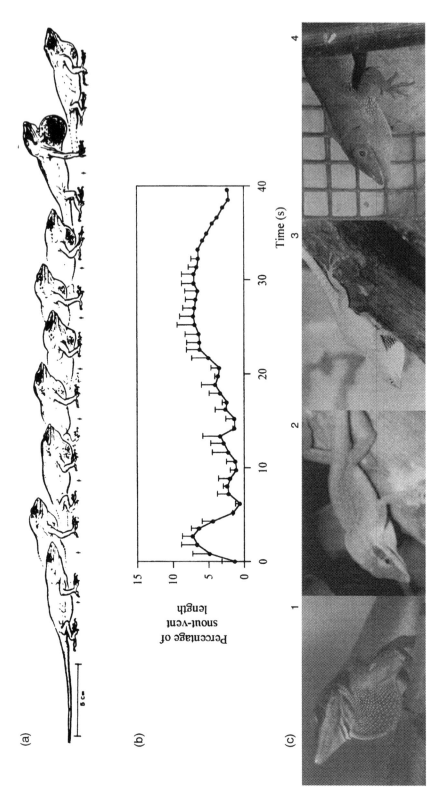

Figure 1. (*a*) *Lateral views of body movements of male Anolis carolinensis in challenge context. Dewlap is fully extended on the sagittal plane as the fore body is fully elevated and head fully backward rotated.* (*b*) *Kinematic profile of shoulder movement against time during a display (Type C, DeCourcy and Jenssen, 1994) of A. carolinensis (mean ± SEM of five DAPs from one male)* (*c*) *The throat is depressed at various levels during the challenge display repertoire in A. carolinensis (1 and 2) and Anolis chlorocyanus (3 and 4).*

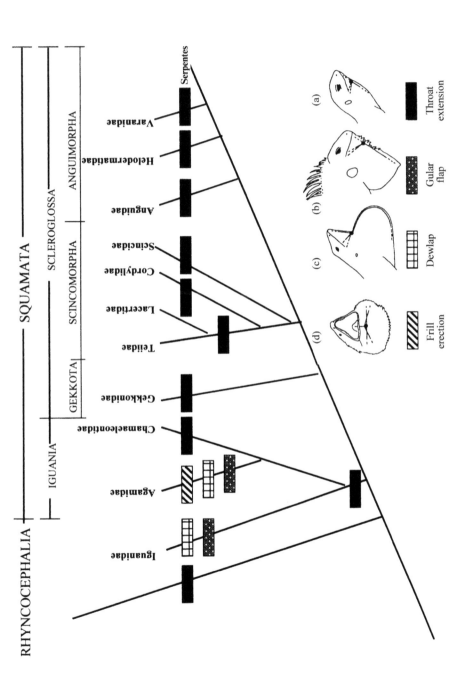

Figure 2. Phylogenetic position of DAPs in lizards (phylogeny is based on Estes et al. (1988)). (a) Throat extension; (b) gular flap; (c) dewlap; (d) frill erection.

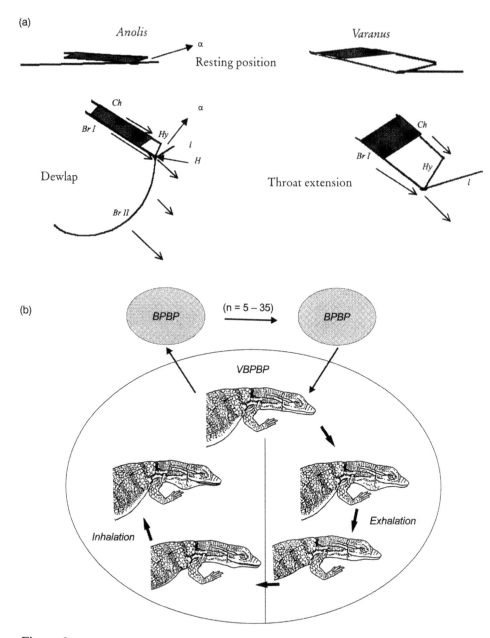

Figure 3. (a) *Comparison between hypothesized mechanisms of dewlap* (Anolis) *and throat extension of VBPBP* (Varanus). *In Anolis, the angle α between the lingual process (l) and the hypohyals (Hy) does not change because these lateral anterior horns are firmly connected to the hyoid body (H) (Bels, 1990). The rectangle connecting ceratohyals (Ch) and ceratobranchials I (Br I) is M. branchiohyoideus. The open arrows indicate movements of the hyoid elements recorded from X-ray films. Their length illustrates the amplitude of the relative displacement of each element. Br II, Ceratobranchials II. (b) Integration of throat and ventilatory cycles during threat display in* Varanus griseus. *BPBP, bucco-pharyngeal breathing pump; VBPBP, ventilatory bucco-pharyngeal breathing pump (Bels et al., 1995).*

that possession of a structurally specialized hyoid has predictive power for the type of throat display. Gular flap and dewlap occur for species with (i) *elongated* ceratobranchials II (gular flap: 54.9 ± 8.9% of the total length; dewlap: 71.1 ± 5.2%), (ii) *associated* cerato-branchials II along their long axis, and (iii) a *narrow* hyoid body (*Figure 4(b)* and *(c)*).

The pattern of mineralization of the ceratobranchials II also differs between lizards displaying gular flap and dewlap (*Figure 4(e)*). In *Iguana iguana* (gular flap), cerato-branchials II contain a complete ring of mineralized matrix, whereas a ventrally opened semicircle of mineralized matrix was observed in *Anolis equestris* and *A. carolinensis*. Preliminary comparison reveals a difference in shape of this ring in these *Anolis* lizards: it is generally more open in small *Anolis* lizards (e.g. *A. carolinensis*, *A. chlorocyanus*). This result could open a discussion on the evolutionary selection of this trait in the context of social radiation in this group of lizards, although an extensive study is needed in a phylogeny of more than 300 species.

2.4 *Motor pattern of throat displays*

The motor patterns of throat displays involve contractions of the throat musculature, which can be roughly divided into hyoid and external lingual muscles. The hyoid musculature can be divided into three groups of muscles: protractor inserted between the mandibulae and hyoid elements (ceratobranchials I and ceratohyals), retractor inserted between hyoid elements and the pectoral girdle, and actuator (M. branchio-hyoideus) of relative movements of the elements (*Figure 4(d)*).

Two conclusions on the motor pattern arose from studies of actions of throat musculature. First, contraction of M. branchiohyoideus plays a key role in this relative movement of ceratobranchials I and ceratohyals, as shown by electrical stimulation of this muscle in agamids (Throckmorton *et al.*, 1985), iguanids (Bels, 1990; Delheusy *et al.*, 1994) and varanids (Bels *et al.*, 1995). Font and Rome (1990) suggested that M. branchiohyoideus acts alone in dewlap extension (called dewlap pulses) in a challenge context, but recognized that throat and dewlap extension may occur simultaneously in a threat (antipredator) context.

Second, contraction of this muscle is probably helped by simultaneous contractions of the other throat muscles. According to Gnanamuthu (1937) and John (1972), dewlap extension in the agamid *Draco dussumieri* may be produced by simultaneous contraction of M. ceratohyoideus and M. mandibulohyoideus II (internus) inserted along the proximal one-third of the ceratobranchials. Throckmorton *et al.* (1985) suggested that other throat muscles (e.g. mm. mandibulohyoideus I and II), and lingual muscles (e.g. mm hyoglossus and genioglossus) contract during frill erection. In this throat display, M. mandibulohyoideus II protracts the hyoid body until the lingual process fits into the mandibular symphysis and rotates the hyoid body because it inserts on the posterior surface of the hyoid body. The ceratobranchials I are thus abducted and rotated around their long axes by the action of mm. mandibulohyoideus I (M. ceratomandibularis externus), hyoglossus, genioglossus and branchiohyoideus (M. ceratohyoideus).

Smith (1986) also reported that simultaneous activity of retractor and protractor muscles occurs in the opening stage of the pharyngeal compression during feeding. This opening stage clearly shares similar behavioural and kinematic features with the depression stage during the throat display in BPBP and VBPBP display-action-patterns of varanids, suggesting a similar mechanism of throat depression in both feeding and displaying contexts (Bels *et al.*, 1995).

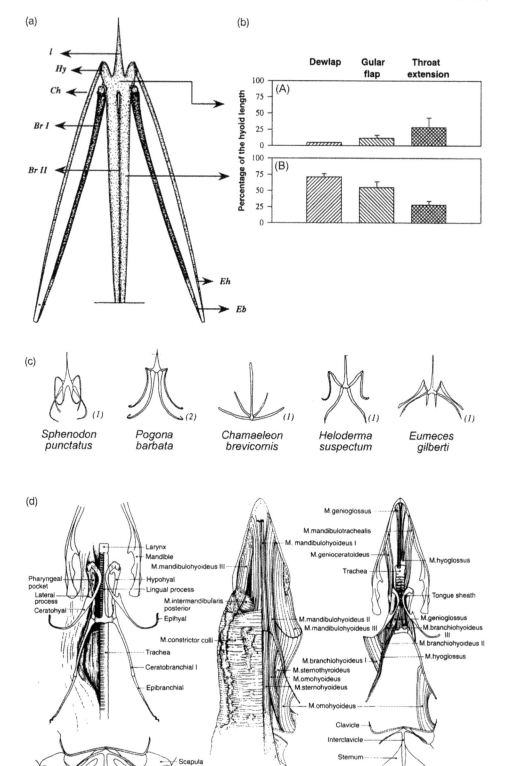

2.5 *A similar mechanism for all throat displays?*

To date, available data on biomechanics of throat display can be used to suggest that a similar mechanism can produce all the throat displays described, which have highly different shapes. This mechanism involves both movements of the hyoid apparatus as a whole and movements of the hyoid elements relative to the hyoid body. The mechanism of all studied throat displays is primarily based on relative movements of both posterior horns. The pivoting movement of the ceratobranchials II is the result of the relative movements of the ceratobranchials and ceratohyals in the dewlap in *Anolis* (Bels, 1990; Font and Rome, 1990; Von Geldern, 1919). The relative movement of both posterior horns of the hyoid apparatus alone produces the throat depression in *Varanus* (Bels *et al.*, 1995) because the ceratobranchials II are absent (*Figure 4(d)*).

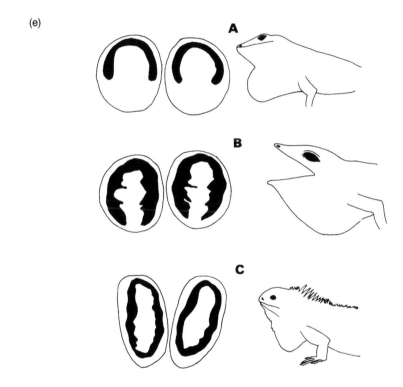

Figure 4. *(a) Ventral view of the hyoid apparatus in* Anolis equestris. *Br I, ceratobranchials I; Br II, ceratobranchials II; Ch, ceratohyals; Eb, epibranchials; Eh, epihyals; H, Hy, hypohyals. (b) Comparison of two structural features of the hyoid apparatus in relationships with DAP. All data were calculated as ratios on the total length of the hyoid apparatus from the tip of the lingual process to the end of the ceratobranchials II. In the case of absence of ceratobranchials II, the total length was considered to be from the tip of the lingual process to the posterior end of the hyoid body. The ratios were therefore greater than one. All data were recorded for one to two specimens of 19 species from Iguania and five species from Scleroglossa. (A) Width of the hyoid body; (B) length of the ceratobranchials II. (c) Shape of the hyoid apparatus shows considerable variability in* Squamata *and the out-group* Rhynchocephalia *(*Sphenodon *sp.): (1) redrawn from Tanner and Avery (1982); (2) redrawn from Throckmorton et al. (1985). (d) Throat musculature of* Varanus griseus *(from Bels et al., 1995, J. Zool. Lond. 235, 95–116) (e) Comparison of mineralization of ceratobranchials II at mid-length in* Anolis carolinensis *(A),* Anolis equestris *(B), and* Iguana iguana *(C).*

According to Throckmorton *et al.* (1985), lateral displacements of the cerato-branchials I relative to the hyoid body and of the ceratohyals relative to the anterior processes (hypohyals) occur in frill erection of *P. barbata*.

Font and Rome (1990) suggested that the hyoid acts as a first-order lever with hypo-hyals being the power arm, the ceratobranchials I–hyoid body joint as the fulcrum, and ceratobranchials the weight arm. This results in a posterior pull of the ceratohyals and immobility of the ceratobranchials I. Throckmorton *et al.* (1995) suggested that abduction and rotation of the ceratobranchials I is larger than that of ceratohyals, with the result that they cross the ceratohyals ventrally (see Figure 5 of Throckmorton *et al.* (1995)). On the basis of X-ray data, Bels (1990) showed that both the ceratobranchials I and the ceratohyals are moved in the dewlap by *A. carolinensis*. The protraction of ceratobranchials I is greater than that of ceratohyals, and acts on the hyoid body, producing vertical displacement of the ceratobranchials II in the sagittal plane. Accordingly, Bels *et al.* (1995) indicated that the angle between the ceratohyal and hypohyal increases as the ceratobranchials I are moved forward and downward during throat depression in *V. griseus* (*Figure 3(a)*).

3. Possible evolutionary scenario of throat display

As suggested in Chapter 2 for 'everyday' behaviours (e.g. locomotion and feeding), a principal component analysis where each principal component represents a combi-nation of the recorded variables (Lauder and Reilly, 1996; Reilly and Lauder, 1992) can provide a preliminary hypothesis on the possible variation among three traits asso-ciated with ritualistic DAPs. The first trait corresponds to the morphology of the hyoid apparatus (shape, length of the horns, pattern of mineralization), the second trait to the throat display exhibited by each species, and the third trait to the motor pattern. From available data (see above), this last trait simply relates to the fact that throat muscles expected to generate the display are contracted or not.

Figure 5 shows that the trajectories followed by the four taxa with respect to each of the three traits are rather complex. At the behavioural level, two taxa, *Anolis* and *Iguana*, share features of DAP (dewlap) produced by extension of the ceratobranchials II on the sagittal plane. Both taxa are therefore grouped together in the principal component analysis of the behavioural trait. *Pogona* and *Varanus* are located in different areas of the multivariate space and are distinct both from each other and from the group *Anolis* and *Iguana*.

At the level of morphology, *Anolis* and *Iguana* have long central posterior horns (ceratobranchials II), and *Pogona* has long lateral posterior horns (ceratobranchials I and ceratohyals) and no ceratobranchials II as *Varanus*. However, relative elongation of the ceratobranchials I is significantly greater in *Pogona* (ratio of approximately 1.05 for *Varanus*, and 1.72 for *Pogona*). Data on the motor pattern may be used to predict that the motor pattern of throat displays in lizards is probably similar in all lizards because M. branchiohyoideus must be contracted to generate the display, probably in association with simultaneous contraction of hyoid protractor and retractor muscles.

A preliminary conclusion can be drawn on the relationship between biomechanical traits and ritualized behaviour in the studied taxa. First, phenotypic different ritualized behaviours can be achieved by similar mechanism and motor pattern, although neuroethological analysis is needed to support this conclusion. Second, the structural

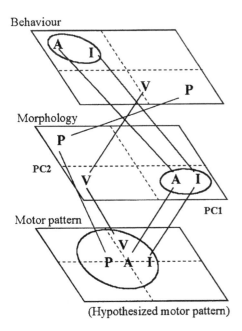

Behaviour

Morphology

PC2

Motor pattern

(Hypothesized motor pattern)

Figure 5. *Visualization of possible pattern of interspecific variation among behavioural, morphological, and motor pattern characters of the DAPs in lizards obtained from results of the principal component analysis of three data sets: behavioural pattern of the DAPs, structural pattern of the hyoid apparatus, and motor pattern established from available data. Each plane represents a plot of principal component analysis 2 (y-axis) versus principal component 1 (x-axis). Lines connect taxa on adjacent planes, and taxa that are not significantly different from each other are enclosed by circles. A, Anolis; I, Iguana; P, Pogona; V, Varanus.*

variations of the hyoid apparatus can be viewed as the major cause of the phenomeno-logically behavioural differences described in the traditional literature of ethology (Carpenter, 1978). Third, the action of one throat muscle plays the primary role in the motor pattern producing the mechanism involving relative movements of the hyoid elements for the different throat social signals in lizards.

4. Throat display and the concept of ritualization

In a preliminary kinematic comparison, Bels (1992) concluded that it was not easy to find the origin of the social throat DAP in pre-existing 'everyday' or 'emotionally motivated' (*sensu* Colgan, 1989) patterns. Bels *et al.* (1995) suggested a possible evolu-tionary hypothesis on the combination of movements of body regions used during 'everyday' behaviours to explain the complex throat display exhibited by *V. griseus* in threat context. Such combination could have been constrained under the evolutionary process of ritualization.

Recent data on buccal pumping reported in *Varanus* and some other lizards by Brainerd (1999) and Owerkowicz *et al.* (1999) in relation to locomotion permit us to explore an alternative hypothesis on the possible origin of throat displays in lizards. Throat display recorded in threat context for *V. griseus* could derive from ventilatory behaviour for two reasons: (i) throat display in threat context is viewed as the simplest type of aggressive display (Carpenter, 1978; *Table 1*); (ii) association of ventilation and throat movements produces an air flow permitting the animal to ventilate during a stressful action and to produce clear visual and auditory messages. According to the injunctive definition of ritualization proposed by Lorenz (1966), three main changes can be postulated during the evolutionary transformation of this 'everyday' behaviour into a signalling behaviour: (i) changes in the threshold; (ii) change in the motor coordination of the pattern of ventilation; (iii) exaggeration of the throat movements related to structural transformation of the hyoid apparatus.

The result for the case study presented here clearly demonstrates the value of integrated study of biomechanics and behaviour in understanding the causal basis of one major ritualized social behaviour in lizards. It clearly opens the discussion on the origin of this behaviour and its evolution. For example, the question about relationship between ventilation and throat movements during display behaviours remains to be investigated in detail within all families of lizards. The body inflation reported for many species (i.e. iguanids and agamids) in social interactions could indicate that some lizards at least could hold their ventilation during throat displays.

5. Conclusion and perspective

Available data for the case study of throat displays in lizards were used to emphasize the role of the integration of biomechanical (e.g. structure, performance, motor pattern) and behavioural (i.e. definition of the behaviour, social context) studies for analysing ritualized behaviours in tetrapods. Description of behaviours, their neuro-motor control, their phylogeny, and their role in the evolutionary process constitute one of the bases of the literature in ethology. For many years, major textbooks in this area have included the structural and physiological basis of behaviours (i.e. pattern of muscle activities) and have involved mechanical approaches of some behaviours. Recent textbooks in comparative and evolutionary anatomy often provide detailed descriptions of behaviours to explain the complex interaction between the structures and their functions (Kardong, 1998). A future integration of comparative and evolutionary study of ritualized behaviour in tetrapods from biomechanical, physiological and neuroethological points of view that include different concepts and methods should promise new valuable data. These data will be necessary for understanding the evolution, development and origin of social behaviours in the context of adaptation, defined as *'any feature that promotes fitness and was built by selection for its current role'* (Gould and Vra, 1982).

Acknowledgements

I thank Tom Jenssen for his very precious comments on a first version of this paper. Support was provided by the CARAH (Ath, Belgium). I also thank the editors and an anonymous referee for useful comments on this paper.

References

Alcock, J. (1998) *Animal Behavior : An Evolutionary Approach*. Sinauer, Sunderland, MA.

Avery, D.F. and Tanner, W.W. (1971) Evolution of the iguanine lizards (Sauria, Iguanidae) as determined by osteological and myological characters. *Brigham Young Univ. Sci. Bull.* **3**: 1–71.

Baerends, G.P. (1958) Comparative methods and the concept of homology in the study of behaviour. *Arch. Neerl. Zool.* **13** (Suppl. 1): 401–417. *Behaviour* **8** (1982): 1–416.

Barlow, G.W. (1977) Modal action patterns. In: *How Animals Communicate* (ed. T.A. Sebeok). Indiana University Press, Bloomington, pp. 98–134.

Bels, V.L. (1986) Analysis of the display-action-pattern of *Anolis chlorocyanus* (Sauria: Iguanidae). *Copeia* **1986**: 963–970.

Bels, V.L. (1990) The mechanism of dewlap extension in *Anolis carolinensis* (Reptilia: Iguanidae) with histological analysis of the hyoid apparatus. *J. Morphol.* 206: 225–244.

Bels, V.L. (1992) Functional analysis of the ritualized behavioural motor pattern in lizards; evolution of behaviour and the concept of ritualization. *Zool. Jb. (Anat.)* 122: 141–159.

Bels, V.L. and Goosse, V. (1987) Evolutionary perspectives in the push-up repertoire of the *Anolis* lizards in the carolinensis group. In: *Proceedings of the 4th Ordinary General Meeting of the Societas Europaea Herpetologica* (eds J.J. Van Gelder, H. Strijbosch and P.J.M. Bergers). Faculty of Sciences, Nijmegen, pp. 71–74.

Bels, V.L., Gasc, J.P., Goosse, V., Renous, S. and Vernet, R. (1995) Functional analysis of the throat display in the sand goanna *Varanus griseus* (Reptilia: Squamata: Varanidae). *J. Zool. Lond.* 235: 95–116.

Blest, A.D. (1961) The concept of ritualization. In: *Current Problems in Animal Behavior* (eds W.H. Thorpe and O.L. Zwangwill). Cambridge University Press, Cambridge, pp. 102–124.

Brainerd, E. (1999) Pumping and sucking: new perspectives on the evolution of vertebrate respiratory mechanisms. *Abstract, SEB Annual Meeting*, Edinburgh.

Carpenter, C.C. (1978) Ritualistic social behaviors in lizards. In: *Behavior and Neurology of Lizards* (eds N. Greenberg and P. McLean). National Institute of Mental Health, Rockville, MD, pp. 253–267.

Carpenter, C.C. and Ferguson, G.W. (1977) Variation and evolution of stereotyped behavior in reptiles. In: *Biology of Reptilia 7. Ecology and Behavior A* (eds C. Gans and D.W. Tinkle). Academic Press, London, pp. 335–554.

Chiszar, D. (1978) Lateral displays in the lower vertebrates: forms, functions, and origins. In: *Contrasts in Behavior: Adaptations in the Aquatic and Terrestrial Environments* (eds E.S. Reese and F.J. Lighter). John Wiley, New York, pp. 105–135.

Colgan, P. (1989) *Animal Motivation.* Chapman and Hall, London.

Cullen, J.M. (1966) Ritualization of animal activities in relation to phylogeny, speciation and ecology: reduction of ambiguity through ritualization. *Philos. Trans. R. Soc. Lond.* 251: 363–374.

Daanje, A. (1950) On locomotory movements in birds and the intention movements derived from them. *Behaviour* 3: 48–98.

DeCourcy, K.R. and Jenssen, T.A. (1994) Structure and use of male territorial headbob signals by the lizard *Anolis carolinensis. Anim. Behav.* 47: 251–262.

Delheusy, V., Toubeau, G. and Bels, V.L. (1994) Tongue structure and function in *Oplurus cuvieri* (Reptilia: Iguanidae). *Anat. Rec.* 238: 263–276.

Estes, R., de Queiroz, K. and Gauthier, J. (1988) Phylogenetic relationships within Squamata. In: *Phylogenetic Relationships of the Lizard Families* (eds R. Estes and G. Pregill). Stanford University Press, Stanford, CA, pp. 119–281.

Font, E. and Rome, L.C. (1990) Functional morphology of dewlap extension in the lizard *Anolis equestris* (Iguanidae). *J. Morphol.* 206: 245–258.

Gnanamuthu, C.P. (1937) Comparative study of the hyoid and tongue of some typical genera of reptiles. *Proc. Zool. Soc.* 107: 1–66.

Gould, S.J. and Vra, E.S. (1982) Exaptation: a missing term in the science of form. *Paleobiology* 8: 4–15.

Guthrie, D.M. (1987) *Aims and Methods in Neuroethology.* Manchester University Press, Manchester.

Heinroth, O. (1911) *Beiträge zur Biologie, insbesondere Psychologie und Ethologie des Anatiden.* Verhandlungen des internationalen Ornithologenkongresses, Berlin.

Hinde, R.A. (1966) *Animal Behaviour. A Synthesis of Ethology and Comparative Psychology.* McGraw–Hill, London.

Huntingford, F. and Turner, A. (1987) *Animal Conflict.* Chapman and Hall, London.

Huxley, J.S. (1914) The courtship-habits of the great crested grebe (*Podices cristatus*) with an addition to the theory of sexual selection. *Proc. Zool. Soc.* 2: 491–562.

Huxley, J.S. (1966) A discussion on ritualized behaviour in animals and man. *Philos. Trans. R. Soc. Lond.* **772**: 247–526.

Immelman, K. and Beer, C. (1989) *A Dictionary of Ethology.* Harvard University Press, London.

Jenssen, T.A. (1977) Evolution of anoline lizard display behavior. *Am. Zool.* **17**: 203–215.

Jenssen, T.A. (1978) Display diversity in anoline lizard and problem of interpretation. In: *Behavior and Neurology of Lizards* (eds N. Greenberg and P. McLean). National Institute of Mental Health, Rockville, MD, pp. 269–285.

Jenssen, T.A. (1979) Display modifiers of *Anolis opalinus* (Lacertilia: Iguanidae). *Herpetologica* **35**: 21–30.

John, K.O. (1972) On the gular mechanism of the South Indian flying lizard *Draco dussumieri* Dum and Bib. *Zool. Anz.* **188**: 12–23.

Jollie, M.T. (1960) The head skeleton of the lizard. *Acta Zool.* **41**: 1–64.

Kardong, K.V. (1998) *Vertebrates. Comparative Anatomy, Function, Evolution.* WCB-McGraw–Hill, Boston, MA.

Kitchener, A.C. (1991) The evolution and mechanical design of horns and antlers. In: *Biomechanics and Evolution* (eds J.M.V. Rayner and R.J. Wootton). Cambridge University Press, Cambridge, pp. 229–254.

Lauder, G.V. and Reilly, S.M. (1996) The mechanistic bases of behavioral evolution: a multi-variate analysis of musculoskeletal function. In: *Phylogenies and the Comparative Method in Animal Behavior* (ed. E.P. Martin). Oxford University Press, New York, pp. 104–137.

Lorenz, K. (1966) The psychobiological approach: methods and results. *Philos. Trans. R. Soc. Lond.* **772**: 273–284.

MacFarland, D. (1981) *The Oxford Companion to Animal Behaviour.* Oxford University Press, Oxford.

MacFarland, D. (1993) *Animal Behaviour.* Longman, Harlow.

Manning, A. (1972) *An Introduction to Animal Behaviour.* Edward Arnold, London.

Mayr, E. (1975) *La Biologie de l'Évolution.* Hermann, Paris.

Morris, D. (1957) 'Typical intensity' and its relation to the problem of ritualization. *Behaviour* **11**: 1–12.

Moynihan, M. (1982) Why is lying about intentions rare during some kinds of contests? *J. Theor. Biol.* **97**: 9–21.

Owerkowicz, T., Farmer, C., Hicks, J.W. and Brainerd, E.L. (1999) Contribution of gular pumping to lung ventilation in monitor lizards. *Science* **284**: 1661–1663.

Payne, R.J.H. and Pagel, M. (1997) Why do animals repeat displays? *Anim. Behav.* **54**: 109–119.

Reilly, S.M. and Lauder, G.V. (1992) Morphology, behavior, and evolution: comparative kinematics of aquatic feeding in salamanders. *Brain Behav. Evol.* **40**: 182–196.

Schleidt, W.M. (1974) How 'fixed' is the fixed action pattern? *Z. Tierpsychol.* **36**: 184–211.

Smith, K.K. (1986) Morphology and function of the tongue and hyoid apparatus in *Varanus* (Varanidae, Lacertilia). *J. Morphol.* **187**: 261–287.

Sondhi, K.C. (1958) The hyoid and associated strutures in some Indian reptiles. *Ann. Zool. Agra* **2**: 157–239.

Stamps, J.A. and Barlow, G.W. (1972) Variation and stereotypy in the displays of *Anolis aeneus* (Sauria: Iguanidae). *Behaviour* **47**: 67–94.

Tanner, W.W. and Avery, D.F. (1982) Buccal floor of reptiles, a summary. *Great Basin Nat.* **42**: 273–349.

Throckmorton, G.S., De Bavay, J., Cheffey, W., Merrotsy, B., Noske, S. and Noske, R. (1985) The mechanism of frill erection in the bearded dragon *Amphibolurus barbatus* with comments on the jacky lizard *A. muricatus* (Agamidae). *J. Morphol.* **183**: 285–292.

Tinbergen, N. (1952) Derived activities: their causation, biological significance, origin, and emancipation during evolution. *Q. Rev. Biol.* **27**: 1–32.

Von Geldern, C.E. (1919) Mechanism in the production of the throat-fan in the chameleon, *Anolis carolinensis. Proc. Calif. Acad. Sci.* **9**: 313–329.

Zahavi, A. (1979) Ritualisation and the evolution of movement signals. *Behaviour* **72**: 77–81.

16

The defence response in crayfish: behaviour and kinematics

Newton Copp and Marc Jamon

1. Introduction

Kinematic studies reveal an integrated picture of a particular behaviour in that they reflect the interplay between the internal control of muscles and the external application of forces that produces the observed movement of the animal. Dynamical studies are required, however, to distinguish those movements that represent active force production by parts of the animal from passive movements imposed by externally applied forces or movements of other parts of the animal's body. The combination of these two approaches produces a detailed and complete picture of the motions that characterize a behaviour. This description can prove instrumental in efforts to understand both the physiological basis and the functional significance of the behaviour because the movements that constitute a behaviour provide the interface between the underlying genetic and physiological control mechanisms and the function served by the act. This is readily grasped for continuing motor activities, but it is also relevant for behaviours characterized primarily by expression of a stable frozen posture. This chapter presents the defence posture and associated movements in crustaceans as an example of such behaviour.

Stereotyped visual displays that threaten an opponent are common among animals but are particularly evident in species with well-developed 'weapons' such as claws, horns and teeth. Defence displays by these animals typically include exhibition of their 'weapons' in the course of harmless manoeuvres directed with respect to the opponent. Decapod crustaceans exhibit a wide variety of defence displays (e.g. Schöne, 1968), but among the most common is one that features presentation of both chelipeds toward an opponent. Lobsters, crayfish and crabs commonly respond to threatening stimuli by elevating the anterior end of the body and spreading the chelipeds (e.g. *Figure 1*). (See Chapter 11 for other uses of the claws.) This meral spread display can be observed in confrontations with potential predators or during agonistic encounters concerning access to food, shelter, or mates.

Biomechanics in Animal Behaviour, edited by P. Domenici and R.W. Blake.
© 2000 Taylor & Francis, Oxford.

Figure 1. Specimen of the crayfish Procambarus clarkii *exhibiting a defence display.*

An effective defence display decreases the probability of attack by a predator or reduces the risk of costly outcomes to agonistic encounters. Obviously, this visual display must be directed with respect to the opponent if it is to have a favourable effect for the displaying animal. The display must be presented not only to be clearly visible to the opponent, but also to position the displaying animal for subsequent attacking or retreating movements as appropriate. Among those decapods that elevate their chelipeds as part of the display, some of the targeting can be accomplished by movement of the chelipeds relative to the body. Typically, however, the chelipeds are raised symmetrically with respect to the body, and the targeting is achieved by positioning the body to bring the opponent within the angle subtended by the elevated chelipeds.

The body turn involves a more or less stereotyped sequence of movements (Copp, unpublished: see also Breithaupt *et al.* (1995)). The first visible motion typically is retraction of the ipsilateral antenna (*Figure 2*), which begins within 70 ms after stimulation and precedes rotation of the body. The antenna often sweeps far enough posteriorly to contact the opponent if it is nearby. Shortly thereafter, the contralateral uropod promotes underneath the telson (*Figure 2*) and the abdomen rotates slightly along its long axis away from the stimulated side of the body. The ipsilateral cheliped extends and remotes toward the opponent while the contralateral cheliped flexes and sweeps across the front of the animal (*Figure 2*). As these latter motions are occurring, the body begins to rotate on the legs toward the opponent. Stepping movements of the legs continue the turn if the angle between the opponent and the mid-body axis of the turning animal is sufficiently large. Rotation of the body brings it in approximate alignment with the target and restores the chelipeds to a symmetrical position about the animal's mid-line. After the animal completes the turn, one of several behaviours may ensue including the meral spread display, advancing toward the aggressor, remaining still, or walking backward. The orientation of the body with respect to the external target therefore positions the animal appropriately for subsequent action.

The turning of the body should thus be considered an important and integral

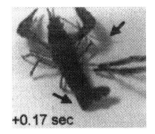

Figure 2. Specimen of P. clarkii *responding to a tactile stimulus with a defence turn. Numbers indicate times relative to the onset of the stimulus. Arrows in the middle panel of the upper row indicate remotion of the ipsilateral antenna and the contralateral uropod. The arrow in the third panel indicates remotion and opening of the ipsilateral cheliped. These motions are characteristic of turning responses.*

component of the overall defence response. Unfortunately, this component has received little attention. Most researchers refer to the 'defence posture' to indicate only the stereotyped meral spread display. We prefer to include under the term 'defence response' both the defence posture and any prior turning motions that position the animal relative to the stimulus that elicited the response. The term 'defence display' is reserved here for the meral spread posture and 'defence turning' will be used to refer to the orienting component of the response. It is important to analyse both components of such a movement from a perspective that integrates behaviour and biomechanics.

Defence displays have been more widely studied than defence turning perhaps because the stereotypy of these behaviours in decapod crustaceans has made them attractive subjects for investigation of the physiological basis of the behaviour, as well as the function and evolution of the behaviour. Physiological studies of decapod defence displays have contributed to our understanding of the neural organization (e.g. Beall *et al.*, 1990; Bowerman and Larimer, 1974; Wiersma, 1961), neurochemical modulation (Beltz and Kravitz, 1983; Livingstone *et al.*, 1980; Mattson and Spaziani, 1986), and sensory control (e.g. Glantz, 1974, 1977) of behaviour. On the other hand, investigations into the function of decapod defence displays have dealt with the information content of animal communication signals (Hazlett and Bossert, 1965), interspecific variation in signalling as it may pertain to the evolution of behaviour (e.g. Schöne, 1968), and social behaviour (e.g. Bovbjerg, 1956; Copp, 1986).

Defence turning behaviour in crustaceans has not been widely studied but is clearly related to other directed turning behaviours, such as prey attack (e.g. Breithaupt *et al.*, 1995; Varju, 1989) and perhaps also to curve walking (e.g. Cruse and Silva Saavedra, 1996; Domenici *et al.*, 1998). In this context, the major focus has been on using kinematic studies of turning motions to extend our understanding of locomotor patterns in arthropods (Copp and Jamon, unpublished data).

Lingering issues regarding the 'how' and 'why' of decapod defence behaviours may be elucidated by kinematic and biomechanical studies, but this effort is only just beginning. In the sections that follow, information about the kinematics of the defence display and defence turning in the crayfish *Procambarus clarkii* is presented along with speculations concerning the implications of these studies for understanding the physiological basis and behavioural function of the defence response in this animal.

2. The defence display

2.1 *Display function*

Agonistic encounters in decapods and other crustaceans may progress through a sequence of stereotyped behaviours characterized by increasing levels of aggressive intensity. Such sequences have been described in crabs (Elwood and Glass, 1981; Glass and Huntingford, 1988), stomatopods (Dingle, 1969), lobsters (Huber and Kravitz, 1995; Scrivener, 1971) and crayfish (Bruski and Dunham, 1987). Frequently, however, agonistic encounters do not lead to actual fighting and damage to one or more of the combatants but are resolved by the defence display. It seems likely that elevation of the chelipeds in a defence display by lobsters, crabs and crayfish provides opportunities for assessment of body size and ability to fight. Garvey and Stein (1993) showed that chela size correlates directly with body size in the crayfish *Orconectes* and also relates directly with success in aggressive encounters and predator defence. Mantis shrimp, *Gonodactylus bredini*, are not decapods, but studies of their behaviour reveal possible communication functions for defence displays in crustaceans. Steger and Caldwell (1983) have argued for an assessment function of meral spread displays in these stomatopods based on an interesting deception practised by newly moulted specimens. A stomatopod residing in its burrow elevates and spreads its formidable raptorial appendages when an opponent advances on the burrow. These displays are 'honest' in that they correlate well with attack. Newly moulted individuals with soft exoskeletons, however, also produce the display despite their inability to fight. In these cases, the display is clearly an act of deception. Steger and Caldwell (1983) proposed that the display provides information about size that an opponent can use to determine the displaying individual's ability and, perhaps, motivation to fight. Indeed, the meral display of lobsters and crayfish allows individuals to determine the dominance status of their opponent in about 80% of the agonistic encounters with conspecifics (Stein *et al.*, 1975).

2.2 *Cheliped kinematics*

Detailed understanding of the motion of the chelipeds is obviously related to the information content of the defence display and so is important for elucidating the signalling function of this behaviour as well as for providing a foundation for comparative studies that may shed light on the evolution of these displays. The kinematics of the defence display in decapod crustaceans has been almost completely ignored, however. One exception is a study by Kelly and Chapple (1990) in which various kinematic parameters of cheliped movement leading to the defence display of the crayfish *Cambarus bartonii* were analysed. These investigators drew a parallel between reaching movements in mammals and the 'unidirectional, multijoint, non-oscillatory movement' of the chelipeds producing the crayfish defence display, and argued from this analogy that similar kinematics techniques could be profitably applied to both types of movement.

Kelly and Chapple (1990) found that the crayfish cheliped comes to a stereotyped end point position in the defence display. Given the prominence with which the chelipeds are featured in this display, it seems reasonable that changes in their end point position may alter the information conveyed by the display. Specimens of *P. clarkii* in the defence display often respond to further provocation by additional spreading of their chelipeds. An important signalling function for cheliped position would be consistent with Kelly and Chapple's finding that the cheliped joint angles at the end point of the reflex were among the least variable parameters that they measured.

Kelly and Chapple's kinematic description of the meral spread display in *C. bartonii* reveals differences from primate reaching movements that have implications for understanding the neural control system underlying crustacean defence displays. Although the duration of the 'reaching' movement is relatively constant in the crayfish and the velocity of the movement increases with the amplitude as occurs in some mammals, the system of co-ordination underlying cheliped movement differs dramatically from that employed in arm movements by primates. A high degree of control over the trajectory of the hand in reaching movements by primates leads to smooth tangential velocity profiles of the hand (Atkeson and Hollerbach, 1985) indicating an underlying system of close co-ordination among the arm joints. In the crayfish, however, movements of the cheliped joints leading to the stereotyped end point are variable in terms of angular velocity and duration and show no discernible inter-joint co-ordination. Asynchronies in joint movement contribute to irregularities in the resulting tangential velocity profiles of the propodite. Furthermore, trajectories of the propodite vary from trial to trial within individual specimens even when the starting and end points are similar. Kelly and Chapple (1990) concluded that the crayfish does not control the trajectory of its propodite, as a primate controls the trajectory of its hand, but that 'the endpoint of the movement simply represents the sum of independent joint activity'. The crayfish apparently lacks the mechanisms for inter-joint co-ordination that would allow control of propodite trajectory in the defence display. Kelly and Chapple speculated that the differences in these two strategies may reflect the fact that the defence display reflex in the crayfish moves the cheliped to a stereotyped position relative to the body whereas reaching movements in primates go to targets that vary in position.

As suggested by the preceding example, our appreciation of the neural control systems underlying defence displays in decapod crustaceans can be enhanced by kinematic and biomechanical analyses of the behaviour. The significance of details and variations of activity in these control systems can be understood only in the context of a kinematic analysis of the behaviour itself. Schaeffer *et al.* (1994), for example, used kinematic descriptions of escape turns in the cockroach to compare tactile and wind-elicited responses, and argued from the kinematic similarity that the two types of response probably utilize the same neural circuitry. On the other hand, kinematic studies of the movement of swimming appendages in the blue crab, *Callinectus sapidus*, revealed that apparently similar rhythms in courtship, sideways walking and backward walking are controlled by different neural systems sharing the same motor units (Wood and Derby, 1995). The latter kinematic studies also set the stage for discovering that neuromodulators can trigger the shift from one pattern of swimming appendage movement to another (Wood *et al.*, 1995).

Kinematic studies might provide a bridge linking the physiological basis of a behavioural act to its function and thus provide a more unified understanding of the behaviour. As described above, the neural control system underlying cheliped movements in the crayfish limits the amount of co-ordination among cheliped joints, leading to significant variability in cheliped motion to produce the meral spread display. What, indeed, is the relationship between the movements of the chelipeds in the meral spread display and the signalling function(s) performed by the display? The most stereotyped feature is the one that appears to be most closely related to a communication function for the display, but this possibility needs further investigation.

3. The defence turn

3.1 *Body movements*

Because turning motions preceding the defence display in decapod crustaceans serve, at least in part, to direct the display towards the threat, it would seem to behove the responding individual to complete the turn and execute the display as rapidly as possible. The average latency between the stimulus and onset of rotation of the body is 130 ± 40 ms when a crayfish is pinched on a leg. This is considerably shorter than the latency measured in specimens of *P. clarkii* turning toward small fish in response to water-borne vibrations (555 ± 220 ms, Breithaupt *et al.*, 1995). The latency to the onset of defence turning in the crayfish is considerably longer, however, than in tailflip escape responses of the crayfish (approximately 20 ms; Wine and Krasne, 1972). Other considerations may mitigate against maximum response speed during defence turning, however. The nature of the stimuli that elicit defence turns in the crayfish apparently call for a response that permits the animal to attack or retreat as circumstances demand. Also, defence turns must be effective under various conditions of substrate composition and water flow, and predators or conspecifics that deliver the threatening stimulus may move while the orienting and display movements are in progress. The need to ensure reasonably accurate orientation to a moveable target may favour a response strategy that sacrifices rapid acceleration or high angular velocity for adaptability of angle.

The conflict between turn speed and accuracy results in a complex mix of translational and rotational components in the defence turning response. The movement of the body during defence turns by specimens of *P. clarkii* approximates rotation in place but includes a translational component as well (*Figure 3(a)* and *(d)*). These two components of the overall motion are difficult to separate. For example, a sideways motion would be pure translation if the lines describing the longitudinal body axis before and after the movement were perfectly parallel. Any slight deviation of these two lines from parallel, however, could be viewed either as a translation superimposed on a rotation about a point on the body or as a pure rotation with the centre of rotation

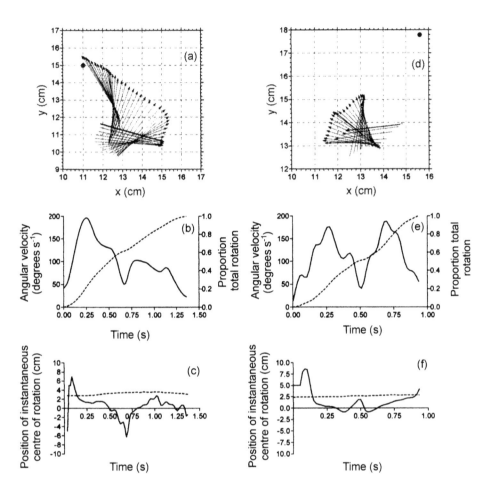

Figure 3. *Two examples (a–c, d–f) of defensive turns by a single specimen of P. clarkii. (a, d) Arrows represent the approximate length of the carapace (arrowhead indicates anterior) and show changes in the orientation of the animal's longitudinal axis in response to a pinch delivered to one of the fifth walking legs.* ●, *position of the stimulated point on the leg at the onset of the tactile stimulus. Time between successive arrows in the figure is 0.033 s. The heavy arrow indicates the orientation of the animal just before it began to rotate. (b, e) Angular velocity of the body (continuous line) and progress toward final orientation angle (dotted line) of the animal producing the turn shown in (a) and (d) respectively. (c, f) Position of the instantaneous centre of rotation (continuous line) relative to the carapace during the turns shown in (a) and (d) respectively. The position of the intersection between the two lines of the long axis of the body in two successive video frames (interval 0.017 s) was determined relative to the carapace in the first of the two video frames. The y-axis represents a line extending along the length of the animal's carapace. Zero on the y-axis indicates the posterior marker on the carapace (between the fourth walking legs). The dotted line indicates the position of the anterior marker at the base of the rostrum. The vertical distance between the x-axis and the dotted line thus indicates the approximate length of the carapace. It changes slightly because the animal was videotaped from directly overhead, and the projected length varies with the tilt of the body.*

positioned well anterior or posterior of the animal, depending on the inclination of the lines to each other. Kinematic analysis of the body rotation allows us to determine the shift in the 'instantaneous' position of the centre of rotation and thus estimate the rotational and translational components of the movement. The position of the centre of rotation at the onset of the response is variable but tends strongly to be in front of the animal or on the carapace near the anterior end (*Figure 3(c)* and (*f*)). If the centre of rotation is not positioned on the carapace at the beginning of a defence turn, it moves posteriorly onto the carapace within approximately 100 ms. It remains somewhere on, or within 1 cm of either end of the carapace for approximately 80% of the duration of the response (*Figure 3(c)* and (*f*)). Therefore, defence turns are mainly a rotational movement, although the mix of rotation and translation, and thus the trajectory of the movement, is highly variable even within a single specimen. As in cheliped movements to the final meral spread position, similar target positions of the body can be achieved by any one of several trajectories.

Freely behaving specimens of *P. clarkii* complete defence turns through approximately 100° at an average rotational rate of 110 ± 23° s^{-1}. This compares reasonably well with the rate measured in specimens of *P. clarkii* turning toward small fish in response to waterborne vibrations (124 ± 48° s^{-1}; Breithaupt *et al.*, 1995). As in their trajectory, however, turning responses vary considerably in duration, average rotation rate, and angular velocity profile (*Figure 3(b)* and (*e*)). The irregularity of the angular velocity profiles of defence turning responses contrasts with the smoothness that characterizes the tangential velocity profiles of reaching motions in primates (Atkeson and Hollerbach, 1985) but resembles the tangential velocity profiles of the propodite and the angular velocity profiles of remotion during cheliped movement in the defence display reflex by the crayfish (Kelly and Chapple, 1990), as well as the angular velocity profiles of crayfish turning toward swimming fish (recomputed from Figure 4 of Breithaupt *et al.* (1995)). Perhaps, as suggested by Kelly and Chapple (1990) for propodite motion, the crayfish does not plan the trajectory of the body in a defence turn. Kinematic descriptions of defence turns thus suggest questions to investigate in the neural control of this directed behaviour.

Despite their overall variability, the angular velocity profiles of the body of a crayfish during a defence turn suggest that the behaviour unfolds in two stages. The initial angular acceleration appears to be sufficiently reliable to describe it as the initial phase of the turning response distinct from the subsequent, more variable phase in which the orienting motion is completed. In general, the rotation rate of the body increases rapidly within the first 260 ± 56 ms of the response to a peak angular velocity, the magnitude of which varies within and between individuals, and then declines in a highly variable way. Additional peaks in angular velocity may occur. A 'trough' in the angular velocity profile frequently occurs about 540 ± 180 ms after the onset of the response and divides the two phases in our proposed scheme.

The proposed two-phase structure of the angular velocity profiles of defence turns in the crayfish may correspond to features of the underlying neural control system. A previous study has shown that two types of control come to play in defence turning of the crayfish (Copp and Watson, 1988). The turn begins under ballistic or open loop control in which the direction and approximate magnitude of the rotation are roughly established by the angle at which the tactile stimulus is delivered to the body. Feedback about motion of the visual field and the appearance of novel objects in the visual field during a turn contribute to determining the magnitude of rotation in closed-loop control fashion. The initial phase of the defence turn may correspond to the period of open-loop control, whereas the variability of the second phase may reflect the influence of sensory feedback in a closed-loop control mode. Some escape behaviours in invertebrates exhibit a two-phase structure in which closed-loop control follows an initial ballistic response (e.g. Camhi and Tom, 1978; Wine and Krasne, 1972), although the same may not always be true for escape responses in fishes (Domenici and Blake, 1997). Breithaupt *et al.* (1995), however, contrasted the continuous motion of turns by the crayfish toward swimming fish with the slower, more discontinuous search behaviour of crustaceans in odour plumes, and suggested that turning towards prey is controlled by an underlying open-loop control system. Further research is needed to determine the nature of the control system governing turning behaviour in the crayfish. If the two-phased control scheme is shown to pertain to defence turns, it would be interesting to speculate that the initial ballistic phase reflects selective pressures for some degree of urgency in the response whereas the subsequent closed-loop control phase permits the animal to retain an effective degree of accuracy despite unpredictable movements of the target.

Despite variability in trajectory, the animal's body actually points at the target with reasonable accuracy, leaving an average error of −10%. The response, on average, is sufficiently accurate to bring the target within the angle subtended by the outstretched chelipeds, that is, approximately 90°. Apparently, selective pressures for more precise orientation in this behaviour are not strong. Subsequent behaviours, such as backward walking or attack, involve further opportunities to orient the body with respect to the target. Perhaps the defence turn should be considered more as a preparatory behaviour than as a complete resolution of a behavioural situation.

3.2 *The walking legs*

The kinematic analysis of body rotation indicates how crayfish produce defence turns that are rapid and accurate, but the motor output also needs to account for biomechanical constraints and adaptability of the neural commands. The defence turn, indeed, relies on the same biomechanical constraints of the motor apparatus and the same sensory–motor circuitry as straight or curve walking behaviour. Indeed, it might be expected that defence turns in the crayfish represent simply a special case of curve walking with a small radius of curvature. Comparisons among these various locomotor behaviours in terms of leg kinematics, with consideration of the relative rotational and translational components of the movement, are helpful in answering this question.

Defence turning by the crayfish features quicker and probably smaller steps than seen in forward walking; the period and stance duration of steps in defence turning are shorter than in forward walking (0.3 and 0.18 s vs. 1 and 0.5 s, respectively; Jamon and Clarac, 1995; see also Clarac (1984) and Klärner and Barnes (1986)). The entire rotation of a defence turn is thus completed with several steps in about 1 s, that is, approximately the same amount of time as the step period in forward walking. The relative timing of the elements of the step cycle in defence turning remains, however, substantially the same as in forward or curve walking.

The legs on the outside of the arc (outer legs) exhibit posteriorly directed stances and the inner legs (legs on the inside of the arc) exhibit anteriorly directed stances, as in forward and backward walking, respectively, but the net trajectory of each leg's stance, defined from the beginning to the end of a stance phase relative to the body, is canted to the body's midline axis such that, collectively, these trajectories describe a circle around the animal (*Figure 4*). In actuality, however, the dactyl of each leg is in a fixed location

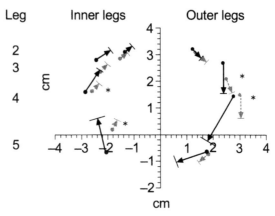

Figure 4. *Mean net stance vectors for initial stances (thick arrows) and all subsequent stances (dashed arrows) by inner and outer legs during defensive turns by P. clarkii. Vertical and horizontal axes indicate body-centred coordinates. The vertical axis indicates the animal's midline (anterior is up). The intersection between the horizontal and vertical axes represents a reference point on the carapace midway between the fourth walking legs. Each vector was determined as the line connecting the average position of a dactyl at stance onset and the average position of the dactyl at the end of the stance. Lines across the tips of the vectors represent 95% confidence intervals about the average vector angles calculated with circular statistics. Asterisks indicate significant differences between the angle of the mean first stance vector and the angle of the mean remaining stance vectors (circular statistics, Watson–Williams test, P < 0.05). n = 25 for all first stance vectors. For the remaining stance vectors, the sample sizes are as follows: I2 = 101; I3 = 98; I4 = 83; I5 = 87; O5 = 90, O4 = 69; O3 = 82; O2 = 93.*

during its stances. This means that the amplitudes and directions shown for each leg in *Figure 4* represent the movement of the body axis relative to that leg during its stance. This more or less circular pattern suggests, therefore, that the crayfish is twisting on the legs, and the centre of rotation is on or near the body axis.

The initial acceleration of a defence turning response occurs when most, if not all, walking legs are in stance as the body rotates on the stable base provided by the legs. The amount of body rotation achieved during the initial phase is directly related to the number of legs that are simultaneously in stance during that phase. The body's angular velocity decreases and the turning response enters its second phase as the walking legs complete their initial stances and are repositioned to begin their next stances. As this transition is made, the kinematics of stepping change. Outer legs three, four and five, and inner legs four and five (numbered from anterior to posterior, with the cheliped on each side labelled as leg number one) direct their stances differently during the initial acceleration of the body than during the second phase of the response (*Figure 4*). The initial steps tend to have longer periods, longer swing and stance durations, and larger stance amplitudes than subsequent steps. The two groups of steps do not differ, however, in relative timing of their various strokes. The difference in stance amplitude between initial and subsequent steps might be related to a change in the translational component of the body movement.

Even within a particular phase of body motion, members of a pair of legs do not behave in the same way. In particular, outer legs three and four exhibit larger amplitude stances than their inner counterparts during the initial phase, and all outer legs produce larger amplitude stances than their inner counterparts during the subsequent steps.

These differences cannot be explained in all instances by accompanying differences in stance duration, step period, or duty factor. They may result from an asymmetry in the distance between the inner and outer legs and the centre of rotation. Perhaps the centre of rotation is moved laterally towards the inner legs. This possibility is underscored by the observation that inner legs three and four take more steps in a defence turning response than their outer counterparts. This difference in stepping also suggests a functional difference between these legs. By taking smaller, more frequent steps, inner legs three and four may be serving more to maintain balance than to produce torque compared with outer legs three and four. A lateral shift of the centre of rotation toward the inner legs could explain the different contributions of inner and outer legs to a defence turn. In this case, inner legs might be producing ground reaction forces of similar magnitude to those for the outer legs, but their contribution to torque of the body would be lower because of their closer proximity to the centre of rotation. Alternatively, the outer legs may actually produce larger forces than the inner legs. Deciding among these options has implications not only for understanding the neural control systems but also for understanding the strategy of turning.

Extrapolating from kinematic studies to physiological or functional consequences for defence turning is a risky venture in the absence of information about the forces involved in the actions of the legs. As discussed by Domenici et al. (1998), it is not possible in purely kinematic studies to separate the active contributions of any particular leg to body motion from the passive effects of forces exerted by other legs. One must either be able to correlate leg motions with accelerations of the body, or relate kinematic studies to force measurements to understand the specific contribution made by each leg to rotation of the body. Measuring forces exerted by the different legs in the course of unrestrained movement presents, however, a difficult challenge with classical force measurement tools such as force platforms. A technique based on photoelastic gelatine has now been developed for invertebrates (Full et al., 1995; Harris, 1978; Harris and Ghiradella, 1980; Jindrich and Full, 1999). This technique has been applied to specimens of P. clarkii during defence turns to measure the forces exerted by the various legs against the substrate. Figure 5 shows a typical example of the timing of forces exerted by the various legs during a defence turn. The time profiles of the ground reaction forces and the body movement (Figure 5(a)–(c)) are consistent with the results of the kinematic analyses reported above. The larger stance amplitudes of outer legs three and four during the initial acceleration and of all outer legs during the second stage of rotation of a defence turning response by specimens of P. clarkii suggest that these legs contribute more than their inner counterparts to the torque that rotates the animal's body. The first apparent increases in force are produced by inner legs three and four and especially leg five, which reacts to the tactile stimulus by extending, and are responsible for the initial lateral translation. The initial sharp increase in the body's angular velocity (to a peak at about 260 ms) is associated with strong forces applied by outer legs three and four and, surprisingly, inner leg two (see also Figure 5(d)). The subsequent decrease in the body's angular velocity is related to the decline in the force applied by the outer legs. This accords well with the statement above that the initial acceleration of the body turn depends on the number of legs in stance at the moment of stimulation. It also suggests that the amplitude of the total force depends on a limited number of legs, as seems to be the case in curve walking (Domenici et al., 1998). In the subsequent steps, the forces exerted by the legs are reduced, as is the angular velocity of the body.

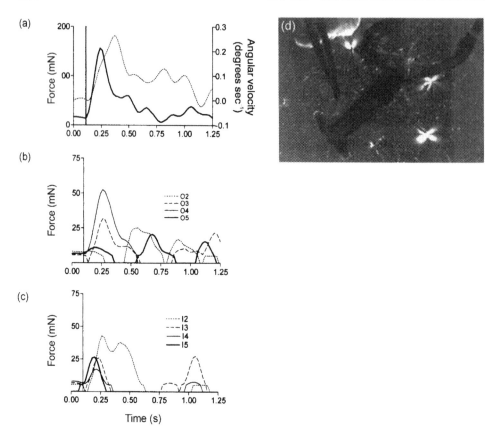

Figure 5. *Forces applied to a photoelastic gelatin create stresses within the gelatin that shift the plane of polarized light passing through it. These stresses can be visualized (d) by suitably arranging one polarizing filter between the light source and the gelatin and another between the gelatin and the recording camera. Appropriate calibration of the gelatin allows quantitative interpretation of the illuminated spots as ground reaction forces. The curves showing the forces have been smoothed. (a) The sum of the forces (in milliNewtons) produced by all legs (continuous line) is shown with the angular velocity profile of the body (dotted line). The vertical line to the right of the y-axis indicates the time of onset of the tactile stimulus that initiated the subsequent defensive turning response. (b) The forces generated by outer legs 2–5 during the turning response shown in (a). (c) The forces generated by inner legs 2–5 during the turning response shown in (a). The vertical axes in (b) and (c) are aligned with the onset of the pinch in (a). (d) Single frame from the response shown in (a) illustrating the forces generated by inner leg 2 (upper left), outer leg 3 (right centre), and outer leg 4 (lower right).*

3.3 *Comparisons with curve walking*

What does this discussion of leg kinematics suggest about the relationship between defence turning and other behaviours that feature a large rotational component? In curve walking, for example, a significant rotational element is added to the forward movement of the body. As in defence turns, the various walking legs of the crayfish make different contributions to the motion of the body in curve walking. The leg kinematics of the two behaviours differ significantly, however. Inner legs three and four step at a higher frequency than their outer counterparts during defence turns, but the situation is reversed in curve walking by freely behaving animals, in which outer legs three and four step with shorter periods than their inner counterparts (Domenici *et al.*, 1998). Also, the degree of co-ordination among legs differs substantially between the two behaviours. Domenici *et al.* (1998) found that the stances of inner legs two and five and outer leg four tend to step as a tripod. The stances of these three legs correlate with increases in the angular velocity of the body during curve walking. Coupling among the legs that would constitute the opposite tripod was found to be much weaker. Domenici *et al.* concluded that the 'tripod' of inner legs two and five with outer leg four is primarily responsible for the rotational component of curve walking. No such co-ordination among legs is evident during defence turning by the crayfish. As might be expected from differences between inner and outer legs in stance amplitude and frequency, the co-ordination among opposing legs is weak at best during defence turns.

4. Discussion

A key element of the typical crustacean defence response is a static meral spread posture. This display, although motionless, is nevertheless the result of behavioural acts that are dependent on the biomechanics of the underlying motor system. A kinematic study of the defence display in the crayfish revealed that the command driving the movement of the chelipeds into the meral spread position is not related to the targeting of the movement toward an opponent but to a given body-referenced position that serves a communication function (Kelly and Chapple, 1990). The turning behaviour that often precedes the defence display is thus an integral part of the defence response because it is crucial for aiming the body at the opponent and maximizing the effectiveness of the signal as well as for positioning the animal appropriately for subsequent action such as attack, retreat, or modulation of the display.

Because the target may move before a defence turn is completed, it may not be possible to maximize simultaneously both response and accuracy. One can imagine a variety of behavioural strategies by which a balance between response speed and accuracy is achieved. It would often be sufficient, for example, for the animal to align its body fairly roughly with respect to the target and then adjust the position of the claws as occurs when a crayfish moves its chelipeds, without repositioning its body, to track small movements of the opponent. A kinematic analysis of defence turning suggests a different strategy, however; that is, that response speed and accuracy are optimized in a two-phase response of the body's movement. The initial, apparently ballistic phase produces rapid angular acceleration of the body, and the subsequent, more variable phase achieves the final orientation, probably with the aid of feedback control. This second phase can be suppressed when it is beneficial to the animal to maximize response speed, as can be seen on occasion when very strong tactile stimuli evoke a whirling turn in which the forces associated with the twist of the body on the legs are sufficiently large to thrust the animal vertically off the substrate and produce a rotation of the body while it settles back down (Copp, personal observation). It will be interesting to determine if the magnitude of rotation in either type of turn is related to the angular position of the stimulus relative to the body or results more simply from biomechanical constraints.

The rotation of the body in a defence turn is driven, of course, by the actions of the legs. Nye and Ritzmann (1992) have shown how a detailed, kinematic understanding of walking legs can reveal adaptive strategies of leg co-ordination in the escape responses of cockroaches. A similar analysis of the walking legs of a crayfish during defence turning responses, however, has not yet uncovered discrete turning strategies. The variability of the leg kinematics may reflect limits on the co-ordination achievable by the underlying control system, as argued above for the cheliped motions leading to the meral spread display. A second possibility is that reduced coupling between contralateral legs may aid the animal in quickly adjusting its turning response according to sensory feedback about the location of the target. The lack of tight co-ordination among the walking legs may also reflect an adaptive strategy by which a rapidly turning and buoyant animal copes with greater instabilities than might attend straight or curve walking. The leg kinematics do show, however, that defence turning behaviour in the crayfish represents neither a simple application of forward or backward stepping patterns to a rotational behaviour nor a direct extension of leg co-ordination patterns observed in curve walking but a distinct locomotor behaviour. The neural control system underlying defence turning behaviour is thus likely to differ significantly from that controlling straight or curve walking. These differences are probably related to specific characteristics of the defence turns, such as the limited translational component of the body's movement, the short duration of the response (about 1 s or the equivalent of one step period in straight or curve walking), and the importance of the orienting component of the movement.

The preceding analysis of defence responses in a crustacean offers glimpses of the questions that can be specified or resolved by conducting kinematic and biomechanical studies of a defence behaviour, questions not only of what the animal does but how and why it behaves in that particular way. Our understanding of this important class of behaviour will be expanded if we utilize these tools more widely.

References

Atkeson, C.G. and Hollerbach, J.M. (1985) Kinematic features of unrestrained vertical arm movements. *J. Neurosci.* **5**: 2318–2330.

Beall, S.P., Langley, D.J. and Edwards, D.H. (1990) Inhibition of escape tailflip in crayfish during backward walking and the defence posture. *J. Exp. Biol.* **152**: 577–582.

Beltz, B.S. and Kravitz, E.A. (1983) Mapping of serotonin-like immunoreactivity in the lobster nervous system. *J. Neurosci.* **3**: 585–602.

Bovbjerg, R.V. (1956) Some factors affecting aggressive behaviours in crayfish. *Physiol. Zool.* **29**: 127–136.

Bowerman, R.F. and Larimer, J.L. (1974) Command fibers in the circumoesophageal connectives of the crayfish. *J. Exp. Biol.* **60**: 95–117.

Breithaupt, T., Schmitz, B. and Tautz, J. (1995) Hydrodynamic orientation of crayfish (*Procambarus clarkii*) to swimming fish prey. *J. Comp. Physiol.* **177**: 481–491.

Bruski, C.A. and Dunham, D.W. (1987) The importance of vision in agonistic communication of the crayfish *Orconectes rusticus*. I. An analysis of bout dynamics. *Behaviour* **63**: 83–107.

Camhi, J.M. and Tom, W. (1978) The escape behaviour of the cockroach *Periplaneta americana* I. Turning response to wind puffs. *J. Comp. Physiol.* **128**: 193–201.

Clarac, F. (1984) Spatial and temporal co-ordination during walking in crustacea. *TINS* **7**: 293–298.

Copp, N.H. (1986) Dominance hierarchies in the crayfish *Procambarus clarkii* (Girard, 1852) and the question of learned individual recognition (Decapoda, Astacidea). *Crustaceana* **51**: 9–24.

Copp, N.H. and Watson, D. (1988) Visual control of turning responses to tactile stimuli in the crayfish *Procambarus clarkii. J. Comp. Physiol.* **163**: 175–186.

Cruse, H. and Silva Saavedra, M.G. (1996). Curve walking in crayfish. *J. Exp. Biol.* **199**: 1477–1482.

Dingle, H. (1969) A statistical and information analysis of aggressive communication in the mantis shrimp. *Anim. Behav.* **17**: 561–575.

Domenici, P. and Blake, R.W. (1997) The kinematics and performance of fish fast-start swimming. *J. Exp. Biol.* **200**: 1165–1178.

Domenici, P., Jamon, M. and Clarac, F. (1998) Curve walking in freely moving crayfish (*Procambarus clarkii*). *J. Exp. Biol.* **201**: 1315–1329.

Elwood, R. and Glass, C. (1981) Negotiation or aggression during shell fights of the hermit crab *Pagurus bernhardus. Anim. Behav.* **29**: 1239–1244.

Full, R.J., Yamauchi, A. and Jindrich, D.L. (1995) Maximum single leg force production: cockroaches righting on photoelastic gelatin. *J. Exp. Biol.* **198**: 2441–2452.

Garvey, J.E. and Stein, R.A. (1993) Evaluating how chela size influences the invasion potential of an introduced crayfish (*Orconectus rusticus*). *Am. Midl. Nat.* **129**: 172–181.

Glantz, R. (1974) Defence reflex and motion detector responsiveness to approaching targets: the motion detector trigger to the defence reflex pathway. *J. Comp. Physiol.* **95**: 297–314.

Glantz, R. (1977) Visual input and motor output of command interneurons in the defence reflex pathway in the crayfish. In: *Identified Neurons and Behaviour of Arthropods* (ed. G. Hoyle). Plenum, New York, pp. 259–274.

Glass, C.W. and Huntingford, F.A. (1988) Initiation and resolution of fights between swimming crabs (*Liocarcinus depurator*). *Ethology* **77**: 237–249.

Harris, J. (1978) A photoelastic substrate technique for dynamic measurements of forces exerted by moving organisms. *J. Microsc.* **114**: 219–228.

Harris, J. and Ghiradella, H. (1980) The forces exerted on the substrate by walking and stationary crickets. *J. Exp. Biol.* **85**: 263–279.

Hazlett, B.A. and Bossert, W.H. (1965) A statistical analysis of the aggressive communication systems of some hermit crabs. *Anim. Behav.* **13**: 357–373.

Huber, R. and Kravitz, E.A. (1995) A quantitative analysis of agonistic behaviour in juvenile American lobsters (*Homarus americanus* L). *Brain Behav. Evol.* **46**: 72–83.

Jamon, M. and Clarac, F. (1995) Locomotor patterns in freely moving crayfish (*Procambarus clarkii*). *J. Exp. Biol.* **198**: 683–700.

Jindrich, D.L. and Full, R. (1999) Many-legged maneuverability: dynamics of turning hexapods. *J. Exp. Biol.* **202**: 1603–1623.

Kelly, T.M. and Chapple, W.D. (1990) Kinematic analysis of the defence response in crayfish. *J. Neurophysiol.* **64**: 64–76.

Klärner, D. and Barnes, W.J.P. (1986) Crayfish cuticular stress detector CSD$_2$. II. Activity during walking and influences on leg coordination. *J. Exp. Biol.* **122**: 161–175.

Livingstone, M.S., Harris-Warrick, R.M. and Kravitz, E.A. (1980) Serotonin and octopamine produce opposite postures in lobsters. *Science* **208**: 76–79.

Mattson, M.P. and Spaziani, E. (1986) Regulation of the stress-responsive X-organ–Y-organ axis by 5-hydroxytryptamine in the crab, *Cancer antennarius. Gen. Comp. Endocrinol.* **62**: 419–427.

Nye, S.W. and Ritzmann, R.E. (1992) Motion analysis of leg joints associated with escape turns of the cockroach, *Periplaneta americana. J. Comp. Physiol.* **171**: 183–194.

Schaeffer, P.L., Varuni Kundagunta, G. and Ritzmann, R. (1994) Motion analysis of escape movements evoked by tactile stimulation in the cockroach *Periplaneta americana. J. Exp. Biol.* **190**: 287–294.

Schöne, H. (1968) Agonistic and sexual display in aquatic and semi-terrestrial brachyuran crabs. *Am. Zool.* **8**: 641–654.

Scrivener, J.C.E. (1971) Agonistic behaviour of the American lobster *Homarus americanus* (Milne-Edwards). *Fish. Res. Board Can. Bull.* **235**: 1–126.

Steger, R. and Caldwell, R.L. (1983) Intraspecific deception by bluffing: a defence strategy of newly molted stomatopods (Arthropoda: Crustacea). *Science* **221**: 558–560.

Stein, L., Jacobson, S. and Atema, J. (1975) Behaviour of lobsters (*Homarus americanus*) in a semi-natural environment at ambient temperatures and under thermal stress. *Woods Hole Oceanogr. Inst. Tech. Rep.* **75-48**: 1–49.

Varju, D. (1989) Prey attack in crayfish: conditions for success and kinematics of body motion. *J. Comp. Physiol.* **165**: 99–107.

Wiersma, C.A.G. (1961) Reflexes and the central nervous system. In: *The Physiology of Crustacea. Vol. II* (ed. T.H. Waterman). Academic Press, New York, pp. 241–279.

Wine, J. and Krasne, F.B. (1972) The organization of escape behaviour in the crayfish. *J. Exp. Biol.* **56**: 1–18.

Wood, D.E. and Derby, C.D. (1995) Coordination and neuromuscular control of rhythmic behaviour in the blue crab, *Callinectes sapidus. J. Comp. Physiol. A* **177**: 307–319.

Wood, D.E., Gleeson, R.A. and Derby, C.D. (1995) Modulation of behaviour by biogenic amines and peptides in the blue crab, *Callinectes sapidus. J. Comp. Physiol. A* **177**: 321–333.

17

Fighting and the mechanical design of horns and antlers

Andrew C. Kitchener

1. Introduction

Horn-like organs have evolved independently in five extant families of mammals (Alvarez, 1992; Geist, 1966; Goss, 1983; Kitchener, 1991; Savage and Long, 1986). A variety of functions have been proposed for horn-like organs, but behavioural observations in recent years have confirmed their use mostly in defence against predators and intraspecific fighting between males to gain access to females in oestrus (Clutton-Brock, 1982; Clutton-Brock et al., 1979; Geist, 1966; Kitchener, 1991). Horns and antlers are most likely to experience the greatest forces during intraspecific fighting, when they often clash directly against each other. Therefore, in this paper I will review the interrelationship between intraspecific fighting and the mechanical design of the horns of bovids and the antlers of cervids, for which published data are available.

Unlike skeletal bones, horns and antlers develop in the complete absence of the forces that they are designed to resist. Therefore, their growth must be linked to the body growth of their owners and must also predict how much force they will eventually resist in intraspecific fighting. Horns and antlers are energetically expensive to grow and carry around, because they may make up to 10% of body mass (e.g. bighorn sheep, *Ovis canadensis*), and the huge antlers of moose, *Alces alces*, and other large deer are cast and regrown annually (Geist, 1971; Goss, 1983). However, the breakage of horns and antlers in fighting may result in severe injury or death, and may affect adversely reproductive success either temporarily or permanently (Geist, 1986). Therefore, we would expect horns and antlers to be minimum-weight structures, which are designed to resist the forces of fighting with minimum incidence of breakage.

Given the costs and benefits of fighting, assessment behaviours before fighting usually ensure equally matched opponents of similar body mass (Clutton-Brock et al., 1979; Spinage, 1982). Fighting often involves what appear to be ritualized movements to allow for the interlocking of horns and antlers of species-specific morphology. Therefore, assessment before fighting and species-specific movements during fighting

Biomechanics in Animal Behaviour, edited by P. Domenici and R.W. Blake.

provide a way of predicting the maximum forces of fighting, which are a function of the body mass of the opponents, often acting in particular directions.

In this paper I consider two aspects of the design of horns and antlers in relation to fighting behaviours. First, the materials from which horns and antlers are made should be appropriate to their use in fighting. Second, the mechanical structure of horns and antlers should be appropriate to the maximum forces of fighting so that they are minimum-weight structures with an appropriate safety factor. Finally, I look at the handful of studies that have attempted to look at the mechanical performance of bovid horns in fighting (e.g. Alvarez, 1990, 1994; Jaslow and Biewener, 1995; Kitchener, 1988; Schaffer, 1968; Schaffer and Reed, 1972), and show how a mechanical assessment of horn and antler structure can be used to predict some aspects of the fighting in extinct species, including the giant deer or Irish elk, *Megaloceros giganteus* (Kitchener, 1987a) and dinosaurs (Alexander, 1989; Farlow and Dodson, 1975; Goodwin *et al.*, 1998).

2. Horns and antlers

Horns and antlers are analogous structures (Goss, 1983). Horns are composite structures that consist of a bony core surmounted by an α-keratin sheath that grows from a germinative epithelium covering the surface of the horn core (Goss, 1983). Each year a new horn sheath grows, pushing the previous year's growth away from the skull. In this way bovids can expand the basal dimensions of their horns as they grow in body size. Therefore, horns grow from their bases and are non-branching permanent structures. Some annual layers of the keratin sheath may be exfoliated in some species, but the only surviving antilocaprid, the pronghorn, *Antilocapra americana*, sheds its branched keratin sheaths annually (O'Gara and Matson, 1975; Petocz and Shank, 1983). The horn core comprises a solid cortex and a spongy core, although in some genera there is a large sinus occupying the interior of the horn core (e.g. *Bos*, *Ovis* and *Capra*) (Schaffer and Reed, 1972).

Antlers are made entirely of bone. Although horns are permanent structures that are added to annually, antlers are deciduous so that they are regrown annually (Goss, 1983; Kitchener, 1991). They grow from pedicles on the frontal bones and while growing are covered in skin (velvet), which is richly supplied with blood vessels. They grow from their tips so that the growing points move away from the skull (Goss, 1983). When completely grown and fully ossified, the skin is shed, leaving a dead bony, often branching structure for use in fighting. Antlers are made of primary Haversian bone (Currey, 1979). In many species the core is spongy (e.g. red deer, *Cervus elaphus*, and Père David's deer, *Elaphurus davidianus*), but in others the antler is completely or almost entirely made of solid bone (e.g. moose, axis deer, *Axis axis*) (personal observations).

The main functions of horns and antlers are in defence from predators and intraspecific fighting between males to gain access to females in oestrus (Clutton-Brock, 1982; Geist, 1966, 1978; Packer, 1983). All female deer except the reindeer, *Rangifer tarandus*, lack antlers, and females of almost half the genera of bovids lack horns (Roberts, 1996). Roberts (1996) has suggested that the female competition hypothesis explains best why some female bovids have horns; to minimize competition for resources between females in larger social groups. However, Packer (1983) analysed the horn shape of male and female bovids and determined that female horns tended to have tips that point outwards for effective defence against predators, whereas males have horn tips that point inwards for more effective intraspecific combat.

3. Fighting

The great variety of horn and antler shapes is reflected in a wide variety of fighting postures (*Figure 1*) (e.g. Kingdon, 1982). Horn and antler shape appears to be related to species-specific fighting, so that it should be possible to predict the direction of forces acting on these structures in fighting (see below).

Geist (1966) has suggested a scheme for the evolution of horns and antlers from the simple spikes of dwarf antelopes and brocket deer, *Mazama* spp., to the complex twists, turns, ridges and branches of most larger species. Species with simple spikes have a small body size, tend to be monogamous and vigorously defend territories for the resources contained therein. Their fighting is less ritualized and is directed against the body in parallel or anti-parallel stances (*Figure 1*) (Geist, 1966; Kingdon, 1982; Spinage, 1986). Species with complex horns and antlers are larger, tend to be polygynous and defend territories containing many females or harems so that mating success can be maximized (Clutton-Brock, 1982; Geist, 1966; Leuthold, 1977). Complex-horned species direct the forces of fighting towards the heads of their opponents, so that horns and antlers must

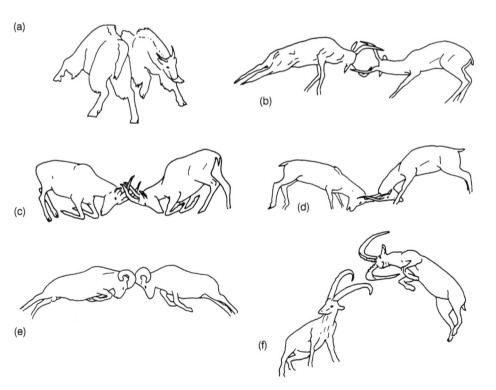

Figure 1. *The diversity of fighting in bovids (after Geist, 1966; Kingdon, 1982). (a) Rocky Mountain goat,* Oreamnos americanus *(stabber); (b) impala,* Aepyceros melampus *(wrestler); (c) topi,* Damaliscus lunatus *(clasher–wrestler); (d) Grant's gazelle,* Gazella granti *(wrestler); (e) bighorn sheep,* Ovis canadensis *(rammer); (f) alpine ibex,* Capra ibex *(clasher).*

resist and transmit the forces of fighting, but horns and antlers may also have an important display function (Geist, 1978) (*Figure 1*). The complex shape of antlers and horns allows them to lock effectively together so that accidental injury as a result of slippage is avoided and so that the fighters can obtain an effective grip to lever their opponents off balance. In the sheep, goats and some other species (e.g. musk-ox, *Ovibos moschatus*), the horns have developed hugely, to transmit very large forces to try and stun or hurt their opponents (Geist, 1971; Schaffer and Reed, 1972). In these species the tips of the horns are often no longer effective in causing physical damage, but forceful butts or clashes against the body may cause considerable trauma (Geist, 1971).

The aims of fighting are to attempt to injure an opponent while avoiding being injured. This results in often highly ritualized behaviour as opponents try to knock each other off balance to obtain an opportunity to stab neck, flank or belly. Fighting may actually deter opponents by maximizing superficial injuries, which may be greatly under-recorded (Geist, 1986). There appear to be rules of fighting, which involve a process of prior assessment of body size and condition. Fighting in defence of territories or harems may occur for up to several months and is energetically very costly. For example, red deer on Rum lose up to 20% of their body weight in the 6-week-long rut (Clutton-Brock *et al.*, 1979). Extra vigilance for opponents and time spent displaying and fighting reduces the time available for feeding and may make males vulnerable to predation. Males that have exhausted their body reserves in the rut may suffer increased winter mortality, especially if the weather is worse than usual (e.g. red deer, Clutton-Brock *et al.* 1979, 1982; chiru, *Pantholops hodgsonii*, Schaller, 1998).

Horns and antlers may be used in agonistic displays, but from the few detailed studies that have been carried out, their size is not an indicator of fighting and reproductive success. In red deer body size and frequency of roaring were the best indicators of fighting success (Clutton-Brock *et al.*, 1979; see also Chapter 11). Spinage (1982) found that larger waterbuck, *Kobus ellipsiprymnus*, had relatively shorter horns because of the wear they had experienced in fighting and horn maintenance (see below). Most species have some form of lateral display, which emphasizes body size. This allows a more sensitive measure of body condition and body size of a potential opponent than does a fixed length of horn or antler.

Kingdon (1982) analysed complex horn structure in terms of different functions. Lundrigan (1996) took this approach one stage further and measured three functional zones on the horns of 21 species of bovids representing 11 of the 12 bovid tribes, along with other measurements. The three functional zones of horns are a straight basal stem, the curved catching arch, which receives the blows from an opponent's horns, and the stabbing zone at the tip, which is designed to injure opponents (*Figure 2*). Lundrigan (1996) also categorized the fighting of these bovids according to the occurrence and frequency of ramming, fencing, wrestling, stabbing and kneeling. This study revealed that particular behaviours and morphological features of the horns are closely linked. For example, the percentage length of the horn occupied by the catching arch is positively correlated with wrestling, and horns with a greater circumference and curvature are positively correlated with ramming. Although Lundrigan's (1996) analysis provides a formal way of examining the functional significance of different horn structures, it does not reveal the limitations on their mechanical performance. To obtain a better understanding of this we need to take a look at the costs and benefits of fighting in terms of reproductive success, risks of injury and death, and consequences of horn and antler failure.

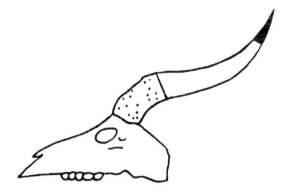

Figure 2. The functional zones of horns as defined by Kingdon (1982) and Lundrigan (1996). Basal stem (stippled), catching arch (white) and stabbing zone (black).

The costs and benefits of fighting for polygynous species are considerable. Successful territory- and harem-holders may gain the highest proportion of matings, but injury, horn or antler breakage and death may seriously limit reproductive success. It was once thought that ritualized fighting between bovids and cervids was intended to avoid any injury. However, data collected over the last 30 years show that injuries and deaths occur in all species that have been studied (mortalities of up to 23% in red deer and 19% in mule deer, *Odocoileus hemionus* (Clutton-Brock *et al.*, 1979, 1982)) (*Table 1*). Geist (1986) has examined the tanned skins of North American cervids and shown the presence of many healed scars from antlers, suggesting that observational records of fighting injuries may be greatly underestimated.

It would be reasonable to assume that bovids and cervids must balance the need to include sufficient materials in their horns and antlers so that they do not break during most intraspecific fighting against the need to minimize horn and antler weight given the high energetic costs involved in growing and carrying them around. Therefore, we might expect cervids and bovids to have evolved minimum-weight structures for effective intraspecific fighting.

4. Horn and antler failure

Horns and antlers could fail in five ways during fighting, representing different mechanical properties of horn structures and/or materials (Kitchener, 1991):

(i) the maximum bending stress during fighting could exceed the bending strength of the horns or antlers, resulting in breakage (i.e. bending strength);
(ii) the maximum shear stress could exceed the shear strength of the horns or antlers (i.e. shear strength);
(iii) the horns or antlers could deflect too much and so would be ineffective at transmitting and receiving the forces of fighting (i.e. stiffness);
(iv) scratches and cracks could act as stress concentrators leading to catastrophic failure of the horn at lower than expected forces of fighting (i.e. notch-sensitivity or toughness);
(v) the energy produced when two animals collide could exceed the amount of energy that horns and antlers can store as strain energy as they bend (i.e. stiffness).

To consider which of the above possible causes of failure are the most critical in the functional design of horns and antlers, there are two important components of the

Table 1. Incidences and annual rates of injury and mortality as a result of intraspecific fighting
(as percentage of the adult male population) in bovids and cervids

Species	Mortality	Injury	Source
Cervidae			
Muntiacus muntjak		+	a
M. reevesi		+	b
Cervus elaphus	5–29	6–23	c
C. canadensis	+	+	d
Alces alces	4	+	e
Odocoileus hemionus		19	c
Rangifer tarandus	+		c
Antilocapridae			
Antilocapra americana	+		f
Bovidae			
Syncerus caffer	+	+	g
Bos gaurus	+		h
Boselapus tragocamelus	+		i
Kobus spp.	+	+	j
Oryx gazella	+		i
Oryx leucoryx	+		k
Alcelaphus buselaphus cokii	+		l
Connochaetes gnou	+		m
Pelea capreolus	+		n
Antilope cervicapra		+	o
Saiga tatarica	+		n
Pantholops hodgsonii	+		p
Oreamnos americanus		+	e
Ovibos moschatus	5–10		c
Ovis canadensis		+	e

a, Barrette (1977); b, Clark (1981); c, Clutton-Brock *et al.* (1979, 1982); d, Leslie and Jenkins (1985); e,
Geist (1971); f, Smith (1990); g, Mloszewski (1983); h, Jarofke *et al.* (1990); i, Cinderey (personal
communication); j, Spinage (1986); k, specimens in National Museums of Scotland; l, Gosling (1974); m,
Boortsma (1979); n, Nowak (1991); o, Ranjitsinh (1989); p, Schaller (1998).

mechanical design of horns and antlers, which need to be examined separately. First, the
mechanical performance of the structure must be appropriate to the forces acting on the
horn or antler. Second, the mechanical properties of the materials that make up horns
and antlers must be appropriate to their mechanical functioning as structures in fighting.

5. Material properties

For horns and antlers to function effectively in fighting, we would expect them to be
made of materials that have mechanical properties appropriate to their use in fighting.
These materials must be stiff and strong enough to resist the forces of fighting, they must
be tough enough to resist the energy of fighting, and must not be weakened by scratches
or cracks, which could act as stress concentrators. Horns and antlers should also be made
of light materials, so that the energetic costs of carrying them around can be minimized.

Not surprisingly, horn sheath keratin, horn core bone and antler bone have
mechanical properties appropriate to structures that experience large bending forces.

Table 2 shows the bending stiffness, rupture modulus (a measure of bending strength, i.e. the maximum force per unit cross-sectional area at bending failure), work of fracture (the total energy required to break the material per unit cross-sectional area) and specific gravity of horn sheath keratin, horn core bone and antler bone. It also shows specific mechanical properties so that comparisons can be made between materials of different specific gravities.

There are some interesting differences and similarities between these three materials. Antler bone is about three times stiffer than horn sheath keratin, which in turn is about three times stiffer than horn core bone (Kitchener, 1991). The work of fracture of horn sheath keratin is about double that of antler bone and 12–15 times that of horn core bone (Kitchener, 1991). 'Bending strengths' of horn sheath keratin and antler bone are about the same, but horn core bone is only about one-fifth as 'strong'. Finally, antler bone is about 50% denser than horn sheath keratin and horn core bone (Kitchener, 1991). By comparing these mechanical properties it is evident that horn sheath keratin rather than horn core bone resists most of the forces of fighting, except at the base of the horn, where forces must be transmitted from the sheath to the core and skull. This accords also with the external distribution of horn sheath keratin to maximize its second moment of area in the cross-sectional shape of the horn (see below). Antler compact bone is denser than horn, but because it is replaced annually and not carried all year round, the overall energetic costs of carrying these apparently relatively heavier structures can be minimized (but see below). This may also explain why antler bone can be less tough than horn sheath keratin; as antlers are deciduous structures, the costs of antler breakage in terms of inability to fight and reproduce successfully thereafter are not permanent as for bovids.

Taking into account the specific gravity of these materials, horn sheath keratin and antler bone are not as stiff as mild steel or wood (*Table 2*) (Kitchener, 1985). However, horn keratin is as tough as mild steel, antler bone is tougher than wood, and both horn keratin and antler bone are four times 'stronger' than mild steel (*Table 2*). Therefore,

Table 2. *A comparison of mechanical properties and specific mechanical properties of horn sheath, horn core, antler bone, wood and mild steel*

	Stiffness (GPa)	Work of fracture (kJ m^{-2})	'Strength' (MPa)	Stiffness/ SG (GPa)	Work of fracture/ SG (kJ m^{-2})	'Strength'/ SG (MPa)	SG
Oryx horn	4.3	19.0	212	3.3	14.2	161	1.32
Waterbuck horn	3.3	20.0	245	2.6	16.0	193	1.27
Sheep horn	4.1	22.0	228	3.2	17.0	178	1.28
Waterbuck horn core	1.3	1.5	42	1.0	1.2	34	1.23
Sika deer antler	13.7	9.0	239	6.9	4.6	121	1.97
Hog deer antler	12.7	10.0	246	5.9	4.6	114	2.16
Mild steel[a]	209.8	99.8	398	26.9	12.8	51	7.80
Wood[a]	14.0	1.2	–	23.3	2.0	–	0.60

(Data from Kitchener (1987c, 1991).[a] Gordon (1976).) SG, specific gravity.

horn sheath keratin and, to a lesser extent, antler bone are light materials with high toughness and strength and moderate stiffness.

Both horn sheath keratin and antler bone are composite materials. In the case of keratin, there are stiff α-keratin fibres set into a weak protein matrix (Fraser and Macrae, 1980; see Kitchener and Vincent (1987) for the modelling of horn keratin stiffness, and Kitchener (1987b) for the effects of water on the viscoelasticity of horn keratin). Antler bone consists of collagen fibres and stiff hydroxyapatite crystals, which are arranged into tubular structures called osteons, which are made of concentric layers of bone (Ascenzi and Bonucci, 1970; Currey, 1979).

The ability of fibrous composites to resist cracks depends on a disparity in the strength of the fibres and matrix, so that as the crack progresses through the material, the crack tip is continually blunted as the matrix fails and directs the crack along the length of the fibres (*Figure 3*). This is known as the Cook–Gordon crack-stopping mechanism (Gordon, 1976). In this way a great deal of energy is expended in creating a crack with a considerable surface area. This is further increased by the need to overcome frictional forces as the fibres pull out of the matrix. By comparing the force deflection curves in three-point bending of horn sheath keratin in its normal state (20% water content) and oven-dried horn (0%), it can be seen that there is a considerable decrease in the energy absorbed by the drier material (*Figure 4*) (Kitchener, 1987c). In normal horn keratin there is considerable plastic yielding of the matrix often far from the crack tip, which accounts for 50–75% of the total work of fracture (Kitchener, 1987c). Therefore, crack-stopping mechanisms at the crack tip (e.g. fibre debonding and pullout, and delamination of keratinocytes) account for only 25–50% of the energy that is absorbed during fracture (Kitchener, 1987c). In the dry horn, the stiffness of the matrix approaches that of the fibres and the Cook–Gordon crack-stopping mechanisms no longer occur, so that little energy is absorbed and the crack tip is not deflected from its path, resulting in catastrophic failure.

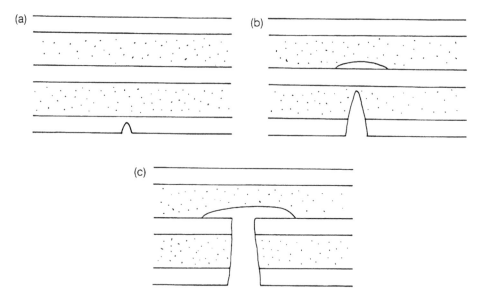

Figure 3. *The Cook–Gordon crack-stopping mechanism of fibrous composites. The crack starts to propagate across the material (a). The stress distribution ahead of the crack tip opens up the weak matrix–fibre interface (b), so the crack is eventually blunted (c).*

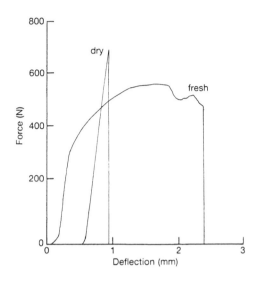

Figure 4. *Force–displacement curves for test-pieces of horn sheath keratin in three-point bending (Kitchener, 1987c). The fresh test piece shows considerable yielding and absorbs much more energy during fracture than the dry test piece.*

Another way of looking at this is to assess the effect of notch length on the strength of horn keratin in tension. If the material is effective at minimizing stress concentrations caused by cracks or notches, we expect to see a proportional reduction in tensile strength as notch length increases, as indeed we do for horn sheath keratin (*Figure 5*) (Kitchener, 1987c). However, when horn keratin is oven-dried, notch length reduces strength more than expected, showing that notches act as stress concentrators where the matrix stiffness approaches that of the fibres (*Figure 5*).

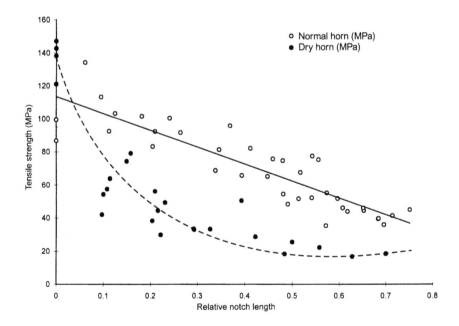

Figure 5. *The effect of notch length on the tensile strength of horn sheath keratin (Kitchener, 1987c).*

What implications does this effect of water content on the fracture properties of horn sheath keratin have for the behaviour of bovids? Inevitably, fighting results in superficial damage to the horn sheath through general wear and tear. In part this is overcome each year in the most forceful fighters (e.g. sheep, goats), because the basal part of the horn is replaced annually as the horn grows. However, despite this in all species there is a risk of cracks and scratches developing that could result in horn failure. In addition, horn sheath keratin is a dead material that is liable to dry out over time (especially in dry tropical climates), so that notch sensitivity could be increased and catastrophic failure would be very likely. It would appear that bovids (and possibly cervids) overcome this problem through horning, which is part of their agonistic repertoire. This behaviour involves the pushing of the horns in wet mud or thrashing them against vegetation, which would allow for the rehydration of the horn keratin material (Anderson, 1980; Child, 1982; Jacobsen, 1972; Leuthold, 1977; Spinage, 1982; Walther, 1966; Walther et al., 1983). From these observations it is possible to see how a maintenance behaviour such as horning could become a highly ritualized agonistic display before fighting.

6. Structures

To understand the mechanical behaviour of horn and antler structures, it is useful to model them as a simple short cylinder fixed at one end (to the skull) and free at the other end to which the force of fighting is applied (Kitchener, 1985, 1991). The maximum stress (σ) in this model horn occurs at its base:

$$\sigma = \frac{FLr}{I}.$$

where F is the maximum force of fighting (N), a function of body mass; L is the moment arm (distance between the base of the horn and the point where the force is acting) (m); r is the radius of the base of the horn (m); I is the second moment of area (m⁴).

The second moment of area is a shape factor, which describes the distribution of material relative to the neutral axis, where there are no net tensile or compressive forces in bending. In our cylindrical horn model, the neutral axis is the diameter of the base of the horn perpendicular to the direction of the force acting on the horn and it is calculated as a fourth-power function of the basal radius of the horn:

$$I = \frac{\pi r^4}{4}.$$

In the most forceful fighting, horns and antlers are engaged at their bases, thereby minimizing the moment arm (L). The maximum force of fighting is a function of body mass, because the most forceful fights involve animals of similar body size (see above). Therefore, we would expect to see second moment of area increase linearly to balance any increase in stress as a result of increase in body weight with age. Unfortunately, there are few data available to see if this relationship holds, although it has been confirmed for red deer, nyala, *Tragelaphus angasi*, and Cretan wild goat, *Capra aegagrus cretica* (Kitchener, 1991; Papageorgiou, 1979; Tello and Van Gelder, 1975).

There is a similar linear relationship between mean second moment of area at the base of horns and antlers and mean body mass in all species of bovids and cervids, but only if the forcefulness of fighting is taken into account (Kitchener, 1985, 1991). I have identified nine linear relationships which show increasing gradients (i.e. greater I for a given body mass) with increasing forcefulness of fighting, including dwarf antelopes (stabbing), duikers (head clashing and stabbing), goat antelopes and deer (wrestling), antelopes (wrestling), cattle (clashing and wrestling), African buffaloes (more forceful clashing and wrestling), goats and ibexes (clashing) and sheep (ramming) (*Figure 6*).

Therefore, horns and antlers appear to have evolved to resist deflection. The deflection (δ) of this model horn is

$$\delta = \frac{FL^3}{3EI}$$

where E is the bending stiffness of horn sheath keratin–horn core bone or antler bone (Pa).

However, there is a consequence of this design solution. As body size increases, so the bending stress increases more in relation to the basal radius, so that we might expect that larger species should break their horns and antlers more frequently than do smaller species, all other factors being equal. Unfortunately, data on natural incidences of horn breakage are scant. Packer (1983) gave some data on incidences of broken horns for various antelope species (*Table 3*), many of which would be included in the wrestling group. As predicted, the incidence of horn breakage increases with body mass, but sample sizes are often small, so it is unclear how representative these data are. Data from Alvarez (1990) show that 7% of Iberian ibex have broken horns, which

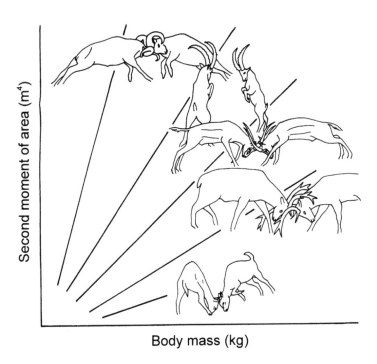

Figure 6. *A diagrammatic representation of the scaling of second moment of area of the base of the horns of bovids (Kitchener, 1991) for different types of fighting. The greater the gradient, the more forceful the fighting. In order of increasing force: duikers, deer, antelopes, goats, sheep.*

Table 3. Percentage broken horns recorded in various bovids grouped according to the fighting categories identified by Kitchener (1985)

	n	% Broken horns	Mean body mass (kg)
Stabbers (dwarf antelopes)			
Kirk's dik-dik, *Madoqua kirkii*	35	5.7	4.5
Steenbok, *Raphicerus campestris*	18	11.1	10.9
Wrestlers–stabbers (goat antelopes–deer)			
Beisa oryx, *Oryx gazella beisa*	6	16.7	188.0
Wrestlers (typical antelopes)			
Thomson's gazelle, *Gazella thomsonii*	1000	0.8	20.0
Bohor reedbuck, *Redunca redunca*	25	0.0	45.0
Bushbuck, *Tragelaphus scriptus*	11	0.0	54.0
Impala, *Aepyceros melampus*	125	5.6	57.0
Grant's gazelle, *Gazella granti*	287	0.4	62.0
Waterbuck, *Kobus ellipsiprymnus*	29	3.5	227.0
Eland, *Taurotragus oryx*	50	6.0	501.0
Clash–wrestle I (hartebeest)			
Topi *Damaliscus lunatus korrigum*[a]	120	20.0	125.0
Topi, *Damaliscus lunatus jimela*	38	5.3	130.0
Kongoni, *Alcelaphus buselaphus cokii*	33	3.0	190.0
Clash–wrestle II (wildebeest)			
Blue wildebeest, *Connochaetes taurinus*	3371	2.8	252.0
Ramming–wrestling (buffaloes)			
Cape buffalo, *Syncerus caffer*	400	3.7	686.0
Clashing (goats)			
Iberian ibex, *Capra pyrenaica*[b]	114	7.0	72.0

Data on broken horns from Packer (1983); [a]Gosling *et al.* (1987); [b]Alvarez (1990).
Body mass data from Kitchener (1985).

is higher than for any of the wrestlers, suggesting that frequency of breakage may be higher in more forceful fighters. Gosling *et al.* (1987) have recorded up to 20% breakage in the horns of topi, *Damaliscus lunatus korrigum*, which form leks, where fighting frequencies are very high. Many more field data on horn and antler breakage could easily be collected by field workers, which would add greatly to our knowledge of the mechanical functioning of horns and antlers.

One final problem remains. Although deer and antelopes have a similar method of fighting (i.e. wrestling), the antelopes have horns with a second moment of area (I) three times greater than that of antlers for deer of similar body weight (Kitchener, 1991). However, antler bone stiffness (E) is approximately three times greater than horn sheath keratin stiffness, so that the structural stiffnesses (EI) of horns and antlers are the same, so that their deflections are also similar (Kitchener, 1991). Therefore, horns and antlers appear to be designed to resist deflections, so as to be as effective as possible in the transmission of and resistance to the forces of species-specific fighting. Bovids with the most forceful fighting methods have horns, which are more resistant to bending, which is in agreement with Lundrigan's (1996) analysis. Now that we have established the mechanical and behavioural bases for fighting in bovids and cervids, let us take a closer look at how the horns of bovids perform mechanically in fighting.

7. The mechanics of fighting

By studying the fighting of bovids and cervids from a mechanical perspective, it is possible to obtain new insights into how horns and antlers are used and the evolution of and constraints on fighting. However, there have been very few studies that have attempted to look at fighting from a mechanical perspective. One of the earliest of these analysed the horns of sheep and goats using curved beam theory (Schaffer, 1968; Schaffer and Reed, 1972). Schaffer (1968) estimated relative forces acting on the horns of a variety of sheep and goat species based on average body masses and descriptions of fighting in the literature. He showed that the skulls of these animals would be subjected to potentially damaging twisting, which is counteracted by the neck muscu-lature. The neck musculature of the goats is more highly developed than in sheep because of the greater torques generated by them in species-specific fighting. Schaffer and Reed (1972) carried out a detailed comparative analysis of horn and skull structure in relation to fighting, including the role of the frontal sinuses in insulating the brain from the forces of fighting, and estimates of ratios of maximum tensile to maximum compressive stresses in the curved horns of sheep and goats.

Alvarez (1994) has suggested that the distribution of bone in and the structure of the antlers of the fallow deer, *Dama dama*, means that they are predisposed to fail at the distal end of the main beam just below the palmation. He believes that this mini-mizes damage to the pedicles, from which the antlers grow, and may leave sufficient antler for males to continue to fight successfully during the rut, if they do break. Alvarez (1990) also carried out an indirect mechanical analysis of fighting in the Iberian ibex, *Capra pyrenaica*, based primarily on behavioural observations. Alvarez (1990) divided the Iberian ibex's various methods of fighting into two broad cate-gories; conventional fighting included all kinds of fighting that involved low forces, including various categories of pushing and butting, whereas the rear clash was the most forceful fighting category. In conventional fighting up to two-thirds of the length of the horns might be used, but the rear clash involved the use of only the proximal third of each horn. Clearly, Iberian ibex minimize the stresses in their horns during the most forceful fighting (i.e. the rear clash) by minimizing the moment arm (see above).

I have analysed film of two bovids that have contrasting fighting methods; the blackbuck, *Antilope cervicapra*, clashes and wrestles like the majority of antelope species (Ranjitsinh, 1989), but the bighorn sheep, *Ovis canadensis*, rears up on its hind legs at some distance away from its opponent, then hops along on its hind legs before dropping onto all fours before a most forceful clash, which may involve both horns or only a single horn (Geist, 1971). Analysis of film allows estimates of fighting forces and energies to be derived at the point of clash for each species, assuming average and equal body masses for both contestants (*Table 4*) (Kitchener, 1988). The application of curved beam theory to the twisted horns of the blackbuck and the curved horns of the bighorn sheep and use of the material properties (*Table 2*) above, allow the bending and shear stresses to be calculated in the horn sheath and horn core of each species at the point of impact when forces are at a maximum (*Table 4*). These show that the horns of both bighorn sheep and blackbuck are most likely to fail in bending. The safety factors in bending (ratio of horn strength to maximum bending stress) were 10.0 and 3.3 for the bighorn sheep and blackbuck, respectively, although this would fall to 5.0 for the bighorn sheep if only one horn were engaged during the clash. These safety factors are consistent with those measured for other biological structures (<10) (Alexander, 1982), so that bovid horns do indeed seem to be optimally designed for resisting bending. Critical crack lengths for both species at maximum stresses were at least 60% of the basal diameter of the horns (*Table 4*). Given the extent of such a crack length, these horns are unlikely to fail catastrophically in fighting. The safety factors may be much higher in the bighorn sheep's horns because the forces acting on its horns may be more unpredictable, as clashes may involve the use of only one horn (Geist, 1971). Therefore, the blackbuck may not need as many materials in its horns, because its fighting is more ritualized (Ranjitsinh, 1989) and produces more predictable forces, so that safety factors are lower.

Table 4. A comparison of the mechanical analyses of fighting in blackbuck and bighorn sheep (Kitchener, 1988)

Parameter	Blackbuck	Bighorn sheep
Fighting	Clash–wrestle	Ram
Body mass (kg)	38	100
Maximum velocity (m s^{-1})	2	6
Maximum force at clash (N)	456	3400
Maximum energy at clash (J)	76	3500
Maximum deceleration at clash (m s^{-2})	−12	−34
Maximum tensile stress (MPa)	11.8 (7.8)[a]	1.4
Maximum compressive stress (MPa)	12.2 (8.2)[a]	4.0
Safety factor in bending (tension/compression)	3.4/3.3 5.1/4.9[a]	28.6/10.0 14.3/5.0[b]
Maximum shear stress (MPa)	0.5	0.4
Safety factor in shear	27.2	34.0
Maximum horn deflection (mm)	3 (2[a])	0.07
Maximum horn strain energy (J)	0.704 (0.212)[a]	0.202
Critical crack length (cm)	2.30 (2.84)[a]	9.11
Muscle mass to absorb energy (kg)	0.48	22.26

[a] Short moment arm.
[b] Only one horn engaged.

However, although these bovid horns do seem to be able to resist the forces of fighting, they may be unable to absorb the energy produced during the clash. For example, a bighorn sheep's horns absorb only about 0.01% of the energy as strain energy as they bend so that the animal must absorb the energy in other ways. I have suggested that the neck and body musculature are able to do this (Kitchener, 1988) because the amount of muscle needed to absorb this energy is up to about 20% of the average body mass of a bighorn sheep (*Table 4*). It is perhaps for this reason that the most forceful fighters have their brains isolated from the rest of the skull by extensive sinuses (Schaffer and Reed, 1972), because the energy of fighting would need to be transmitted via the skull to the neck and body musculature to avoid brain injury.

Rather than taking the theoretical approach to the forces and stresses of fighting, Jaslow and Biewener (1995) have measured the stresses in the horn core and skull of the domestic goat, *Capra hircus*, during fighting by implanting strain gauges. They were able to compare the theoretical stresses estimated by curved beam theory with *in vivo* measurements, and found that the latter were much higher than expected, thereby reducing safety factors greatly (*Table 5*).

8. Extinct fighters

Although there is much more work to be done on the mechanical analysis of fighting in bovids, cervids and other horned mammals, the techniques described above can also be applied to extinct species to test hypotheses on whether horn-like organs were functional in fighting and, indeed, they may be used to reconstruct the behaviour of the most forceful encounters.

My studies of living bovids and cervids have shown that horns and antlers are indeed able to mechanically resist the forces of fighting. However, Gould (1974) had suggested that the Irish elk or giant deer, *Megaloceros giganteus*, could not have used its enormous antlers for fighting, but instead used them in displays to maximize its reproductive success. Although it is clear how the large palms on the 3.75 m span of bony antlers could be used effectively in displays, it is difficult to see how an individual would back down in a display contest without the final deterrent of fighting.

First of all I analysed the fighting of other living cervids and was able to show that in New World deer (to which the moose, *Alces alces*, and reindeer, *Rangifer tarandus*, belong), the antlers are engaged towards their bases in the most forceful encounters with heads held almost parallel to each other (*Figure 7*). However, in Old World deer

Table 5. A comparison of the stresses in and safety factors of the horns during fighting in domestic goats, Capra hircus, as estimated by curved beam theory and as measured by strain gauges in vivo during fighting (Jaslow and Biewener, 1995)

	Curved beam theory	Strain gauges *in vivo*
Horn core compressive stress (MPa)	5.75	9.30
Compressive:tensile stress (horn core)	1.43	2.76
Safety factor	7.0	4.3

(a)

(b)

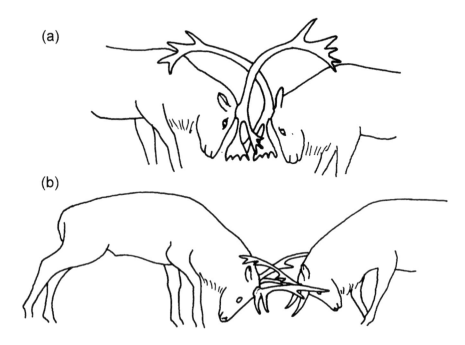

Figure 7. *The fighting of New World (a) and Old World deer (b).*

(to which red deer and giant deer belong), the antlers engage at their bases when the heads are held often almost parallel to the ground with noses pointing away from each opponent (*Figure 7*). If the skulls of giant deer are held in this way, the antlers engage just above the second tine at the distal end of the short beam (Kitchener, 1987a) (*Figure 8*). In this position the branched brow tines at the base of the antler are in the best position to parry any accidental blows aimed towards the eyes. I then sectioned the beam of a giant deer antler and calculated the stresses in the beam for a display function (i.e. the force of gravity) and a fighting function (two males at 500 kg each colliding at 3 m s^{-1} and decelerating in 0.1 s) assuming that the antler bone had the same mechanical properties as those of modern cervids. This simple analysis showed that giant deer antlers were ludicrously overdesigned for display, with a safety factor of more than 200, whereas they seemed optimally designed for fighting (with a safety factor of two; see *Table 6*) (Kitchener, 1987a). Moreover, the beam of the antler was not cylindrical in cross-section as might be expected for a display function, but was elliptical with the mean second moment of area ($n = 62$) 15% greater in the fighting axis than if it were a circular cross-section (Kitchener, 1987a).

    Being so-called subfossils, giant deer antlers from Ireland are not mineralized, so that it was possible to use a novel technique to obtain an indirect measure of the mechanical properties of giant deer antler bone. Obviously, after more than 12 000 years in lake deposits below a peat bog, it would be unrealistic to measure the mechanical properties of the antler bone directly. Neutron diffraction can be used to measure the degree of preferred orientation of the c-axes of the hydroxyapatite crystals in bone (Bacon *et al.*, 1979). The c-axes are aligned parallel to the maximum stresses that act normally on that bone. Therefore in a powder with a random

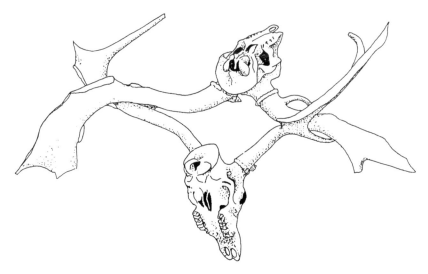

Figure 8. *The antlers of the extinct Irish elk interlock just above the second tine when the skulls are held in the Old World deer fighting posture (Kitchener, 1987a). Dorsal view.*

arrangement of crystals, the degree of preferred orientation would be one in all directions, but in a human femur, the ratio is more than three measured along the long axis of the bone, which also experiences the greatest stresses (Bacon and Griffiths, 1985). We used neutron diffraction to measure the degree of preferred orientation of the c-axes of the hydroxyapatite crystals along the length of a giant deer antler beam and compared it with that along the length of the antler of a living cervid, the hog deer, *Axis porcinus*, which is known to use its antlers in fighting (Kitchener *et al.*, 1994).

The mean degree of preferred orientation of hydroxyapatite crystals in the hog deer antler was 5.5, which was consistent with maximum longitudinal stresses as in the human femur, but the mean value for the giant deer was even higher (6.5), which is consistent with higher longitudinal stresses and totally inconsistent with an antler used only for display (Kitchener *et al.*, 1994). However, there were subtle variations in the degree of preferred orientation of hydroxyapatite crystals in different quadrants of

Table 6. *A comparison of the safety factors for the antlers of the Irish elk for fighting and display (after Kitchener, 1987a)*

	Function	
	Display	Fighting[a]
Maximum force (N)	200	15000
Maximum stress at antler base (MPa)	3	125
Safety factor[b]	80.0	1.92

[a] Assuming body mass 500 kg, collision velocity 3 m s^{-1}, deceleration time 0.1 s.
[b] Using bending strength of antler bone of 240 MPa.

the roughly circular cross-section of the antler beam of the two species of deer. The ratio of the values in the anterior to posterior quadrants of the hog deer antler was 1.115, compared with 1.2 for the dorsal:ventral quadrants and 1.015 for the anterior: posterior quadrants of the giant deer antler (Kitchener *et al.* 1994). There is a similar disparity in hydroxyapatite orientation in the human femur, with the anterior face of the femur having a higher degree of orientation than the posterior (a ratio of 1.2) (Bacon and Griffiths, 1985). The differences may be due to the mechanical properties of osteons with different degrees of orientation of hydroxyapatite. Ascenzi and Bonucci (1970) found that osteons with a high degree of longitudinal fibres (and hence hydroxyapatite) crystals were better at resisting tensile forces, whereas osteons with a lower degree of longitudinal fibres were more resistant to compressive forces. Therefore, the antler bone of hog deer and giant deer appear to be adapted to taking bending forces in the antero-posterior and dorso-ventral axes of the antler beam, respectively. This corresponds to our knowledge of fighting in the hog deer and matches that predicted for the giant deer. Therefore, comparative behaviour and morphology, beam theory and neutron diffraction were used in a complementary way to test the hypothesis that the extinct giant deer used its antlers in fighting. These approaches also allowed us to reconstruct its most forceful fighting, a behaviour not seen for 10 000 years.

This approach could be taken further by analysing whether the horns of ceratopsian dinosaurs could have been used in intraspecific fighting and perhaps analysing how they were used. There is circumstantial evidence of fighting in *Triceratops* based on healed injuries to the skull, which suggests that its horns were functional (Forster and Sereno, 1997). Farlow and Dodson (1975) have distinguished three possible fighting methods for ceratopsian dinosaurs based on morphological types of ceratopsians, which can be equated with Geist's (1966) hypothesis for the evolution of complex horns in bovids. Therefore, the earliest ceratopsians (e.g. *Protoceratops*) probably used their small nasal horns against the flank of their opponents. The later ceratopsians divided into two groups depending on the length of their neck frills. The long-frilled group (e.g. *Torosaurus*) probably used their prominent brow horns in frontal engagements to wrestle like many bovids and cervids, and used their neck frills in prominent displays. The short-frilled group (e.g. *Monoclonius*) may have used their larger nasal horns in frontal encounters like rhinoceroses. However, there has been little or no mechanical analysis of the horns of ceratopsians to date.

Alexander (1989) has carried out a very simple analysis by comparing the ratio of the cross-sectional area of the base of the brow horns to the body mass of *Triceratops* with the scaling relationship derived by Packer (1983) for the horn cores of bovids. It would appear that *Triceratops* had much thinner horns than a bovid of similar size would be expected to have. However, Packer's (1983) scaling data mask at least six fighting types of bovids (*Figure 9*). If the wrestling antelopes are compared alone with *Triceratops*, its horn cores has a second moment of area only slightly less $(1.12 \times 10^{-4} \, \text{m}^4)$ than the predicted value $(1.44 \times 10^{-4} \, \text{m}^4)$ for an antelope of that size. My own measurements of second moment of area in *Triceratops* range from $5.68 \times 10^{-5} \, \text{m}^4$ to $1.83 \times 10^{-4} \, \text{m}^4$, which includes the predicted value from bovids. Therefore, it seems likely that ceratopsians had a similar wrestling type of fighting to many antelopes as reconstructed by Farlow and Dodson (1975) and Alexander (1989).

I have carried out a preliminary analysis of the scaling of second moment of area of the nasal and brow horns of ceratopsian dinosaurs. This analysis includes both long-

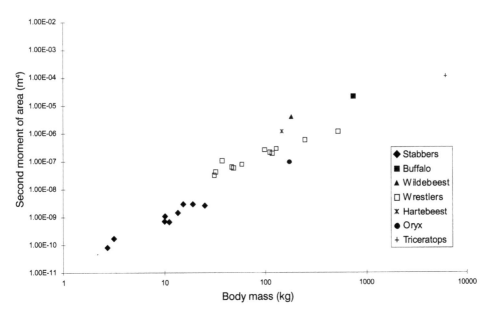

Figure 9. *A reanalysis of Packer's (1983) data for the scaling of horn core thickness and cross-sectional area. The data have been recalculated as second moments of area and assigned to the fighting categories of Kitchener (1985). Data for* Triceratops *are included from Alexander (1989).*

and short-frilled species, and many putative *Triceratops* species, which have recently been re-identified as variable males and females of a single species, *T. horridus* (Ostrom and Wellnhofer, 1986). Therefore, this analysis is slightly different from the one I have carried out for bovids and cervids (Kitchener, 1985), because we are dealing with individual specimens and mostly intraspecific variation, rather than means and only interspecific variation. Estimates of body mass were unavailable for each specimen, so I have used cubed skull length instead to examine scaling relationships. *Figure 10(a)* reveals that the second moment of area of the nasal horn does not scale with size for most short-frilled ceratopsians and for females of long-frilled animals, including *Torosaurus*, *Pentaceratops* and the short-frilled *Triceratops*. However, male long-frilled ceratopsians and *Triceratops* do show a linear relationship between nasal horn *I* and size as would be expected. This suggests that short-frilled ceratopsians and females of long-frilled animals do not use their horns for horn-to-horn fighting, but instead use them in attacks on softer flanks and possibly against predators (see Farlow and Dodson, 1975). The relationship between brow horn *I* and size shows a similar dichotomy in scaling (see *Figure 10(b)*), although this is less obvious, because there is some increase in I with size in female long-frilled and short-frilled animals. Neck frill area also shows some division into subgroups, although the data are less clear (see *Figure 10(c)*). It would seem that ceratopsians do indeed have horns that are adapted to fighting similar to that of bovids as suggested by Farlow and Dodson (1975), although there is clearly still much interesting research to pursue.

Another group of dinosaurs, the pachycephalosaurs, had a very thick cranium (180 mm thick in a 280 mm long skull of *Stegoceras*; Alexander (1989)), which they

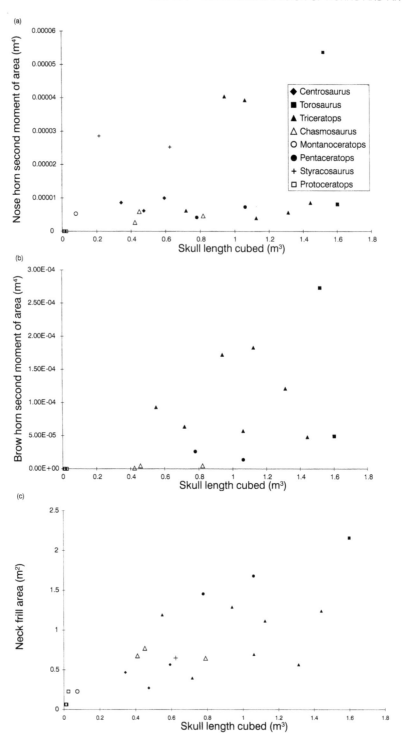

Figure 10. (a) *The scaling of the nasal horn second moment of area of ceratopsian dinosaurs.* (b) *The scaling of the brow horn second moment of area of ceratopsians.* (c) *The scaling of neck frill area in ceratopsians.*

were supposed to have used to clash together with considerable force in intraspecific combat like modern wild sheep. However, there is no sinus to protect the brain as found in modern sheep and goats, so it is unclear how damage to the brain was avoided. Alexander (1989) has estimated that one species, *Stegoceras*, probably produced a force of about 9000 N on impact, which he considered would easily be resisted by the deformation of the spongy bone of the skull. However, pachycephalosaurs were said to be able to lock their vertebrae together like a rod behind the skull and in line with the force of fighting. As Alexander (1989) pointed out, this arrangement would not have allowed the energy produced by the impact to be dissipated. Instead, he suggested that the necks of pachycephalosaurs folded up, absorbing energy in this way. However, I wonder whether the focus for the head blows of pachycephalosaurs was against soft body tissues rather than against hard heads, rather like the fighting of male giraffes, *Giraffa camelopardalis* (Dagg and Foster, 1982). This would surmount the problem of protecting the brain from impact and may make the shock absorption problem of the vertebral column unnecessary, as the soft tissues of the opponent would absorb most of the energy. Recently, Goodwin *et al.* (1998) have suggested that there were varying fighting techniques for different morphological groups of pachycephalosaurs, ranging from head-shoving in the flat-headed pachycephalosaurs (e.g. *Homalocephale*) to flank-butting in the high-domed pachycephalosaurs (e.g. *Stygimoloch*). However, this is mere speculation and emphasizes that there is much still to be discovered about fighting in both extinct and extant species.

9. Conclusion

Mechanical analyses of the fighting of horned mammals may provide useful insights into the functional design of horn-like organs and fighting, and the mechanical or developmental constraints that affect horn and antler design. In this paper I have provided a brief overview of the interaction between fighting, horn and antler structure, material properties, mechanical design of horns and antlers, and their mechanical functioning in fighting. I have shown also how these techniques may also be applied to extinct species, including the dinosaurs. Although I have been able to present a broad approach to the application of mechanics to fighting, this area of research is very much in its infancy and there is considerable scope for further fruitful research.

Acknowledgements

I am grateful to Paolo Domenici for inviting me to give this paper at the SEB meeting on Biomechanics and Behaviour. I am most grateful to many people who have helped me in various ways in this work, including Julian Vincent, Bev Halstead, Mike Watkins, Andrew Jackson, George Jeronomidis, George Bacon, Glenn Storrs, John Ostrom, Nicholas Hotton and Eugene Gaffney. Part of this work was carried out under a Short Term Visitor Program of the Smithsonian Institution—I am grateful to Chris Wemmer for assisting in this and for his kind hospitality at Front Royal. I am also very grateful to Evelyn Bowyer and John Pease, who looked after me so well while I measured dinosaurs in New York.

References

Alexander, R.McN. (1982) *Optima for Animals.* Edward Arnold, London.

Alexander, R.McN. (1989) *Dynamics of Dinosaurs and Other Extinct Giants.* Columbia University Press, New York.

Alvarez, F. (1990) Horns and fighting in male Spanish ibex, *Capra pyrenaica. J. Mammal.* **71**(4): 608–616.

Alvarez, F. (1992) Animal weapons and sexual selection. *Ric. Biol. Selvaggina (Suppl.)* **19**: 415–437.

Alvarez, F. (1994) Bone density and breaking stress in relation to consistent fracture position in fallow deer antlers. *Acta Vertebr. Donaña* **21**(1): 15–24.

Anderson, J.L. (1980) The social organisation and aspects of behaviour of the nyala, *Tragelaphus angasi* Gray, 1849. *Z. Säugetierkunde* **45**: 90–123.

Ascenzi, A. and Bonucci, E. (1970) The mechanical properties of the osteon in relation to its structural organisation. In: *Chemistry and Molecular Biology of the Intercellular Matrix, Vol. III* (ed. E.A. Balazs). Academic Press, London, pp. 1341–1359.

Bacon, G.E. and Griffiths, R.K. (1985) Texture, stress and age in the human femur. *J. Anat.* **143**: 97–101.

Bacon, G.E., Bacon, P.J. and Griffiths, R.K. (1979) The orientation of apatite crystals in bone. *J. Appl. Crystallogr.* **12**: 99–103.

Barrette, C. (1977) Fighting behaviour of muntjac and the evolution of antlers. *Evolution* **31**: 169–176.

Boortsma, M. (1979) Intertwining of horns of fighting black wildebeeste bulls. *J. S. Afr. Vet. Assoc.* **50**(2): 122.

Child, K.N. (1982) Moose antlers. *Wildlife Rev.* **10**(4): 17–20.

Clark, M. (1981) *Mammal Watching.* Hamlyn, London.

Clutton-Brock, T.H. (1982) The functions of antlers. *Behaviour* **79**: 108–124.

Clutton-Brock, T.H., Albon, S.A. and Guinness, F.E. (1979) The logical stag: adaptive aspects of fighting in red deer. *Anim. Behav.* **27**: 211–225.

Clutton-Brock, T.H., Guinness, F.E. and Albon, S.D. (1982) *Red Deer: The Behaviour and Ecology of Two Sexes.* Chicago University Press, Chicago, IL.

Currey, J.D. (1979) Mechanical properties of bone tissues with greatly differing functions. *J. Biomech.* **12**: 313–319.

Dagg, A.I. and Foster, J.B. (1982) *The Giraffe. Its Biology, Behavior and Ecology.* Krieger, Malabar.

Farlow, J.O. and Dodson, P. (1975) The behavioral significance of frill and horn morphology in ceratopsian dinosaurs. *Evolution* **29**: 353–361.

Forster, C.A. and Sereno, P.C. (1997) Marginocephalians. In: *The Complete Dinosaur* (eds J.O. Farlow and M.K. Brett-Surman). Indiana University Press, Bloomington, IN, pp. 317–329.

Fraser, R.D.B. and Macrae, T.P. (1980) Molecular structure and mechanical properties of keratins. In: *The Mechanical Properties of Biological Materials* (34th Symposium of the Society of Experimental Biology) (eds J.F.V. Vincent and J.D. Curry). Cambridge University Press, Cambridge, pp. 211–246.

Geist, V. (1966) The evolution of horn-like organs. *Behaviour* **27**: 175–214.

Geist, V. (1971) *Mountain Sheep.* Chicago University Press, Chicago, IL.

Geist, V. (1978) On weapons, combat, and ecology. In: *Aggression, Dominance and Spacing* (eds L. Kramer, P. Pliner and T. Alloway). Plenum, New York, pp. 1–30.

Geist, V. (1986) New evidence of high frequency of antler wounding in cervids. *Can. J. Zool.* **64**: 380–384.

Goodwin, M.B., Buchholtz, E.A. and Johnson, R.E. (1998) Cranial anatomy and diagnosis of *Stygimoloch spinifer* (Ornithischia: Pachycephalosauria) with comments on cranial display structures in agonistic behavior. *J. Vertebr. Paleontol.* **18**(2): 363–375.

Gordon, J.E. (1976) *The New Science of Strong Materials*, 2nd Edn. Penguin, London.

Gosling, L.M. (1974) The social behaviour of Coke's hartebeest. In: *The Social Behaviour of Ungulates and its Relation to Management*, Vol. 2 (eds V. Geist and F. Walther). IUCN, Morges, pp. 488–511.

Gosling, L.M., Petrie, M. and Rainy, M.E. (1987) Lekking in topi: a high cost, specialist strategy. *Anim. Behav.* **35**(2): 616–618.

Goss, R.J. (1983) *Deer Antlers*. Academic Press, London.

Gould, S.J. (1974) The origin and function of 'bizarre' structures: antler size and skull size in the 'Irish elk', *Megaloceros giganteus. Evolution* **28**: 191–220.

Jacobsen, N.H.G. (1972) Distribution, home range and behaviour patterns of bushbuck in the Lutope and Sengwa Valleys, Rhodesia. *J. S. Afr. Wildlife Manage. Assoc.* **4**: 75–93.

Jarofke, D., Klös, H.-G. and Frese, R. (1990) Todesursachen der Gaure (*Bos gaurus*) in Zoologischen Gärten (Zuchtbuchauswertung). *Erkrank. Zootiere* **32**: 349–351.

Jaslow, C.R. and Biewener, A.A. (1995) Strain patterns in the horncores, cranial bones and sutures of goats (*Capra hircus*) during impact loading. *J. Zool. Lond.* **235**: 193–210.

Kingdon, J. (1982) *East African Mammals: An Atlas of Evolution in Africa*, Vol. IIIC Bovids. Academic Press, London.

Kitchener, A. (1985) The effect of behaviour and body weight on the mechanical design of horns. *J. Zool.* **205**: 191–203.

Kitchener, A. (1987a) Fighting behaviour of the extinct Irish elk. *Mod. Geol.* **11**: 1–28.

Kitchener, A. (1987b) Effect of water on the linear viscoelasticity of horn sheath keratin. *J. Materials Sci. Lett.* **6**: 321–322.

Kitchener, A. (1987c) Fracture toughness of horns and a reinterpretation of the horning behaviour of bovids. *J. Zool.* **213**: 621–639.

Kitchener, A. (1988) An analysis of the forces of fighting of the blackbuck (*Antilope cervicapra*) and the bighorn sheep (*Ovis canadensis*) and the mechanical design of the horns of bovids. *J. Zool.* **214**: 1–20.

Kitchener, A.C. (1991) The evolution and mechanical design of horns and antlers. In: *Biomechanics and Evolution* (eds J.M.V. Rayner and R.J. Wootton). Cambridge University Press, Cambridge, pp. 229–253.

Kitchener, A. and Vincent, J.F.V. (1987) Composite theory and the effect of water on the stiffness of horn keratin. *J. Materials Sci.* **22**: 1385–1389.

Kitchener, A.C., Bacon, G.E. and Griffiths, J.F.V. (1994) Orientation in antler bone and the expected stress distribution, studied by neutron diffraction. *Biomimetics* **2**(4): 297–307.

Leslie, D.M. and Jenkins, K.J. (1985) Rutting mortality among male Roosevelt elk. *J. Mammal.* **66**(1): 163–164.

Leuthold, W. (1977) *African Ungulates*. Springer, Heidelberg.

Lundrigan, B. (1996) Morphology of horns and fighting behavior in the Family Bovidae. *J. Mammal.* **77**(2): 462–475.

Mloszewski, M.J. (1983) *The Behaviour and Ecology of the African Buffalo*. Cambridge University Press, Cambridge.

Nowak, R.M. (1991) *Walker's Mammals of the World*, Vol. II. 5th Edn. Johns Hopkins University Press, Baltimore, MD.

O'Gara, B.W. and Matson, G. (1975) Growth and casting of horns by pronghorns and exfoliation of horns by bovids. *J. Mammal.* **56**: 829–846.

Ostrom, J.H. and Wellnhofer, P. (1986) The Munich specimen of *Triceratops* with a revision of the genus. *Zitteliana* **14**: 111–158.

Packer, C. (1983) Sexual dimorphism: the horns of African antelopes. *Science* **211**: 1191–1193.

Papageorgiou, N. (1979) Population energy relationships of the Agrimi (*Capra aegagarus cretica*) on Theodorou I., Greece. *Mammal Depicta No. 11*. Paul Parey, New York.

Petocz, R.G. and Shank, C.C. (1983) Horn exfoliation in Marco Polo sheep *Ovis ammon poli* in the Afghan Pamir. *J. Mammal.* **64**: 136–138.

Ranjitsinh, M.K. (1989) *The Indian Blackbuck.* Naraj, Dehradun.

Roberts, S.C. (1996) The evolution of hornedness in female ruminants. *Behaviour* **133:** 399–442.

Savage, R.J.G. and Long, M.R. (1986) *Mammal Evolution.* British Museum (Natural History), London.

Schaffer, W.M. (1968) Intraspecific combat and the evolution of the Caprini. *Evolution* **22:** 817–825.

Schaffer, W.M. and Reed, C.A. (1972) The co-evolution of social behaviour and cranial morphology in sheep and goats (Bovidae, Caprini). *Fieldiana Zool.* **61:** 1–62.

Schaller, G.B. (1998) *Wildlife of the Tibetan Steppe.* Chicago University Press, Chicago, IL.

Smith, D.R. (1990) Two pronghorn antelope found locked together during the rut in west central Utah. *Great Basin Nat.* **50**(3): 287.

Spinage, C.A. (1982) *The Uganda Waterbuck: A Territorial Antelope.* Academic Press, London.

Spinage, C.A. (1986) *The Natural History of Antelopes.* Croom Helm, Beckenham.

Tello, J.L.P.L. and Van Gelder, R.G. (1975) The natural history of the nyala *Tragelaphus angasi* (Mammalia, Bovidae) in Mozambique. *Bull. Am. Mus. Nat. Hist.* **155:** 319–386.

Walther, F. (1966) *Mit Horn und Huf.* Paul Parey, Berlin.

Walther, F., Mungall, E.C. and Grau, G.A. (1983) *Gazelles and their Relatives.* Noyes, Park Ridge, NJ.

18

Co-evolution of behaviour and material in the spider's web

Fritz Vollrath

1. Introduction

Many animals make nests using materials they collect. In some cases, such as hamsters, nesting material is collected in the process of ordinary foraging for food; but storage in the sleeping chamber would turn it into nest material. However in other cases, such as birds, nesting materials cannot be eaten (and never were during evolution) but are collected using specific criteria, which presumably differ from the criteria used in foraging for food objects. Moreover, the collected sticks or objects are not just piled on top of each other but interwoven with specifically adapted behavioural patterns. Indeed, the collections can be specific to a particular phase in the building process. Further, in addition to collecting the appropriate object, the animal might shape it to provide an even better fit; for instance, the caddis fly cuts a leaf to suit the changing shape of its tube. Here the action patterns of the combined collection and integration behaviour have been augmented by shaping–cutting behaviour patterns. Many animals add further value to the collected items by gluing them together using a bodily excretion (such as spit, mucus, silk or faeces). Here specific wiping, tapping and pulling behaviour patterns have evolved to combine the collected items with the excretion, which is massaged into or spread over the collection with specific behaviour patterns. Finally, a few animals no longer collect objects for their nests but construct them entirely from excretions. Here the collection behaviour has been dropped but the 'gluing together' behaviour has been refined. Spiders are the most prominent examples of this evolutionary pinnacle of nest building (*Figure 1*).

In the following, using the orb web as a specific example, I shall show how the interaction of behaviour and material is forever being optimized by evolution with respect to both the behaviour and the material. I consider it appropriate to use the term co-evolution for this kind of concerted evolution. In the traditional use, co-evolution denotes the interaction between the genotype fitnesses of separate but closely interacting species, generally within a very specific meaning (Futuyma, 1998). The arms race between a

Biomechanics in Animal Behaviour, edited by P. Domenici and R.W. Blake.

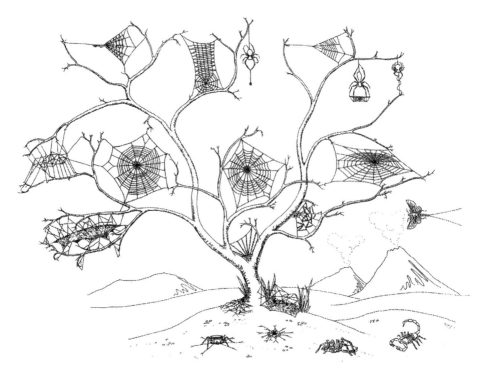

Figure 1. *A selection of orb weavers on a tree (emphatically not a phylogenetic tree) to demonstrate the various web types. On the ground we see distant ancestors and relatives (scorpion, mygalomorph, trapdoor spider and amaurobiid). Further ancestral relatives (eresid and agelenid) have built their webs on the base of the tree, a dictynid web spans the fork of the tree. The right-hand branch contains (in order from its base) the webs of the cribellates* Stegodyphus *followed by the orb-weaving hackled orb weaver* Uloborus *and the triangle spider* Hypotiotes *as well as the gladiator spider* Deinopis *and the line lasso spider* Miagrammopes. *The centre branch holds an ecribellate orb web which might be by the garden cross spider* Araneus. *The left-hand branch holds another ecribellate orb web, which might be by* Meta *or* Tetragnatha; *this branch also supports (on the extreme left) three-dimensional* Theridiion *(above) and* Linyphia *(below) type webs as well as (upper left to right) ecribellate orb webs by* Theridiosoma *and* Scoloderus, *and finally the bolas spider* Mastophora *with its glue-drop line web (from Vollrath, 1988a). (For more details on web types, see Stowe (1986) and Eberhard (1990), and for recent hypothesis on cladistic relationships of orb spiders, see Hormiga* et al. *(1995).)*

specialist predator and its prey would be an example of co-evolution, as would be the shape of a flower and the anatomy of its pollinator. Co-evolution would not typically be used for denoting the interactive evolution between the parts of an individual. But by employing this term in the context of the co-optimization of behaviour and material I aim to draw attention to the fact that these two are encoded in totally different genes and subject to rather different selection factors. The genes for the building behaviour are evolved versions of genes encoding for modules of sensor, motor and decision behaviour. The genes encoding for the building material are evolved versions of genes encoding for silk protein sequence and silk production pathways. I can see no element where the genes for material and those for behaviour might overlap even within the same individual. Because of this separation of functional gene complexes coupled to the parallel but highly interactive evolution of the two, I feel that the use of the term co-evolution is justified.

2. The unit of spider and web

For our garden cross spider, *Araneus diadematus*, a healthy bee (of 100 mg) in full flight (travelling at, say, 40 km h^{-1}) constitutes a fine parcel of food. But this potential prey item also contains a fair amount of kinetic energy that must, somehow, be absorbed by the web and its strands. If this energy were not absorbed, then the bee would break the threads and fly straight through the trap. Or, if the strands were very strong and elastic, then the insect would be flung back out, trampoline fashion, whence it came. Neither is a desired effect. The web has been built to feed the spider, and for that it has to efficiently and effectively catch the bee (or fly, or grasshopper) by intercepting it in mid air. How is this done? Some of the energy is taken up by the silk, some by the structure of the web, and some by the interaction between the web and the environment (Lin *et al.*, 1995). Once the energy is taken up, the spider itself springs into action to prevent the escape of the struggling prey.

The web does not function without the spider (Witt *et al.*, 1968). First the spider has to find the right location (Vollrath, 1992a). Then it has to build the structure—ideally with the appropriate design features tailored to the climatic conditions of that location and the specific prey expected here (Witt, 1963a; Vollrath *et al.*, 1997). Then the trap has to be operated: the spider must quickly run to the prey and overpower it, drag it back to the lair, wrap it up and prepare for another prey (Robinson, 1969; Eberhard, 1989). If debris falls into the web, then the spider must clean it out because, first, debris interferes with the vibration–information transmission through the web (Witt, 1963b) and, second, it might render the trap visible to potential prey (Craig and Freeman, 1991). Finally, the spider regularly takes the web down and replaces it with a new structure, if the site was rewarding (Vollrath, 1985). If it was not, then the spider must decide whether to pack up and leave, or whether to stay and give it another try; after all, who is to know whether a new site would be better (Vollrath and Houston, 1986). Moreover, there are dangers en route. In many species the web is recycled, that is, it is eaten and the amino acids (Witt *et al.*, 1968) as well as any water on the silk (Edmonds and Vollrath, 1992) are absorbed by the spider.

Some spider silks are virtually unbeatable for material qualities (*Figure 2*). Far tougher than steel, they equal or even surpass the best of man-made fibres in strength, elasticity and toughness (Denny, 1980). Yet silk is produced by the spider at ambient temperatures and pressures. Clearly, evolution has led to a biopolymer that allows the spider to build daring constructions. Or is it that the spider's drive to build these aerial masterworks has led to the evolution of this extraordinary material? What came first, the silk or the behaviour?

In the following I shall give an overview of (i) the building material silk (reasonably well understood if only at a beginner's level), (ii) the engineering of the web (hardly understood even at that level), (iii) the building behaviour of the spider (well understood at all levels), and, finally, (iv) the co-evolution of all three (highly hypothetical).

3. The spider's building material

Let us examine first the silk. What kind of material is it? How is it made? How does it work? Silk is a protein, a fibroprotein to be more precise (for a good review of recent work on silk, see Kaplan *et al.* (1994a). Spider silk is similar in chemistry and, possibly, also molecular structure, to the silk of the commercial silkworm. The silkworm is an

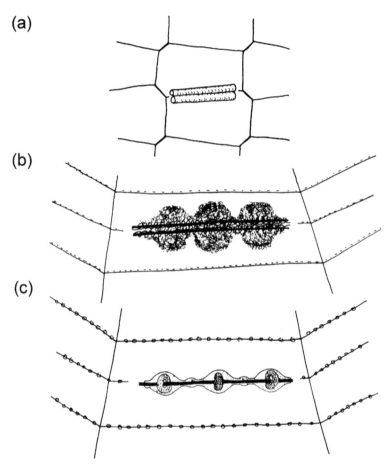

Figure 2. *Sketches of adhesive capture spiral thread. (a) Dry dragline thread is used for both thread types (radial and spiral) that make the long-lasting capture web of* Cyrtophora *spiders. (For more information on this type of silk and web, see Lubin (1973).) (b) Cribellate hackled capture thread here shown in a composite picture for a hypothetical spider based on scanning electron microscopy of threads of* Deinopis spp. *and* Uloborus spp. *The axial threads (diameter* ~ 0.2 μm*) are generally thinner than the crimped threads (* ~ 0.5 μm*), and both are very much thinner than the hackled threads (* ~ 200 nm*). (For more information on hackled capture threads and webs, see e.g. Opell (1993) or Peters (1987) and Eberhard (1988a).) (c) Ecribellate coated capture thread (Vollrath, 1994). A composite picture based on Nomarsky and scanning electron microscopy of threads from* Araneus diadematus, Zygiella x-notata *and* Meta segmentata. *The coat is mostly water, and the rings consist of glycoproteins. They are highly adhesive and can slide on the thread. Thus they avoid exerting localized force on the core threads while at the same time accumulating on the prey strengthening points of attachment. (For more information on coated capture threads, see Vollrath and Edmonds (1989) and Vollrath and Tillinghast (1991).) The two types of thread are drawn roughly to scale.*

insect, the larva of a moth and not the only insect that makes silk. Spider silk is generally much stronger and more elastic than insect silk, not least because of its function as a high impact absorber. Strictly speaking, silk is not a simple material but a composite consisting of protein crystals and fibrils embedded in a protein matrix— sub-microscopic wattle and daub might serve as a rough analogy. The crystals are wavy sheets of interlinked amino acids, neatly stacked into heaps. They give the silk thread its strength whereas the matrix gives it resilience. The great strength of a composite when compared with a homogeneous material derives from its ability to deflect a crack (Gordon, 1976). For a material under tension, a crack can be fatal because at its tip the stress lines that normally run parallel within the thread now bunch up, creating a greater localized load and thus further propagating the crack with ever increasing speed and force. In a composite material the growing crack encountering an obstacle, be it crystal or wattle, has to detour around it, losing energy in the process. A crack meeting obstacle after obstacle in its path is quickly stopped for good.

One model to explain the material properties of silk uses a crystal–matrix image (Gosline et al., 1994); it is based on studies of silkworm silk and may be true for some spider silks. Another model may apply to the ultrastructure of other silks, specifically the dragline silk of the golden silk spider Nephila clavipes (Vollrath et al., 1996). Recent light and electron microscopy studies show that the Nephila dragline thread has a highly organized microstructure that could explain the extraordinary tensile strength of this, as well as comparable silks. Here, a thread seems to be composed of many of the finest micro-fibrils wound around a thin core, reminiscent of a climbing rope (Vollrath et al., 1996). A multiple-stranded thread always has greater tensile strength than a single-stranded fibre of the same material diameter, as it, too, is better at resisting the propagation of cracks. Whatever the correct model, it is obvious that spider silks have evolved complex microstructures that enhance the mechanical properties of the material (Shao et al., 1999).

It seems that the spider can modify these mechanical properties and adapt them to a task at hand (Vollrath and Köhler, 1996). We may assume that the animal uses small adaptations in a direct and reflex-like response; for example, to meet an increased demand for strength on a windy day the spider may simply vary the speed at which it pushes the silk through the nozzle of the spigot. Greater changes of material quality would seem to require more fundamental alterations, for example, adjustments in the chemical composition of the chains. Starvation, for example, seems to lead to modifications in silk mechanics (Madsen et al., 1999) as well as, not surprisingly, amino acid compositions (C. Craig, personal communication).

To satisfy greater and impromptu needs for special silks, the spider relies on different glands producing different raw materials (Vollrath, 1992b). The raw liquid silk is made in special glands in the spider's abdomen. It is stored in sacs and, when needed, pressed or pulled through a duct and out into the open through fine openings at the tops of small turrets, the spools or spigots (Tillinghast and Townley, 1994). Spigots aggregate at the tips of short limbs, the spinnerets, which are found on the spider's abdomen, near the apex. The common garden spider has seven kinds of silk (Kovoor, 1987). Inside its storage sac all silks are liquid and all but one kind harden when pressed through the spinning duct because here the crystals align and interlink (Knight and Vollrath, 1999; Vollrath and Knight, 1999). Only the viscid, sticky silk that covers the capture spiral of the web remains a liquid. It is a very special silk, hardly deserving the name 'silk' for it consists mostly of water enriched by a few protein compounds that help to collect more water from the air (Vollrath et al., 1990). Other constituent chemicals prevent bacteria from growing in this nutritive medium (Schildknecht et al., 1972) and yet other chemicals make this silk sticky (Vollrath and Tillinghast, 1991).

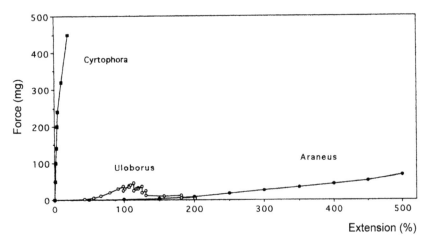

Figure 3. Stress–strain curve for averaged values of dry, hackled and coated spiral threads stretched until they broke: Plain Cyrtophora thread (typical for dragline, radii and spiral), hackled Uloborus capture thread (made up of core fibres and fine cribellum threads) and coated Araneus thread (made up of core fibres and aqueous cover containing many compounds). It should be noted that the dry thread is rather stiff, that the coated thread stretches most and that the hackled thread has sudden losses of stress suggesting the subsequent breaking of fibre components (based on data from Köhler (1992)). (For more information on the technique and data aquisition, see Köhler and Vollrath (1995), Madsen and Vollrath (2000) and Vollrath and Edmonds (1989).)

Each silken thread is very thin indeed, a hundred times thinner than a human hair. Because it is so thin it appears to be weak when pulled by a human hand. But far from being weak, spider silk is in fact extremely strong (*Figure 3*), weight for weight much stronger than nylon or steel and equal to Kevlar in toughness (Kaplan *et al.*, 1994b). Some spider silks are as tough as the best man-made fibre mainly because silk is not only strong but also very elastic (Vollrath and Edmonds, 1989). Some spider silks, especially the sticky silk of the garden spider's orb web, are much more elastic than any commercial rubber known. These capture silks can be extended to many times their own length without breaking. After each extension they easily and immediately contract back again to their former lengths, and even far beyond this point to a small fraction of the original lengths. Such silks are 'more' than composite materials; they are complex mechanical systems in their own right, albeit on a minute scale. A sticky capture thread consists of a pair of core fibres surrounded by a coat of highly viscous liquid (more of a syrup than a fluid). This coat, although applied as a continuous film of water-attracting chemicals, quickly separates into little droplets swelled by water taken from the atmosphere (Edmonds and Vollrath, 1992). Moreover, this water creeps into the silken core fibres, altering their mechanical behaviour. Whereas dry silk is rather stiff, and breaks at extensions of about 30%, waterlogged silk becomes floppy and elastic, and breaks at extensions of 100% or more (Shao and Vollrath, 1998). The coated sticky spiral can be extended to many times its own length ($\sim$ 400%) before it breaks (Vollrath and Edmonds, 1989). At the initial phase, little force is needed to stretch the thread but suddenly, just before breaking, it begins to resist further stretching (Vollrath and Edmonds, 1989). Thus, although the sticky thread is soft and appears to require very little force to stretch, it is actually strong, although its real strength is shown only at the very end of its extension. Thus the sticky thread has the strength to retain the hapless prey without acting as a trampoline or vibration damper and it has the toughness, by its extensibility, to absorb huge amounts of kinetic energy.

How does the capture thread do it? Ordinary, dry silk sags when suddenly relaxed from a stretched position and only very slowly 'creeps' back to its former length (Denny, 1980). The sticky threads snap back into shape instantaneously and, even more astonishing, can be relaxed beyond their initial length without ever sagging. Amazing as it seems, the droplets act as tiny windlasses and cable drums, with the surface tension of the watery coat powering the reeling-in of the core fibres (Vollrath and Edmonds, 1989). Stretching and relaxing a sticky thread in rapid succession will wind and unwind the core fibres inside their droplets. Thus the entire system of threads and coat is always under tension, maintaining the sticky thread taut. The energetic forces applied from the outside, by buffeting winds or blundering insects, are absorbed not just by the silken core fibres but also by the windlass system. The kinetic energy of the wind or of an insect suddenly stopped in its tracks is converted into heat and distributed in the water-logged threads. And, in a clever twist, this absorbed energy actually strengthens the thread because rubber, unlike other materials, develops greater rather than lesser strength when hot and wet silk is *de facto* a rubber (Gosline *et al.*, 1984). Thus, the kinetic energy of the hapless insect actually helps the felon rather than the victim.

The spider's wet capture thread is nature's tinkering at its best. But then, what do we expect from 180 million years (or so) of time spent on parallel 'research and development' in many million 'laboratories'? It has been shown that modern orb spiders first emerged in the Cretaceous, *circa* 180 million years ago (Selden, 1989). Spider silk, however, is much older, having evolved about 400 million years ago in tarantula-like arachnids (Shear *et al.*, 1989). During this time, spider silk evolved from a relatively simple protein ribbon to the broad range of highly specialized silks found in modern orb weavers. As a highly advanced orb spider *Araneus* has had millions of generations in which to fine tune each and every one of its silks, turning them into veritable show-pieces for the adaptedness of a material and its economic use. As we have seen, simply by applying a watery coat an ancestral orb spider has turned a great and potentially fatal disadvantage of silk (namely, that it becomes floppy when wet) into a great advantage (that it is soft where required). This wet silk, in combination with 'normal' dry silk, makes a most effective trap. The uncoated, dry, stiff and nonsticky threads form the orb spider's 'highways', signal lines, and firm stays; whereas the coated, wet, soft and sticky threads form its mesh of fetters. Although a most amazing material, silk is not the only great invention of the ancestors of our common garden cross spider *A. diadematus*. The engineering of the web employs another and very different trick to convert a potentially constraining feature into an advantage. This trick, the mini-windlass, makes further use of the extreme thinness of the threads as I shall demonstrate later. But first let us examine some architectural and engineering aspects of the trap.

4. Architecture of the structure

There are two aspects to an orb web's architecture. One concerns the structure in splendid isolation, the other its 'behaviour' in the natural environment. This natural function provides some exciting insights into constraints acting on the spider's orb web as a completed structure (rather than on its sections, the threads). The concept of constraints in biology is an old one (generally used in morphology), which has recently received new and wider attention as it illustrates well the force of natural selection and the extraordinary flexibility of the evolutionary process (Maynard Smith *et al*. 1985).

The structural engineering of the orb web is far from simple (Witt *et al*., 1968). The web, as I have shown, incorporates silks of different chemical compositions and with different mechanical properties, connected in a polar network that has both firm and soft sections. The firm sections, the spokes or radii, transmit the vibrations that signal the presence of prey; they also form pathways on which the spider traverses its web. In addition, they are crucial for the arrest of the incoming insect and the rapid distribution of its kinetic energy away from the point of impact. The radii support the capture spiral, which forms the soft sections, sticky and extraordinarily elastic. The softness prevents the capture spiral from interfering with the vibration transfer along the radii, which would be dampened by a tightly strung connecting spiral. In addition, as argued earlier, its very softness is necessary to arrest the insect's flight without catapulting it back out or, once caught, providing purchase for its struggling legs.

Measurements of radial geometry have shown that all radii have some pre-tension, and that strategically placed radii can be pre-tensed more than others that are less crucial (Wirth and Barth, 1992). As each radius is connected to its two neighbours by 20–40 spiral threads, the radii do not really act independently, and their interaction is governed partly by the resilience of these interconnecting threads (Lin *et al*., 1995). Moreover, these spiral threads (super-contracting under their watery coat) form tightening 'rings' around the web's hub and thus pull the radials inward, creating a gradient of tension along each, which increases towards the web centre (Wirth and Barth, 1992). At present, we do not fully understand whether this is a neutral, desired or unwanted side effect of the spiral threads.

The degree of radial pre-tension determines how firmly a web is strung, and thus how 'bouncy' it is (Craig, 1987). Further, the tensile strength of these threads determines to some degree the strength of the whole structure. However, as we have seen, both elasticity and strength themselves are also strongly affected by the material properties of the capture spiral threads. These capture threads with their tiny beads of water convey another interesting and unexpected property to the web, that of aerodynamic damping (Lin *et al*., 1995). Consider that the threads are extremely thin, even those bearing droplets. In this region of small size and low Reynolds number strange things happen to the physics of everyday life (McMahon and Bonner, 1983). For example, water behaves like treacle and air like water. Waterfleas and other minute creatures suffer (or benefit) from these features as much as do spider's webs. For the latter, this means that the strands of a web, as they are pushed through the air by the impact of an insect, are 'braked' by air resistance. This explains why an orb web in a second or two can totally dissipate the kinetic energy of even a large insect in full flight. Of course, the spider, however ingenious, cannot 'have its cake and eat it'. Aerodynamic damping also means that wind has an extraordinarily strong effect (Lin *et al*., 1995): the billowing of a web in only moderate breezes can force the spider to take down its structure and retreat into a crevice. This observation illustrates well the usual conundrum of a biological constraint: benefits are typically bought at some price.

5. Building behaviour

The spider's web-building behaviour combines orientation in space with the manipulation of threads. Rules of manipulation determine points where threads are joined

(Peters, 1970; Eberhard, 1982, 1988b; Vollrath, 1988a,b, 1992a,b) and rules of orientation guide the animal's path to the next joint (Gotts and Vollrath, 1991; Krink and Vollrath, 1997, 1998a,b). The construction of the web's 'rim' (its frame) and its spokes (the radii) are crucial in laying down the engineering of the web because here the animal fashions the working platform on which to build the rest of the web in the most efficient way (Witt *et al.*, 1968). Frame and radius construction are followed by the construction of a temporary auxiliary (or scaffolding) spiral running from the hub outwards. This, in turn, is followed by a permanent capture spiral running from the periphery inwards which is an important construction aid (Zschokke, 1995). In the process of building the capture spiral the spider cuts, step by step, the auxiliary spiral. Finally, orb spiders tune their web by tensioning threads on the hub (Eberhard, 1981). Distorting webs during construction do not seem to affect radius placement, which indicates that tensions are not crucial for the spider's orientation (Krieger, 1992). However, tensions would affect the engineering and we might assume that the final tuning at the hub will take care of such inadequacies or inconsistencies arising from distortions during construction (Eberhard, 1987). Moreover, the mechanical properties of the silks used in web construction also affect web performance, and tuning the web to suit the silk presumably allows the spider to optimize the trap's effectiveness. But efficiency in prey capture is only one of the web's traits; the other is survival at least until the prey has made contact.

Weather affects a web and can severely limit its longevity. Quatremére-Disjonval (1797) during his long sojourn as the King's guest in the Bastille recorded and understood this; others have followed in his footsteps (albeit under more liberal conditions) by proving that spiders adapt their webs to climatic conditions (Wolff and Hempel, 1951). In addition to weather, and often in conjunction with it, the spider must consider the structural parameters of a site (Heiling and Herberstein, 2000). Like some other spiders (Ades *et al.*, 2000), orb webs of the garden cross spider *A. diadematus* can be adapted in shape to fit into highly eccentric web supports (Krink and Vollrath, 1999, 2000). Moreover, they can be re-engineered with sudden changes in wind, temperature and humidity (Vollrath *et al.*, 1997). A sudden drop in the spider's supply of silk can be accommodated by modifications of the structure (Vollrath *et al.*, 1997), as can long periods of starvation with the accompanying decrease in silk raw material (Vollrath and Samu, 1997). Even slightly windy conditions (0.5 m s⁻¹) during web building cause spiders to build smaller and rounder webs with fewer capture spirals and larger distances between capture spiral meshes (Vollrath *et al.*, 1997); stronger winds will eventually lead to the cessation of all building activity. A fall in temperature from 24° to 12°C will cause the capture spiral to have fewer but wider spaced meshes, not altering overall capture area but reducing the length of capture spiral threads laid down (Vollrath *et al.*, 1997). Decreased humidity (from 70 to 20% RH) has the effect of reducing web and capture spiral size, the latter by reducing mesh number while keeping mesh spacing constant (Vollrath *et al.*, 1997). Webs built in unnaturally rapid succession by the same spider (four webs in 24 h when one is the norm) become sequentially smaller, have fewer radii and shorter capture spirals, and are wider meshed (Eberhard, 1988c; Vollrath *et al.*, 1997; Zschokke, 1997). Starvation leads to an increase in the frequency of web-building activity as well as a higher variability in web mesh and capture area (Vollrath and Samu, 1997).

Conditions such as climate or silk supply affect not only a web's superficial geometry but also its engineering and thus its performance. The studies of the effect of wind on web engineering have shown (Hieber, 1984; Lin, 1997) that the overall architecture of an orb web is affected by the wind blowing while the spider is building, rather than just on the finished structure, i.e. the pre-stress forces in the radii or the mechanical properties of the radius silk (although those can also be affected). The net effect of the web design modified to suit windy conditions is an increase in the tilt of the capture area. This results in a dynamic asymmetric deformation, which allows those webs to withstand higher wind velocities before structural failure.

Clearly, the interaction of the building behaviour (which determines the shape of the structure) and the building materials (which co-determine the engineering of the structure) is crucial for the performance of the finished web. Moreover, already during construction the building materials interact with the behaviour and, presumably, guide it into certain pathways (*Figures 4* and *5*).

6. Capture behaviour

Architecture is important not only for a web's function as a snare but also for its function as a working platform (Witt and Reed, 1965). After all, a web functions as a trap because spider and web interact: the web intercepts and snares the insect, the spider rushes in and overpowers the struggling prey. If the spider does not interfere, or takes a long time to interfere, then the prey has a fair to very good chance of escaping. Therefore we must not overlook the function of a web as a conductor of information when we examine the engineering of the structure. Moreover, not only would the spider benefit from acquiring information about the prey but it also needs to act rapidly on that information. The spider must be able to run quickly in its web, ideally in a straight line towards the prey. Then the spider must be able to deal with the prey, fight it into submission if necessary, ideally without terminal damage to the integrity of the web.

For both intelligence gathering and predatory action, the design of a web is near perfect, with its stiff dry radiating spokes (the telegraph wires and runways) and its very soft capture spiral (the snares between the runways). The radii are tough draglines, designed to hold the spider's weight and its prey. The combination of many radii dividing a web into many small sections reinforces the strength of this design and prevents rips from running across a web. The windlass system in the droplets of the capture spiral allows those sticky threads to be supersoft while at the same time being eminently extensible and strong. Thus they hardly interfere with vibration transfer along the radii. Moreover, because of the windlass system, the capture spiral will always be taut and thus even the large distortions of a spider running along a radius do not lead to entanglements in the evenly spaced capture threads. Finally, local damage hardly ever affects the function of the rest of the web because of the modular construction of the orb into structural sections that are reasonably independent.

7. Co-evolution of material and behaviour

Behaviour is thought by many to be the pacemaker of evolutionary change (Tinbergen, 1951). The orb web might be a good example for this hypothesis. Changes in orb-web design during its evolutionary history were drastic: such webs started with web sectors radiating off a retreat in a substrate, continued via the free swinging radial orb and ended in the reduced single-line or glue-droplet webs. The initial pre-orb led, probably about 180 million years ago, to a minor shift into the new adaptive zone of fully aerial

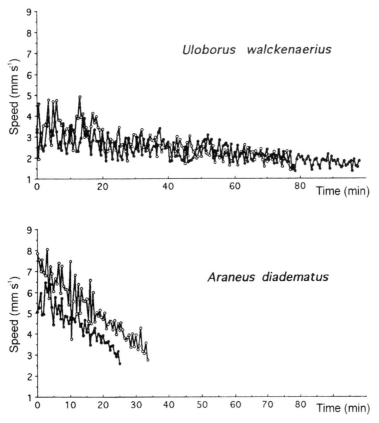

Figure 4. *Speed of capture spiral construction in* Araneus diadematus *and* Uloborus walckenaerius. *For each species, building speeds are shown for two different capture spirals. Both spirals coil from the periphery inwards. The slope of the speed curve is an indication of the spider's building speed when approaching the hub, where the thread sections between radii become successively shorter. It should be noted that the* Uloborus *spiral has a rather shallow slope; this indicates that combing of the silk takes up much of the building time. The* Araneus *spiral has a much steeper slope; this indicates that joining threads takes up much of the building time. One can estimate the time it takes the spider to connect the spiral to a radius and the speed of the 'pure' silk production, by drawing the regression lines. The intersection of a regression line with the y-axis gives the average time it takes the spider to attach the spiral to a radius. The inverse of the slope would be the speed of the silk production (based on data from Zschokke and Vollrath (1995a,b)).*

two-dimensional traps (Vollrath, 1992a). A further, more recent step to open up this adaptive zone was the invention (for want of a better word) of a new type of glyco-protein glue for the capture threads (Tillinghast and Kavanagh, 1979; Vollrath and Tillinghast, 1991) combined with the new mini-windlass system of energy dispersal (Vollrath and Edmonds, 1989). This reduced the costs of such orb webs drastically, and increased the attraction (i.e. the selective advantage) of the new adaptive zone (Opell, 1998). In this new functional environment the orb web experienced rapid adaptive radiation into many niches, and today we can observe a wide range of specializations in orb and post-orb web architecture (Shear, 1986).

    Explaining the invention of the new glue presents us with a set of hypotheses. First, the glue itself was either a totally new invention or a transformation from another type

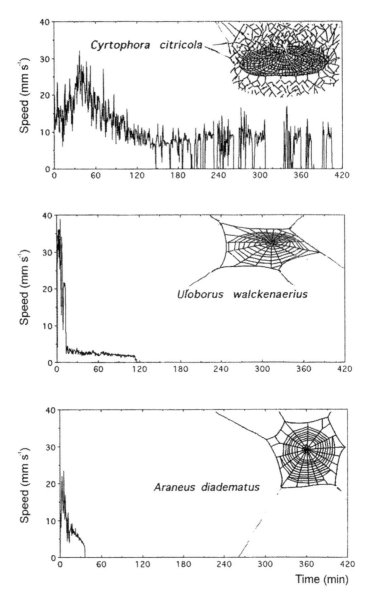

Figure 5. Activity patterns of the spiders during the construction (radii to capture spiral) of the webs shown in the upper right-hand corner. (a) Cyrtophora citricola's space web takes a long time and the spider may add sections of the capture 'spiral' on subsequent days (longer periods of inactivity indicate lapsed days). The spider moves fast but takes much time for the many firm connections between spiral and radii augmented by constant running out to the periphery to connect additional radii; thus we note that radial and spiral construction is intimately intercorrelated. This makes for a very fine and strong mesh. (b) Uloborus walckenaerius also takes a long time over its capture spiral; it walks slowly because it has to hackle (comb out) the capture silk with its hind legs. It should be noted that the construction of radials and auxilary spiral takes up only a small proportion of the total time of web building. This is very different for (c) Araneus diadematus, which is by far the fastest builder, and which gains most time during capture spiral construction. Note that the lengths of the capture spirals and equivalents in these three examples is ~ 50 000 cm for Cyrtophora, ~ 350 cm for Uloborus and ~ 1000 cm for Araneus.

of glue. Second, the web structure that now relies on the new glue was either a new invention or the transformation of a similar type of web. Before expanding on this statement, let me explain the two types of stickiness found in the web spiders. One type of stickiness relies on dry hackled silk (Opell, 1997, 1998). This type is found in cribellate spiders (Eberhard, 1988; Foelix, 1996). Hackled silk consists of thousands of the finest threads that originate each in a tiny gland, with all glands aggregated underneath a spinning plate, the cribellum (leading to the name 'cribellated' for the spiders that carry these glands). The threads are drawn out and hackled on top of core support fibres with a comb on the hind legs of the spider. This action fluffs them up and makes them sticky, possibly by charging them electrostatically (Vollrath, 1994). The other type of stickiness relies on aqueous glycoprotein glue and is found in ecribellate spiders only. This type, as outlined above, consists of droplets of water surrounding a glycoprotein core (Vollrath and Tillinghast, 1991). These droplets and glue self-assemble after a thick fluid has been applied to the core support fibres as they are drawn out (Edmonds and Vollrath, 1992). The highly hygroscopic fluid is produced by two pairs of specialized glands (glandulae aggregatae; Foelix, 1996); and spiders with these glands always lack the cribellum (hence the term 'ecribellate'). Application of the hackled silk is costly and time consuming, whereas application of the aqueous glue is cheap and instantaneous. Clearly, invention of the glue (which is phylogenetically younger than the hackled silk) bestowed energetic benefits on the construction process. Further, it seems (Opell, 1998) that glycoprotein glue is as sticky and lasting as the hackled silk (if, indeed, not stickier), so there would be no hidden costs (such as more frequent renewal of the capture threads).

Both types of silk are found in orb webs, where they provide stickiness to the capture spiral. The capture spirals of these two silk types are very similar. Indeed, web building in both types of spiders, cribellate and ecribellate orb weavers, is very similar, albeit with important temporal differences (*Figures 4* and *5*). The cribellate type of stickiness is the more ancestral, as are other features of these spiders, and it has been argued that the cribellate orb web is also the more ancestral of the two, whereas the ecribellate orb web is the more derived form (Coddington, 1986). This is an interesting argument but we do not know how one type of web could have been transformed into the other assuming linear, serial evolution (see below). Still, we must remember that such hypotheses have a fair amount of uncertainty in their assumptions and conclusions (Eberhard, 1982; Shear, 1986; Vollrath, 1992a; Opell, 1994).

Let us assume, as some do, that the cribellate orb web evolved straight into the ecribellate orb web (Coddington, 1986). There is very strong evidence that the ecribellate glue-gland (the aggregate gland) did not evolve from the cribellate hackled silk glands (Glatz, 1973). So, theoretically the two glands could coexist in one spider in the transition from cribellate to ecribellate. However, the two silks would not mix. We must assume that a hackled silk would not function in its normal way when wet. Thus the coexistence of the two types of stickiness in one capture spiral is an unrealistic hypothesis and the scenario of a smooth transition from hackled to wet spiral should be ruled out. Could there have been a transition period when the spiders built an orb web without a capture spiral? This is a realistic hypothesis; indeed, the present-day *Cyrtophora* builds an orb web without the typical capture spiral (typically running inwards) but instead with a permanent auxiliary spiral (running outwards) covering the whole web (Zschokke and Vollrath, 1995b). Basically, this web has (indeed is) one large hub that extends to the periphery of the web. The web catches prey by first knocking it out of the air with strong trip- and anchor-threads above the main sheet of a web and then catching the tumbling prey on its trampoline of a net where the spider lurks (Lubin, 1973). *Cyrtophora* shows us that a totally unsticky orb web with only radials and a hub can capture prey. Neoteny thus becomes a possibility to explain how spiders in the transition period between the two types of stickiness might have fed themselves. Very young instars of cribellate orb weavers do not have the cribellum and thus lack hackled silk (Hajer, 1988). In some species they nevertheless build an orb web, but one without spirals. To compensate, their webs consist of very many, very fine radii (Szlep, 1961), which apparently stick to the prey simply by adhesion (probably electrostatic) without hackling or coating with glue. One could imagine that a fully cribellate orb weaver was selected to reduce the number of instars (i.e. became adult as a very early instar without a web). There is no reason to assume that a strong drive towards neoteny cannot happen in spiders; indeed, there is evidence that such a strategy might have compelling advantages if growth-related mortality is high (Hutchinson *et al.*, 1997). Much later in the phylogeny of this group such a selection for neoteny might have been reversed and replaced by selection for large size. However, what if by then these spiders had lost the ability to make cribellate silk? Would they not have to either reinvent the cribellum or invent another type of glue? In either case, the extended hub scenario and the neoteny scenario would retain, in the transforming species, the basic web-building algorithm of beginning a web with a wheel of spokes held together in the hub by a minimal or an extended mesh of thread that is laid down in a spiral fashion. Such webs would not, however, have had a 'proper' capture spiral, because such a spiral is necessary only if it contains some type of glue. We recall that in both the cribellate and ecribellate orb web the capture spirals are built from the outside inwards after the radial wheel has been bound together by the auxiliary spiral winding from the hub outwards. The *Cyrtophora* web shows that a glueless auxiliary spiral can very well replace a glueless capture spiral, albeit with a slightly different mode of action. Building a dry and glueless capture spiral in addition to a dry and glueless auxiliary spiral would be an extra cost that would be unnecessary (and thus presumably selected against). On the other hand, one could build a web without a secondary (capture) spiral but with a sticky primary (auxiliary) spiral, such as found in the orb-like web of the cribellate spider *Fecinia* sp. (Robinson and Lubin, 1979). This web is interesting as it shows that there are possibilities of building a two-dimensional orb-type web following totally different rules of construction (Zschokke and Vollrath, 1995b).

Now, if the adhesiveless webs of the ancestral bridge species had no capture spiral, then this raises an interesting question of evolutionary thinking. I have just argued that during the long capture-spiralless generations, the ability to make the original hackled adhesive might have been lost, that is, that the genes responsible for this material were lost or jumbled. Why would we assume that the original rules of capture spiral construction would escape such a fate? I see no reason why they should have survived unaltered, as the genes for behaviour are not inherently more buffered against mutations than are the genes for morphology. Unless, of course, it were that those genes are multi-functional, meaning that a gene or set of genes encoded a modal action pattern that is used in many behaviour patterns. In that case, only the one or very few rules that are responsible only and exclusively for capture spiral construction (and participate in no other behaviour pattern) would have had the opportunity to be lost during the 'time in the wilderness'. If there are no such highly specific construction rules, then any capture spiral is like any other. If there are specific rules, then the question arises whether the cribellate and ecribellate capture spirals are one and the same behaviourally speaking (Vollrath, 1992a). Maybe there is only one good, efficient way of making a secondary capture spiral although there are two ways of making the adhesive. And if there is only one way, then the two spirals would look identical (but for the glue) yet they could very well have evolved independently without us ever being able to find out unless we can find telltale differences in the two web-building algorithms (Krink and Vollrath, 1997).

It is certainly true that a gland–duct–spinneret system would require many genes to encode for the various components and their relative positions. And it is certainly true that a behaviour, however complex, would be able to draw on reflexes, modal action patterns and full behaviour patterns already existing for totally different purposes. All the organism has to do is to switch the behaviour on and direct it. Thus all that would have to evolve would be the pairing of the first stimulus–response action, maybe followed by a few further stimulus–response pairings. Consequently, one might assume that behaviour is both more buffered against loss of function through mutations and more flexible in its response to selection than is morphology. Thus behaviour could at the same time be more conservative and more responsive to selection than morphology. Further, this means that the direction of evolution could be reversed easily for a behavioural trait—such as is shown for highly specific behaviour patterns in duck courtship, where hybrids show the ancestral pattern still evident in distant relatives but lost in both parent species (Lorenz, 1941). Such total and functional reversal (re-emergence of a lost trait) is a difficult if not an impossible feat for the evolution of morphological traits (Ridley, 1998) where atavistic traits are typically non-functional.

8. Conclusions

By applying the concept of co-evolution to the concerted evolution of (1) building material for the web and (2) web-building behaviour we can clearly see the close interaction of the two separately encoded yet functionally interlinked character traits, silk and movement. The spider's web, that is, its construction, engineering and evolution, is an ideal example to show the importance of studying in detail the material and technological properties of animal structures. We cannot understand the origin and evolution of the building behaviour without a good understanding of the building material. Without such understanding we can comprehend even less the economics (and thus, ultimately, the evolution) of the web-building and prey capture behaviour. In short, manufacture and operation of the web depend on the inherent qualities (constraints as well as opportunities) of the material, and so does the evolution of the entire araneid clade, which has made so much of what was once a humble fibrous protein gel that once encased eggs.

Surely, in many other animal groups, too, only a detailed knowledge of the mechanical properties of the materials integrated into structures will enable us to understand the evolution from simple to complex behaviour patterns. Caddis flies, wasps and termites (to name but three other animal clades known for their constructions) rely on the interaction of behaviour and building material as much as spiders do. It would seem that estimating the true economics of a behaviour requires as much an understanding of the principles of mechanics and materials (e.g. Alexander, 1983; Vincent, 1982; Vogel, 1988) as an understanding of modelling the costs and benefits in the interaction of ecology and behaviour (e.g. Krebs and Davis, 1997; Lendrem, 1989).

Acknowledgements

I thank my students, collaborators and friends Tamara Köhler, Samuel Zschokke, Thiemo Krink and Bo Madsen for their dedication to data acquisition and rigorous analysis. Data used in some of the figures derive from their work.

Dedication

This essay is dedicated to the fond memory of Peter Witt, a father of modern behavioural arachnology and a friend *par excellence* to spiders and humans alike.

References

Alexander, R.M. (1983) *Animal Mechanics*. Blackwell Scientific, Oxford.

Coddington, J. (1986) The monophyletic origin of the orb web. In: *Spiders, Webs, Behaviour and Evolution* (ed. W.A. Shear). Stanford University Press, Stanford, CA, pp. 319–363.

Craig, C.L. (1987) The ecological and evolutionary interdependence between web architecture and web silk spun by orb web weaving spiders. *Biol. J. Linnean Soc.* 30, 135–162.

Craig, C.L. and Freeman, C.R. (1991) Effects of predator visibility on prey encounter. A case study on aerial web weaving spiders. *Behav. Ecol. Sociobiol.* 29: 249.

Denny, M.W. (1980) Silks—their properties and functions. In: *The Mechanical Properties of Biological Materials* (eds J.F.V. Vincent and J.D. Currey). *Symp. Soc. Exp. Biol.* 34: 245–271.

Denny, M.W. (1994) Elastomeric network models for the frame and viscid silks from the orb web of the spider *Araneus diadematus*. In: *Silk Polymers. Materials Science and Biotechnology* (eds D. Kaplan, W.W. Adams, B. Farmer and C. Viney). American Chemical Society, Washington, DC, pp. 328–341.

Eberhard, W.G. (1981) Construction behaviour and the distribution of tensions in orb webs. *Bull. Br. Arachnol. Soc.* 5, 189–204.

Eberhard, W.G. (1982) Behavioural characters for the higher classification of orb-weaving spiders. *Evolution* 36: 1067–1095.

Eberhard, W.G. (1987) Hub construction by *Leucauge mariana* (Araneae, Araneidae). *Bull. Br. Arachnol. Soc.* 7: 128–132.

Eberhard, W.G. (1988a) Memory of distances and directions moved as cues during temporary spiral construction in the spider *Leucauge mariana* (Araneae: Araneidae). *J. Insect Behav.* 1: 51–66.

Eberhard, W.G. (1988b) Combing and sticky silk attachment behaviour by cribellate spiders and its taxonomic implications. *Bull. Br. Arachnol. Soc.* 7: 247–251.

Eberhard, W.G. (1988c) Behavioural flexibility in orb web construction: effects of supplies in different silk glands and spider size and weight. *J. Arachnology* 16: 295–302.

Eberhard, W.G. (1989) Effects of orb web orientation and spider size on prey retention. *Bull. Br. Arachnol. Soc.* 8: 45–48.

Eberhard, W.G. (1990) Function and phylogeny of spider webs. *Annu. Rev. Ecol. Syst.* 21: 341–372.

Edmonds, D.T. and Vollrath, F. (1992) The contribution of atmospheric water vapour to the formation and efficiency of a spider's capture web. *Proc. R. Soc. Lond.* 248: 145–148.

Foelix, R. (1996) *Biology of Spiders*. Oxford University Press, Oxford.

Futuyma, D. (1998) *Evolution*, 2nd edn. Sinauer, Sunderland, MA.

Glatz, L. (1973) Der Spinnapparat der Orthognatha (Arachnida, Araneae). *Z. Morphol. Tiere* 75: 1–50.

Gordon, J.E. (1976) *The New Science of Strong Materials*, 2nd Edn. Pelican, London.

Gosline, J., Denny, M. and DeMont, M. (1984) Spider silk as rubber. *Nature* 309: 551–552.

Gosline, J.M., Pollak, C.C., Guerette, P., Cheng, A., DeMont, M.E. and Denny, M.W. (1994) Elastomeric network models for the frame and viscid silks from the orb web of the spider *Araneus diadematus*. In: *Silk Polymers. Materials Science and Biotechnology* (eds D. Kaplan, W.W. Adams, B. Farmer and C. Viney). Washington: American Chemical Society, Washington, DC, pp. 328–341.

Gotts, N.M. and Vollrath, F. (1991) Artificial intelligence modelling of web-building in the garden cross spider. *J. Theor. Biol.* **152**: 485–511.

Hajer, J. (1988) The study of the ontogenesis of the spinning apparatus in Cribellatae spiders and its utilising in solving phylogenetic and taxonomic problems. *Fauna Bohem. Septentr.* **13**: 109–126.

Heiling, A.M. and Herberstein, M.E. (2000) The role of experience in web-building spiders (Araneidae). *Anim. Cognition* (in press).

Herberstein, M.E. and Heiling, A.M. (2000) Asymmetry in spider orb-webs: a result of physical constraints? *Anim. Behav.* (in press).

Hieber, C.S. (1984) Orb-web orientation and modification by the spiders *Araneus diadematus* and *Araneus gemmoides* (Araneae: Araneidae) in response to wind and light. *Z. Tierpsychol.* **65**: 250–260.

Hormiga, G., Eberhard, W.G. and Coddington, J.A. (1995) Web-construction behaviour in Australian *Phonognatha* and the phylogeny of nephiline and tetragnathid spiders (Araneae: Tetragnathidae). *Aust. J. Zool.* **43**: 313–364.

Hutchinson, J.M., McNamara, J.M., Houston, A.I. and Vollrath, F. (1997) Dyar's rule and the investment principle: the consequences of a size-dependent feeding rate when growth is discontinuous. *Philos. Trans. R. Soc. Lond.* **352**: 113–138.

Kaplan, D.L., Adams, W.W., Viney, C. and Farmer, B.L. (1994a) *Silk Polymers. Materials Science and Biotechnology*. American Chemical Society, Washington, DC.

Kaplan, D., Adams, W.W., Farmer, B. and Viney, C. (1994b) Silk: biology, structure, properties and genetics. In: *Silk Polymers. Materials Science and Biotechnology* (eds D. Kaplan, W.W. Adams, B. Farmer and C. Viney). American Chemical Society, Washington, DC, pp. 2–16.

Knight, D.P. and Vollrath F. (1999) Liquid crystals in a spider's silk production line. *Proc. R. Soc. Lond.* **266**: 519–523.

Köhler, T. (1992) The mechanical properties of different threads of the orb webs of *Araneus diadematus* and *Uloborus walkenaerius*. Diploma Thesis, University of Basel, Switzerland.

Köhler, T. and Vollrath, F. (1995) Thread biomechanics in the two orb weaving spiders *Araneus diadematus* (Araneae, Araneidae) and *Uloborus walkenaerius* (Araneae, Uloboridae). *J. Exp. Zool.* **271**: 1–17.

Kovoor, J. (1987) Comparative structure and histochemistry of silk-producing organs in arachnids. In: *Ecophysiology of Spiders* (ed. W. Nentwig). Springer, Heidelberg, pp. 160–186.

Krebs, J.R. and Davies, N.B (1997) *Behavioural Ecology: An Evolutionary Approach*. Blackwell Scientific, Oxford.

Krieger, M. (1992) Radienbau im Netz der Radnetzspinne. Ph.D. thesis, University of Basel.

Krink, T. and Vollrath, F. (1997) Analysing spider web-building behaviour with rule-based simulations and genetic algorithms. *J. Theor. Biol.* **185**: 321–331.

Krink, T. and Vollrath, F. (1998a) Using a virtual robot to model the use of regenerated legs in a web spider. *Anim. Behav.* **57**: 223–241.

Krink, T. and Vollrath, F (1998b) Emergent properties in the orb web of the garden cross spider *Araneus diadematus*. *Proc. R. Soc. Lond.* **265**: 2051–2055.

Krink, T. and Vollrath, F. (1999) Spatial orientation of virtual spider robots by object-oriented rule-based simulations. *IEEE Intelligent Systems* **14**: 77–84

Krink, T. and Vollrath, F. (2000) Optimal area use in orb webs of the garden cross spider. *Naturwiss* **87** (2): 90–93.

Lendrem, D. (1989) *Modelling Animal Behaviour*. Chrom Helm.

Lin, L. (1997) The biomechanics of spider silks and orb webs. Ph.D. thesis, Oxford University.

Lin, L., Edmonds, D. and Vollrath, F. (1995) Structural engineering of a spider's web. *Nature* 373: 146–148.

Lorenz, K. (1941) Vergleichende Bewegungsstudien an Anatiden. *Suppl. J. Ornithol.* 89: 194–294.

Lubin, Y.D. (1973) Web structure and function: the non-adhesive orb-web of *Cyrtophora moluccensis* (Doleschall) (Araneae, Araneidae). *Forma Functio* 6: 337–358.

Lubin, Y.D. (1986) Web building and prey capture in the Uloboridae. In *Spiders: Webs, Behavior and Evolution* (ed. W. A. Shear). Stanford University Press, Stanford, CA, pp. 132–171.

Madsen B. and Vollrath, F. (2000) Mechanics and morphology of silk reeled from anaesthetised spiders. *Naturwiss* 87 (3): 148–153.

Madsen, B., Shao, Z. and Vollrath, F. (1999) Variability in the mechanical properties of spider silks on three levels: interspecific, intraspecific and intraindividual. *Int. J. Biol. Macromol.* 24: 301–306.

Maynard Smith, J., Burian, R., Kauffman, S., Alperch, P., Campbell, J., Goodwin, B., Lande, R., Raub, D. and Wolpert, L. (1985) Developmental constraints and evolution. *Quart. Rev. Biol.* 60: 265–287.

McMahon, T.A. and Bonner, J.T. (1983) *On Size and Life.* Freeman, New York.

Opell, B.D. (1993) What forces are responsible for the stickiness of spider cribellar prey capture threads? *J. Exp. Zool.* 265: 469–476.

Opell, B.D. (1994) Increased stickiness of prey capture threads accompanying web reduction in the spider family Uloboridae. *Functional Ecol.* 8: 85–90.

Opell, B. (1997) A comparison of capture thread and architectural features of deinopoid and araneoid orb-webs. *J. Arachnol.* 25: 295–306.

Opell, B. (1998) Economics of spider orb-webs: the benefits of producing adhesive capture thread and of recycling silk. *Functional Ecol.* 12: 613–624.

Peters, H.M. (1987) Fine structure and function of capture threads. In: *Ecophysiology of Spiders* (ed. W. Nentwig). Springer, Berlin, pp. 187–202.

Peters, P.J. (1970) Orb web construction: interaction of spider *Araneus diadematus* (Cl.) and thread configuration. *Anim. Behav.* 18: 478–484.

Quatremére-Disjonval, D.B. (1797) *L'Aranéologie.* J.J. Fuchs, Paris.

Ridley, M. (1998) *Evolution.* Blackwell Scientific, Oxford.

Robinson, M.H. (1969) Predatory behaviour of *Argiope argentata* (Fabricius). *Am. Zool.* 9: 161–173.

Robinson, M.H. and Lubin, Y.D. (1979) Specialists and generalists: the ecology and behavior of some web-building spiders from Papua New Guinea. II *Psechrus argentatus* and *Fecinia* sp. (Araeae: Psechridae). *Pacific Insects* 21: 133–164.

Schildknecht, H., Munzelmann, P., Krauss, D. and Kuhn, C. (1972) Über die Chemie der Spinnwebe. *Naturwissenschaften* 59: 98–99.

Selden, P.A. (1989) Orb-web weaving spiders in the early Cretaceous. *Nature* 340: 711.

Shao, Z. and Vollrath, F. (1998) The effect of solvents on the contraction and mechanical properties of spider silk. *Polymers* 40: 2493–2500.

Shao, Z., Wen Hu, X., Frische, S. and Vollrath, F. (1999) Heterogeneous morphology in spider silk and its function for mechanical properties. *Polymers* 40: 4709–4711.

Shear, W.A. (1986) The evolution of web-building behavior. In: *Spiders: Webs, Behavior and Evolution* (ed. W. A. Shear). Stanford University Press, Stanford, CA, pp. 364–400.

Shear, W.A., Palmer, J.M., Coddington, J.A. and Bonamo, P.M. (1989) A Devonian spinneret: early evidence of spiders and silk use. *Science* 246: 479–481.

Stowe, M.K. (1986) Prey specialisation in the Araneidae. In: *Spiders: Webs, Behavior and Evolution* (ed. W. A. Shear). Stanford University Press, Stanford, CA, pp. 101–131.

Szlep, R. (1961) Developmental changes in the web spinning instinct of Uloboridae: construction of the primary type web. *Behavior* 17: 60–70.

Tillinghast, E.K. and Kavanagh, E.J. (1979) Fibrous and adhesive components of the orb webs of *Araneus trifolium* and *Argiope trifasciata*. *J. Morphol.* 160: 17–32.

Tillinghast, E.K. and Townley, M.A. (1994) Silk glands of araneid spiders: selected morpho-logical and physiological aspects. In: *Silk Polymers. Materials Science and Biotechnology* (eds D. Kaplan, W.W. Adams, B. Farmer and C. Viney). American Chemical Society, Washington, DC, pp. 29–44.

Tinbergen, N. (1951) *The Study of Instinct.* Oxford University Press, London.

Vincent, J. (1982) *Structural Biomaterials.* John Wiley, New York.

Vogel, S. (1988) *Life's Devices.* Princeton University Press, Princeton, NJ.

Vollrath, F. (1985) Web spider's dilemma: a risky move or site dependent growth. *Oecologia* 68: 69–94.

Vollrath, F. (1988a) Untangling the spider's web. *Trends Ecol. Evol.* 3: 331–335.

Vollrath, F. (1988b) Spiral orientation of *Araneus diadematus* orb webs built during vertical rotation. *J. Comp. Physiol.* 162: 413–419.

Vollrath, F. (1992a) Analysis and interpretation of orb spider exploration and web-building behavior. *Adv. Stud. Anim. Behav.* 21: 147–199.

Vollrath, F. (1992b) Spider webs and silk. *Sci. Am.* 266: 70–76.

Vollrath, F. (1994) General properties of some spider silks. In: *Silk Polymers. Materials Science and Biotechnology* (eds D. Kaplan, W.W. Adams, B. Farmer and C. Viney). American Chemical Society, Washington, DC, pp. 17–28.

Vollrath, F. (1999) Biology of spider silk. *Int. J. Biol. Macromol.* 24: 81–88.

Vollrath, F. and Edmonds, D. (1989) Modulation of the mechanical properties of spider silk by coating with water. *Nature* 340: 305–307.

Vollrath, F. and Edmonds, D.T. (1992) The contribution of atmospheric water vapour to the formation and efficiency of a spider's capture web. *Proc. R. Soc. Lond.* 248: 145–148.

Vollrath, F. and Houston, A. (1986) Previous experience and site tenacity in the orb spider *Nephila clavipes. Oecologia* 70: 305–308.

Vollrath, F. and Knight, D. (1999) The silk press of the spider *Nephila edulis. Int. J. Biol. Macromol.* 24: 243–249.

Vollrath, F. and Köhler, T. (1996) Mechanics of silk produced by loaded spiders. *Proc. R. Soc. Lond.* 263: 387–391.

Vollrath, F. and Samu, F. (1997) The effect of starvation on web geometry in an orb-weaving spider. *Bull. Br. Arachnol. Soc.* 10: 295–298.

Vollrath, F. and Tillinghast, E.K. (1991) Glycoprotein glue inside a spider web's aqueous coat. *Naturwissenschaften* 78: 557–559.

Vollrath, F., Fairbrother, W.J., Williams, R.J.P., Tillinghast, E.K., Bernstein, D.T., Gallagher, K.S. and Townley, M.A. (1990) Compounds in the droplets of the orb spider's viscid spiral. *Nature* 345: 526–528.

Vollrath, F., Holtet, T., Thogersen, H. and Frische, S. (1996) Structural organization of spider silk. *Proc. R. Soc. Lond.* 263: 147–151.

Vollrath, F., Downes, M. and Krackow, S. (1997) Design variables in web geometry of an orb weaving spider. *Physiol. Behav.* 62: 735–743.

Vollrath, F., Wen Hu, X. and Knight, D. (1998) Silk production in a spider involves acid bath treatment. *Proc. R. Soc. Lond.* 263: 817–820.

Wirth, E. and Barth, F.G. (1992) Forces in the spider orb web. *J. Comp. Physiol. A* 171: 359–371.

Witt, P.N. (1963a) Environment in relation to the behaviour of spiders. *Arch. Environ. Hlth* 7: 4–12.

Witt, P.N. (1963b) The web as a means of communication. *Biosci. Commun.* 1: 7–23.

Witt, P.N. (1965) Do we live in the best of all worlds? Spider webs suggest an answer. *Persp. Biol. Med.* 8: 475–487.

Witt, P.N. and Reed, C.F. (1965) Spider web-building: measurement of web geometry identifies components in a complex invertebrate behaviour pattern. *Science* 149: 1190–1197.

Witt, P.N., Reed, C.F. and Peakall, D.B. (1968) *A Spider's Web: Problems in Regulatory Biology.* Springer, Heidelberg.

Wolff, D. and Hempel, U. (1951) Versuche über die Beeinflussung des Netzbaues von *Zilla x-notata* durch Pervitin, Scolopamin und Strychnin. *Z. Vergl. Physiol.* **33**: 297–528.

Zschokke, S. (1995) The coiling of the spirals in the orb web of *Araneus diadematus. Newslett. Br. Arachnol. Soc.* **74**: 9–10.

Zschokke, S. (1997) Factors influencing the size of the orb web in *Araneus diadematus. 16th Eur. Arachnol. Congr.*, Siedlice, Poland.

Zschokke, S. and Vollrath, F. (1995a) Unfreezing the behaviour of two orb spiders. *Behav. Physiol.* **58**: 1167–1173.

Zschokke, S. and Vollrath, F. (1995b) Web construction patterns in a range of orb-weaving spiders (Araneae). *Eur. J. Entomol.* **92**: 523–541.

Index